Asian Paleoanthropology

Vertebrate Paleobiology and Paleoanthropology Series

Edited by

Eric Delson
Vertebrate Paleontology, American Museum of Natural History,
New York, NY 10024, USA
delson@amnh.org

Eric J. Sargis
Anthropology, Yale University
New Haven, CT 06520, USA
eric.sargis@yale.edu

Focal topics for volumes in the series will include systematic paleontology of all vertebrates (from agnathans to humans), phylogeny reconstruction, functional morphology, Paleolithic archaeology, taphonomy, geochronology, historical biogeography, and biostratigraphy. Other fields (e.g., paleoclimatology, paleoecology, ancient DNA, total organismal community structure) may be considered if the volume theme emphasizes paleobiology (or archaeology). Fields such as modeling of physical processes, genetic methodology, nonvertebrates or neontology are out of our scope.

Volumes in the series may either be monographic treatments (including unpublished but fully revised dissertations) or edited collections, especially those focusing on problem-oriented issues, with multidisciplinary coverage where possible.

For other titles published in this series, go to
www.springer.com/series/6978

Asian Paleoanthropology

From Africa to China and Beyond

Edited by

Christopher J. Norton

Department of Anthropology, University of Hawaii at Manoa, Honolulu, USA

David R. Braun

Department of Archaeology, University of Cape Town, Rondebosch, South Africa

Springer

Editors
Christopher J. Norton
Department of Anthropology
University of Hawaii at Manoa
Honolulu
USA
cjbae@hawaii.edu

David R. Braun
Department of Archaeology
University of Cape Town
South Africa
david.braun@uct.ac.za

ISBN 978-90-481-9093-5 e-ISBN 978-90-481-9094-2
DOI 10.1007/978-90-481-9094-2
Springer Dordrecht Heidelberg London New York

Library of Congress Control Number: 2010933110

Cover illustration: Skullcap of Sambungmacan 3 (reproduced with permission from Eric Delson), Skull of Zhoukoudian *Homo erectus* (reproduced with permission from Ian Tattersall), and handaxe from Chongokni (reproduced with permission from Kidong Bae). Cover design by Chaeho Lee (University of Hawaii at Manoa)

Printed on acid-free paper

Springer is part of Springer Science+Business Media (www.springer.com)

Preface

One of the most talked about topics in paleoanthropology in the last few years is the nature of early hominin dispersals out of Africa and into Asia. Although it was once thought that central Asia played a primary role in early human evolution, it is generally accepted nowadays that hominins evolved in Africa and dispersed into Asia. In all likelihood these dispersals were multiple events that began occurring sometime after 2 million years ago. Nevertheless, the timing and nature of these dispersals are currently poorly understood. Fortunately, a growing number of paleoanthropologists are currently developing models, conducting multi-disciplinary field and laboratory projects, and importantly increasing the number of available datasets to conduct more systematic research. Paleoanthropologists working in various regions of Asia are now able to address the who, what, where, why, and how types of questions. The increasing number of studies is quickly filling in the gaps across space (from western Asia to eastern Asia) and time (from the Pliocene to the Late Pleistocene) and facilitating cross-comparative studies between Asia, Africa and Europe.

This volume is the primary result of a symposium that we [along with Jack Harris (Rutgers University)] organized at the Association of American Physical Anthropologists (AAPA) conference in Philadelphia, Pennsylvania in 2007. Most of the participants of the AAPA symposium contributed papers. To increase the breadth of coverage, we invited specialists from regions that often do not receive as much attention in this debate to contribute chapters. The goal of this edited volume is to add to the current state of knowledge on this topic, by focusing on particular recent developments in Asian paleoanthropology. The particular areas covered in this volume are the nature of Early and Middle Pleistocene hominin dispersals into particular regions; the evidence for the earliest hominin occupations in certain regions; and hominin morphological and behavioral variation. Importantly, almost all of the contributors have active field and laboratory research projects. The primary data of these projects formed the foundation of many of the chapters.

We sincerely thank Eric Delson for the initial invitation to contribute this volume to the Springer series *Vertebrate Paleobiology & Paleoanthropology* that he and Ross MacPhee (and now Eric Sargis) edit. Eric provided many helpful comments and support on the entire project. Eric and Alison Brooks, who served as Discussants during the AAPA symposium, provided much constructive feedback on the presentations. We also express our appreciation to the contributors and more than 20 external reviewers who took the time and effort to contribute constructive comments on the papers that appear in this volume. The reviewer's thoughtful reviews improved the quality of the papers tremendously. We are grateful to Steven Heritage, Rachel Hoerman, and Robert Whalley for helping out with the index. We thank Judith Terpos and her group from Springer for support and help during the entire process. This project was started while Christopher J. Norton was a research fellow in the Institute of Vertebrate Paleontology and Paleoanthropology (IVPP) in Beijing, China. Christopher J. Norton sincerely thanks the following IVPP researchers: Wu Xinzhi, Gao Xing, Liu Wu, Jin Changzhu, Pei Shuwen, Feng Xingwu, Wu Xiujie, Zhang Yingqi, and Wang Yuan. Without the sincere and strong support of these colleagues from the IVPP during the editing this book, it would have been very difficult to complete. Christopher J. Norton also appreciates the continued support from the Department of Anthropology, University of Hawaii at Manoa. David R. Braun was an NSF International Research Post-Doctoral Fellow at the outset of this project and acknowledges the support of that program and the University of Cape Town URC-ERP program. As always, Christopher J. Norton is especially grateful to Jennie JH Jin, who has been by his side every step of the way while this project was developing and helped to see it through to completion. David R. Braun acknowledges the support of his family, especially Robert, Lynda, Jenifer and Kathryn, and now Henry.

Contents

Contributors

Sheela Athreya
Department of Anthropology, Texas A&M University, 4352 TAMU,
College Station, TX 77843, USA
athreya@tamu.edu

Fachroel Aziz
Geological Museum, Geological Survey Institute, Jl. Diponegoro No. 57, Bandung, 40122, Indonesia; University of Wollongong, Wollongong, NSW 2522, Australia
azizf@bdg.centrin.net.id

Karen L. Baab
Department of Anatomical Science, Stony Brook University Medical Center,
Stony Brook, NY 11794, USA
klbaab@yahoo.com

Hisao Baba
Department of Anthropology, National Museum of Nature and Science,
3-23-1 Hyakunincho, Shinjuku-ku, Tokyo, 169-0073, Japan
baba@kahaku.go.jp

Kidong Bae
Department of Anthropology, Hanyang University, Ansan, Kyunggi-do, 425-791, South Korea
Bkd5374@chol.com

Miriam Belmaker
Department of Anthropology, Peabody Museum, Harvard University,
11 Divinity Ave, Cambridge, MA 02138, USA
belmaker@fas.harvard.edu

David R. Braun
Archaeology Department, University of Cape Town, Rondebosch 7701, South Africa
david.braun@uct.ac.za

Parth R Chauhan
226 Centennial Avenue, Cranford, NJ 07016, USA
parthrchauhan@gmail.com

Robin William Dennell
Department of Archaeology, University of Sheffield, Sheffield, S1 4 ET, UK
r.dennell@sheffield.ac.uk

Xing Gao
Institute of Vertebrate Palaeontology and Paleoanthropology, Chinese Academy of Sciences, Beijing, 100044, People's Republic of China
gaoxing@ivpp.ac.cn

Michelle M Glantz
Department of Anthropology, Colorado State University, Fort Collins, CO 80524, USA
Mica.Glantz@colostate.edu

John W.K. Harris
Department of Anthropology, Rutgers University, 131 George Street,
New Brunswick, NJ 08901, USA
jwharris@rci.rutgers.edu

Etty Indriati
Laboratory of Bio- and Paleoanthropology, Gadjah Mada University, Yogyakarta, 55281,
Indonesia
ettykurtz@idola.net.id

Changzhu Jin
Institute of Vertebrate Paleontology and Paleoanthropology, Chinese Academy of Sciences,
142 Xizhimenwai Street, Beijing, 100044, The People's Republic of China
jinchangzhu@ivpp.ac.cn

Yousuke Kaifu
Department of Anthropology, National Museum of Nature and Science,
3-23-1 Hyakunincho, Shinjuku-ku, Tokyo, 169-0073, Japan;
Department of Biological Science, The University of Tokyo, 6-3-1 Hongo,
Bunkyo-ku, Tokyo, 113-0033, Japan
Kaifu@kahaku.go.jp

Iwan Kurniawan
Geological Museum, Geological Survey Institute, Jl. Diponegoro No. 57,
Bandung, 40122, Indonesia

Wu Liu
Institute of Vertebrate Paleontology and Paleoanthropology,
Chinese Academy of Sciences, 142 Xizhimenwai Street, Beijing, 100044,
The People's Republic of China
liuwu@ivpp.ac.cn

Kazuto Matsufuji
Archaeology Lab, Faculty of Letters, Doshisha University, Kyoto, Japan, 602-8580
kmatsufu@mail.doshisha.ac.jp

Christopher J. Norton
Department of Anthropology, University of Hawaii, 2424 Maile Way,
Saunders Hall 346, Honolulu, HI 96822, USA
cnorton@hawaii.edu

Michael D. Petraglia
Research Laboratory for Archaeology and the History of Art, University of Oxford, Dyson
Perrins Building, South Parks Road, Oxford, OX1 3QY, UK
Michael.petraglia@rlaha.ox.ac.uk

Chen Shen
Department of World Cultures, Royal Ontario Museum, Toronto, ON M5S 2C6, Canada
chens@rom.on.ca

Ceri Shipton
Garreglefain, Porthyrhyd, Llanwrda, Sir Gar, SA19 8PE, UK
c.shipton@human-evol.cam.ac.uk

Yuan Wang
Institute of Vertebrate Paleontology and Paleoanthropology, Chinese Academy of Sciences, 142 Xizhimenwai Street, Beijing, 100044, The People's Republic of China
Graduate School, Chinese Academy of Sciences, Beijing, 100039, The People's Republic of China
power_xzw@163.com

Qi Wei
Institute of Vertebrate Palaeontology and Paleoanthropology, Chinese Academy of Sciences, 142 Xizhimenwai Street, Beijing, 100044, People's Republic of China
weiqigood@163.com

Xiujie Wu
Institute of Vertebrate Paleontology and Paleoanthropology, Chinese Academy of Sciences, 142 Xizhimenwai Street, Beijing, 100044, The People's Republic of China
wuxiujie@ivpp.ac.cn

Yingqi Zhang
Institute of Vertebrate Paleontology and Paleoanthropology, Chinese Academy of Sciences, 142 Xizhimenwai Street, Beijing, 100044, The People's Republic of China
arvicolid@gmail.com

Chapter 1
Asian Paleoanthropology: An Introduction

Christopher J. Norton and David R. Braun

Introduction

In the days of Ernst Haeckel, Eugene Dubois, Henry Fairfield Osborn, and Roy Chapman Andrews, Asia was often considered the center where major events in human evolution occurred. Since the middle of the twentieth century, however, the focus of paleoanthropology shifted to Africa, due (at least in part) to the many significant hominin fossils found there (Dennell 2001). It is now generally accepted that most of the major human evolutionary events during the late Neogene and the early Early Pleistocene took place in Africa, rather than Asia.

Currently it is now generally thought that sometime after 2 Ma (millions of years ago) hominins dispersed (likely more than once) from Africa into Asia, reaching Dmanisi, Georgia, by ~1.7 Ma and perhaps the eastern Asian sites of Yuanmou, Majuanggou, Mojokerto, and Sangiran by ~1.8–1.6 Ma (Swisher et al. 1994; Gabunia et al. 2001; Larick et al. 2001; Zhu et al. 2003, 2004, 2008; Anton and Swisher 2004; Anton 2007; but see Hyodo et al. 1993, 2002). This is commonly known as the Out of Africa I hypothesis. In order to reach the sites in northern and southern China and Indonesia from Africa, hominins would have been forced to skirt the Himalayan-Tibetan Plateau which would have formed a formidable dispersal barrier by at least the Late Miocene (Fort 1996; Dennell 2009). The possible dispersal corridors would have included a more northerly route from Dmanisi, through Mongolia, and reaching the Nihewan Basin in northern China. However, there is currently a paucity of evidence that Early and possibly even Middle Pleistocene hominins dispersed into north-central Asia (Glantz 2010). The other plausible hominin dispersal would have followed a more southern route. The evidence for an Early Pleistocene southern dispersal corridor is not much stronger (Chauhan 2010); though a few southern Asian sites may serve as possible evidence of hominin movements (e.g., Pabbi Hills, Isampur) (Dennell 2001, 2009). Although these purported 'early' sites are beginning to shape a picture of southern hominin dispersals, further detailed investigation will determine their importance for the story of ancient human dispersals.

The nature of the early hominin dispersals into Asia is only slowly becoming clearer as paleoanthropologists are currently concentrating on who, what, where, why, and how types of questions. The goal of this edited volume is to contribute to the current state of knowledge on this topic, by focusing on particular recent developments in Asian paleoanthropology. The idea of this volume came about while the two editors (along with John W.K. Harris) were sitting in the back of a minivan visiting over the course of several days various sites in Anhui Province in central-east China in May 2006. The result of this discussion was a proposal to organize a symposium that focused on the current state of Asian paleoanthropology. The resulting symposium entitled "The nature of the earliest Asian hominin lifeways: The current state of evidence" was held at the American Association of Physical Anthropologists 2007 meeting in Philadelphia, Pennsylvania, USA. In order to develop a comprehensive synthesis of the current state of Asian paleoanthropology, we invited contributors who are working in very diverse areas of research (e.g., hominin paleontology, archaeology) and in different areas of Asia. Most of the symposium participants agreed to contribute papers to this volume. To augment this selection of contributions we invited scholars from other areas of Asia (e.g., China, Russia, Korea, and Japan) who were unable to attend the conference in Philadelphia, but had important information that we felt should be disseminated more widely. Since the start of the *Asian Paleoanthropology* project, we organized another symposium at the 2008 Society for American Archaeology (SAA) conference in Vancouver. The SAA symposium resulted in the recent publication of a special issue of *Quaternary International* (Norton and Jin 2010). Although the content of the *Quaternary International* volume overlaps a little with the *Asian Paleoanthropology*

C.J. Norton (✉)
Department of Anthropology, University of Hawaii, 2424 Maile Way, Saunders Hall 346, Honolulu, HI 96822, USA
e-mail: cjbae@hawaii.edu

C.J. Norton and D.R. Braun (eds.), *Asian Paleoanthropology: From Africa to China and Beyond*, Vertebrate Paleobiology and Paleoanthropology, DOI 10.1007/978-90-481-9094-2_1, © Springer Science+Business Media B.V. 2010

volume, the former covers eastern Asia (including East and Southeast Asia) and Australasia and includes contributions that focus on the Late Pleistocene and Early Holocene. Thus, the two volumes dovetail each other nicely.

Contents of This Volume

We divided the chapters of this volume along two themes. The first section consists of papers that focus more on theoretical approaches, expectations, and re-evaluations of existing data. The second section comprises papers that were more regional surveys of the paleoanthropological records of those respective regions. We tried to provide complete coverage of the different topics related to the nature of early hominin dispersals, but some of the areas that are left out of this volume include Siberia, Mongolia, and Southeast Asia (except for the important Indonesian sites and materials), in addition to more theoretical studies like the timing of early hominin dispersals. We will provide coverage of these important topics and regions in more detail in future works.

Theoretical Approaches, Expectations, and Re-Evaluations

The chapters in the first half of this volume cover theoretical approaches toward deciphering the Asian paleoanthropological record, and expectations and re-evaluations of the current state of the Asian record. Chapter 2, by Robin Dennell, is intended to give an alternative to the Out of Africa I hypothesis. Expanding on earlier works (e.g., Dennell 2003, 2004; Dennell and Roebroeks 2005), Dennell presents a detailed counter-hypothesis to the Out of Africa I model. Dennell suggests that hominins may have followed the savannas that expanded into central Asia during the late Neogene. Then from central Asia, hominins evolved (either *Homo ergaster* or *H. erectus*) and then dispersed into other regions of Asia. He also suggests it may be possible that hominins expanded back into Africa from Asia, rather than the reverse (see also Dennell and Roebroeks 2005). Even though there is currently very little support for Dennell's hypothesis, it is important because alternative models for the expansion of early hominins throughout the Old World should be sought. From this perspective, Dennell's contribution is very important.

The next four chapters contribute to the growing number of inter-regional cross-comparative studies involving sites and materials from Asia and Africa. Chapter 3, by Miriam Belmaker, is an interesting comparative study of the species diversity present at an array of different sites across Early Pleistocene Eurasia. One of the interesting conclusions that Belmaker draws from her study is that even though there is evidence of a less forested environment beginning with the Pliocene, the Asian grasslands were probably different from the savannah-type environment that was present in equatorial Africa. Thus, Belmaker concludes that early hominins would have been dispersing into a novel environment and would have required the ability to adapt to these new environmental pressures.

Chapters 4 and 5, by David Braun and colleagues, and Ceri Shipton and Michael Petraglia, draw regional comparisons of the Early Paleolithic lithic tool industries present in East Asia. In Chapter 4 Braun et al. conduct a preliminary comparative study of the Mode 1 (or Oldowan) stone tool industries between East Africa and the Nihewan Basin, northern China. Because the Mode 1 lithics from China have not been studied and/or published in the same detail as the African materials, it was not feasible to conduct a comprehensive comparative analysis of the stone tool industries between the two regions. However, Braun et al. lay the groundwork for a series of hypotheses that can be tested as their research further develops. In Chapter 5 Shipton and Petraglia focus their lithic analysis on more recent materials, namely the bifacial implements east and west of the Movius Line. Their piece provides an important contribution to this volume by compiling a fairly extensive dataset from a diversity of sources. The conclusion that Shipton and Petraglia draw is that the morphological variability of the handaxes from Korea, China, India, and East Africa overlap; thus, they cannot be distinguished (see also Petraglia and Shipton 2009). These analyses are the subject of current debate (see Norton and Bae 2009). This debate will surely contribute to more detailed multivariate studies of the lithic technologies east and west of the Movius Line as this research progresses.

In Chapter 6 Karen Baab studies the regional variation of *Homo erectus* crania by performing a multivariate analysis of the hominin fossils from Zhoukoudian Locality 1 and Indonesia (e.g., Sangiran, Sambungmacan). Baab found that time and space could best explain the morphological variation between the northern Chinese and Southeast Asian hominin fossils. What makes Baab's study particularly interesting is the application of three dimensional geometric morphometrics methods, an innovative approach that has been widely successful in deducing patterns in fossil material in other regions of the world (e.g., Harvati 2001; Frost et al. 2003; McNulty 2003). Application of these types of methodologies is contributing to a deeper understanding of the nature of the hominin morphological record in eastern Asia.

In Chapter 7, Christopher Norton and colleagues review the Quaternary vertebrate paleontological record in East Asia, focusing on China and Korea. Their primary conclusion drawn from their synopsis is that the traditional boundary separating the Palearctic and Oriental biogeographic zones is fluid. In fact, the fluctuating boundary actually represents climatic changes with colder taxa appearing south of the Yangtze River during stadials and warmer faunas appearing in northern China (e.g., Zhoukoudian) and Korea during interstadials. Another point that Norton et al. note is the region east of the Qinling Mountain Range which is low-lying, has no natural boundaries. Norton et al. hypothesize that this region would have facilitated the movement of fauna between the two different biogeographic zones. Norton et al. suggest that hominin movements across the landscapes of East Asia would likely have been similarly influenced by climatic changes.

The Current State of the Asian Paleoanthropological Record

The remaining chapters in this volume focus on synthesizing the current state of research in a particular region. One of the strengths of the following chapters is that each of these authors has an active research program, so they are able to contribute primary data from each of these respective regions.

In Chapter 8, Michelle Glantz reviews the central Asian paleoanthropological record. Glantz's primary conclusion is that the evidence for Early Pleistocene hominin occupation of the region is weak and that the evidence currently suggests late Middle Pleistocene hominins (e.g., Neandertals) interspersed throughout the region. Glantz also notes that her analysis of the Obi-Rakhmat and Teshik-Tash Neandertal fossils from Uzbekistan indicates admixture with hominin groups moving into the region from North and East Asia.

Chapters 9 and 10 focus on the sites and materials from India. In Chapter 9, Parth Chauhan reviews the lithic tool industries in Peninsular India. In particular, Chauhan compares the distribution of Acheulean and the non-Acheulean sites throughout the region. One of the primary conclusions that Chauhan draws from his review is that most of the non-Acheulean sites are younger in age. In addition, most of the lithics from these sites are surface collected or derive from questionable contexts. In Chapter 10, Sheela Athreya conducts a multivariate analysis of the Middle Pleistocene hominin cranium from Hathnora,[1] India. Athreya determines that the Hathnora cranium displays a mosaic of sub-Saharan African and Southeast Asian hominin morphological features. Athreya suggests the reason for this is because India is at a biogeographic crossroads between Africa and Southeast Asia; thus, the Hathnora cranium displays a combination of African and Asian features.

In Chapter 11, Yousuke Kaifu and colleagues conduct a detailed analysis of the cranial remains from the important Sangiran site in Indonesia. They study between-group and within-group variation using the penecontemporaneous hominin fossils from Dmanisi and East Africa. Kaifu et al. find little support for the hypothesis that the Sangiran hominin assemblage should best be described as super-robust. They determine a substantial amount of variability in the Sangiran hominin fossil assemblage that includes robust and gracile morphotypes. Kaifu et al. also present the morphological description of a poorly known hominin occipital fragment from Sangiran: Bu 9604.

Chapters 12 and 13 review the current state of paleoanthropological research in China. In Chapter 12, Christopher Norton and colleagues synthesize the hominin fossil and Paleolithic archaeological records of central-east China, a region they argue could have served as an early hominin migration corridor. Some of the sites described by Norton et al. (e.g., Longgupo, Renzidong, Gongwangling, Hexian) are fairly well known to the paleoanthropological scientific community. However, this chapter also describes several lesser known localities (e.g., Yiyuan, Huanglongdong). Norton et al.'s synthesis provides the opportunity to develop hypotheses that are currently being tested through the fieldwork program in this region (e.g., see Norton et al., Table 2). In Chapter 13, Chen Shen and colleagues review the history and current state of research of the important Early and Middle Pleistocene Nihewan Basin localities in northern China. Interestingly, Shen et al. note that they identified use-wear on some of the stone implements that were recovered from the Early Pleistocene Xiaochangliang site. Shen et al. are expanding on earlier research in the Nihewan Basin in the late 1980s and early 1990s that was led by Wei Qi from the Institute of Vertebrate Paleontology and Paleoanthropology in Beijing and the late Desmond Clark and the UC Berkeley team.

Chapters 14 and 15 review the current state of evidence of early hominin occupation of the Korean Peninsula and the Japanese Archipelago. In Chapter 14, Kidong Bae describes recent developments in the study of the Chongokni site, which is best known for the presence of Acheulean-like bifacial implements. Bae suggests that the earliest hominin occupations of Chongokni may date to 350–300 ka (see also Norton et al. 2006; Norton and Bae 2009) and that other sites on the Korean Peninsula (e.g., Komunmoru, Jangsanni) may actually predate Chongokni (see also Norton 2000). Bae concludes that the earliest hominin occupation of the Korean Peninsula is likely to be at least the Middle Pleistocene. In Chapter 15, Kazuto Matsufuji reviews the Japanese paleoanthropological record and discusses his multi-disciplinary research at the Kanedori site. Matsufuji draws the conclusion that Kanedori is currently the oldest site in Japan and dates to

[1]The Hathnora cranium is also commonly known as the Narmada cranium because it was found in the Narmada valley.

Marine Isotope Stage ("MIS") 5 (~127–71 ka). Matsufuji further suggests that because sea levels would have been higher during MIS 5, that Japanese researchers should be looking in deposits that date back to MIS 6. Although there is currently no evidence to support this hypothesis, it is worth investigating (Norton and Jin 2009).

Discussion

It is generally agreed that the Asian record, vis-à-vis Africa, is still poorly known despite being much larger geographically. Probably one of the most parsimonious explanations for this is that hominin population densities in Pleistocene Asia were much lower than in coeval Africa (Movius 1948; Wolpoff et al. 1984; Schick 1994; Lycett 2007; Norton et al. 2006; Lycett and Norton 2010). However, another equally plausible explanation is the paucity of systematic multi-disciplinary fieldwork in many regions of Asia, a point emphasized by many of the contributors of this volume (e.g., Chauhan, Dennell, Glantz, Norton). The increase of collaborative multi-disciplinary field and laboratory work in areas like India and China are serving to increase our knowledge about the paleoanthropological records of these regions.

During the latter part of the 1800s and early 1900s, Ernst Haeckel and Henry Osborn suggested Asia was the cradle of hominin evolution. Although Haeckel and Osborn may have been mistaken in their view of the paleoanthropological record, what should be evident from the chapters in this volume is that a diversity of new ideas, innovative research approaches, and data related to deciphering the nature of the Asian paleoanthropological record is slowly contributing to our knowledge of how the Asian landmass was populated during the Quaternary (see also Norton and Jin 2009, 2010).

Without a doubt the paleoanthropological record of Asia is the fastest changing part of our field. Two decades ago the earliest movement off the African continent was assumed to be no older than 1.4 Ma, and it was largely accepted that Early Pleistocene hominins were restricted to tropical environments. These assumptions have been proven false. Fieldwork on the Asian continent consistently continues to redraw the picture of human evolution with each new discovery. What the immediate future (next few decades) holds for further research and discoveries in Asia will help to clarify the position of the Asian record when compared to the better-known African and European records. There is no doubt that a number of significant hominin fossils and Paleolithic artifacts will be found by the research teams dauntless enough to venture into parts of Asia largely unknown to the broader scientific community….

Acknowledgments We thank Eric Delson for comments on an earlier draft of this manuscript.

References

Anton, S. C. (2007). Climatic influences on the evolution of early Homo? *Folia Primatol, 78*, 365–388.

Anton, S. C., & Swisher, C. C., III. (2004). Early dispersals of Homo from Africa. *Annual Review of Anthropology, 33*, 271–296.

Chauhan, P. R. (2010). Core-and-flake assemblages in India. In C. J. Norton, D. Braun (Eds.), *Asian paleoanthropology: from Africa to China and beyond. Vertebrate paleobiology and paleoanthropology series.* Dordrecht, The Netherlands: Springer (in press).

Dennell, R. W. (2001) From Sangiran to Olduvai, 1937–1960: The quest for "centres" of hominid origins in Asia and Africa. In R. Corbey, W. Roebroeks (Eds.), *Studying human origins: Disciplinary history and epistemology* (pp. 45–66). Amsterdam: University Press.

Dennell, R. (2003). Dispersal and colonization, long and short chronologies: How continuous is the Early Pleistocene record for hominids outside East Africa? *Journal of Human Evolution, 45*, 421–440.

Dennell, R. (2004). Hominid dispersals and Asian biogeography during the Lower and early Middle Pleistocene, c. 2.0–0.5 Mya. *Asian Perspectives, 43*, 205–226.

Dennell, R. (2009). *The Palaeolithic settlement of Asia.* Cambridge: Cambridge University Press.

Dennell, R. W., & Roebroeks, W. (2005). An Asian perspective on early human dispersal from Africa. *Nature, 438*, 1099–1104.

Fort, M. (1996). Late Cenozoic environmental changes and uplift on the northern side of the central Himalaya: a reappraisal from field data. *Palaeogeography, Palaeoclimatology, Palaeoecology, 120*, 123–145.

Frost, S. R., Marcus, L. F., Bookstein, F. L., Reddy, D. P., & Delson, E. (2003). Cranial allometry, phylogeography, and systematics of large-bodied Papionins (Primates: Cercopithecinae) inferred from geometric morphometric analysis of landmark data. *The Anatomical Record Part A, 275*(2), 1048–1072.

Gabunia, L., Anton, S. C., Lordkipanidze, D., Vekua, A., Justus, A., & Swisher, C. C. (2001). Dmanisi and dispersal. *Evolutionary Anthropology, 10*, 158–170.

Glantz, M. M. (2010) The history of hominin occupation of Central Asia in review. In C. J. Norton, D. Braun (Eds.), *Asian paleoanthropology: From Africa to China and beyond. Vertebrate paleobiology and paleoanthropology series.* Dordrecht, The Netherlands: Springer (in press).

Harvati, K. (2001) *The Neanderthal problem: 3-D geometric morphometric models of cranial shape variation within and among species.* Ph.D. Dissertation, City University of New York.

Hyodo, M., Watanabe, N., Sunata, W., Susanto, E. E., & Wahyono, H. (1993). Magnetostratigraphy of hominid fossil bearing formations in Sangiran and Mojokerto, Java. *Anthropological Science, 101*(2), 157–186.

Hyodo, M., Nakaya, H., Urabe, A., Saegusa, H., Xue, S. R., Yin, Y. Y., et al. (2002). Paleomagnetic dates of hominid remains from Yuanmou, China and other Asian sites. *Journal of Human Evolution, 43*, 27–41.

Larick, R., Ciochon, R. L., Zaim, Y., Sudijono, S., Rizal, Y., Aziz, F., et al. (2001). Early Pleistocene 40Ar/39Ar ages for Bapang Formation hominins, Central Jawa, Indonesia. *Proceedings of the National Academy of Sciences of the USA, 98*, 4866–4871.

Lycett, S. J. (2007). Why is there a lack of Mode 3 Levallois technologies in East Asia? A phylogenetic test of the Movius-Schick hypothesis. *Journal of Anthropological Archaeology, 26*, 541–575.

Lycett, S. J., & Norton, C. J. (2010). A demographic model for Palaeolithic technological evolution: the case of East Asia and the Movius Line. *Quaternary International, 211*, 55–65.

McNulty, K. P. (2003). *Geometric morphometric analyses of extant and fossil hominoid craniofacial morphology*. Ph.D. Dissertation, City University of New York.

Movius, H. L. (1948). The Lower Paleolithic cultures of southern and eastern Asia. *Transactions of the American Philosophical Society, 38*, 329–420.

Norton, C. J. (2000). The current state of Korean paleoanthropology. *Journal of Human Evolution, 38*, 803–825.

Norton, C. J., & Bae, K.D. (2009). Erratum to "The Movius Line sensu lato (Norton et al. 2006) further assessed and defined" J. H. Evol. 55 (2008) 1148–1150. *Journal of Human Evolution, 57*, 331–334.

Norton, C. J., & Jin, J. (2009). The evolution of modern humans in East Asia: Behavioral perspectives. *Evolutionary Anthropology, 18*, 247–260.

Norton, C. J., & Jin, J. (2010). Hominin morphological and behavioral variation in eastern Asia and Australasia: Current perspectives. *Quaternary International, 211*, 1–3.

Norton, C. J., Bae, K. D., Harris, J. W. K., & Lee, H. Y. (2006). Middle Pleistocene handaxes from the Korean Peninsula. *Journal of Human Evolution, 51*, 527–536.

Petraglia, M.D., & Shipton, C. (2009). Erratum to "Large cutting tool variation west and east of the Movius Line" J. H. Evol. 55 (2008) 962–966. *Journal of Human Evolution, 57*, 326–330.

Schick, K. D. (1994). The Movius Line reconsidered: Perspectives on the Earlier Paleolithic of Eastern Asia. In R. Corruchini & S. Ciochon (Eds.), *Integrative Paths to the Past: Paleoanthropological Advances in Honor of F* (pp. 569–594). New Jersey: Clark Howell. Prentice Hall.

Swisher, C. C. III, Curtis, G. H., Jacob T., Getty, A. G., Suprijo A., & Widiasmoro (1994). Age of the earliest known hominids in Java, Indonesia. *Science, 263*, 1118–1121.

Wolpoff, M. H., Wu, X. Z., & Thorne, A. G. (1984). Modern Homo sapiens origins: A general theory of hominid evolution involving the fossil evidence from East Asia. In F. H. Smith & F. Spencer (Eds.), *The Origins of Modern Humans* (pp. 411–484). New York: Liss.

Zhu, R. X., An, Z. S., Potts, R., & Hoffman, K. A. (2003). Magnetostratigraphic dating of early humans in China. *Earth-Science Reviews, 61*, 341–359.

Zhu, R. X., Potts, R., Xie, F., Hoffman, K. A., Deng, C. L., Shi, C. D., et al. (2004). New evidence on the earliest human presence at high northern latitudes in northeast Asia. *Nature, 431*, 559–562.

Zhu, R. X., Potts, R., Pan, Y. X., Yao, H. T., Lu, L. Q., Zhao, X., et al. (2008). Early evidence of the genus Homo in East Asia. *Journal of Human Evolution, 55*, 1075–1085.

Chapter 2
The Colonization of "Savannahstan": Issues of Timing(s) and Patterns of Dispersal Across Asia in the Late Pliocene and Early Pleistocene

Robin William Dennell

Abstract This paper examines current weaknesses in the Out of Africa 1 model concerning the earliest hominin dispersals into Asia. It proposes first that the development of grasslands in Late Pliocene East Africa was the final part of a process of eastward expansion of grasslands across Asia that began in the Miocene; and secondly that early *H. erectus* in East Africa was not particularly distinctive relative to its contemporaries. It then reviews assessments that the Dmanisi hominins may have been ancestral to *H. erectus* in both East Africa and East Asia, and argues that the current fossil vertebrate record of Southwest Asia cannot demonstrate that hominins were absent before 1.8 Ma. Some alternatives are explored, of which the most parsimonious is that hominins may have dispersed into Southwest Asia before 2.0 Ma, and perhaps shortly after 2.6 Ma when stone toolmaking became routine. Regarding the direction of dispersal, hominins probably dispersed southwards towards Java, and northwards via Central Asia to North China. Because of the climatic and vegetational changes that affected Asia during the Pliocene and Pleistocene, hominin populations would have expanded and contracted in tune with these fluctuations. Out of Africa 1 was therefore not an isolated, continental-level colonization event shortly after 1.8 Ma, but probably a process of numerous, small-scale latitudinal and longitudinal dispersals that began before 2 Ma.

Keywords Dmanisi • Grasslands • *Homo erectus* • Monsoon • Nihewan • Out of Africa • Sangiran

Introduction

The "Out of Africa 1" model has proved a useful way of using a very small amount of Asian evidence to explain a great deal about human evolution outside Africa in the Late Pliocene and Early Pleistocene. According to this model, the genus *Homo*, and the ability to flake stone tools and butcher large mammals, and the development of other skills needed to survive in the grasslands[1] that were expanding in East Africa during the Late Pliocene all originated in this region. Under the increasingly seasonal and arid conditions of the Late Pliocene and Early Pleistocene, *H. erectus* s.l. (sensu lato) emerged in East Africa as a hominin uniquely adapted to take advantage of these new conditions, and at some point after 2 Ma colonised much of Asia (see Fig. 2.1). Two pieces of information often highlighted in this model are the Nariokotome *H. erectus* s.l. skeleton WT 15000, dated to 1.53 ± 0.03 Ma[2] (Brown and McDougall 1993: 19), and the re-dating by Swisher et al. (1994) of the earliest hominins at Mojokerto[3] and Sangiran in Java to 1.81 ± 0.04 and 1.66 ± 0.04 Ma, respectively. The body size and proportions as well the brain size of the Nariokotome skeleton seemed to provide an explanation of how and why Asia was colonised; as an obligate biped with long limbs and modern body proportions, it was better adapted for life in the grasslands that were expanding in East Africa at that time than its immediate predecessors (Ruff and Walker 1993: 262). Its thermoregulatory efficiency in sweating and cooling would have helped it cope with the hot and dry conditions in the savannah grasslands of

R.W. Dennell (✉)
Department of Archaeology, University of Sheffield,
Sheffield S1 4 ET, UK
e-mail: r.dennell@sheffield.ac.uk

[1] "Grasslands" are used here in a generic sense to denote ones with a short growing season, characterized by strong seasonal contrasts in temperature and precipitation, and a prolonged dry season of ≥3 months. Precipitation can be either in summer (as in monsoonal grasslands) or in winter and spring, as in Central and Southwest Asia. Edaphic (i.e. wetland grasslands) are excluded. See Harris (1980) for a selection of papers on past and present usage.

[2] This estimate is not affected by the more recent dates now available for the tuffs over and under this specimen. The revised ages (McDougall and Brown 2006) for the KBS, Chari and Lower Nariokotome Tuffs are indistinguishable from the previous ones, although the revised estimates for the age of the Morutut Tuff are slightly younger (1.607 ± 0.019 Ma instead of 1.64 ± 0.03 Ma). As the age of the Nariokotome specimen is derived from estimates of sedimentation rates between dated tuffs, the original estimate of 1.53 ± 0.05 Ma is not significantly affected.

[3] The Mojokerto hominin now appears to be ≤1.49 Ma (Morwood et al. 2003).

C.J. Norton and D.R. Braun (eds.), *Asian Paleoanthropology: From Africa to China and Beyond*, Vertebrate Paleobiology and Paleoanthropology, DOI 10.1007/978-90-481-9094-2_2, © Springer Science+Business Media B.V. 2010

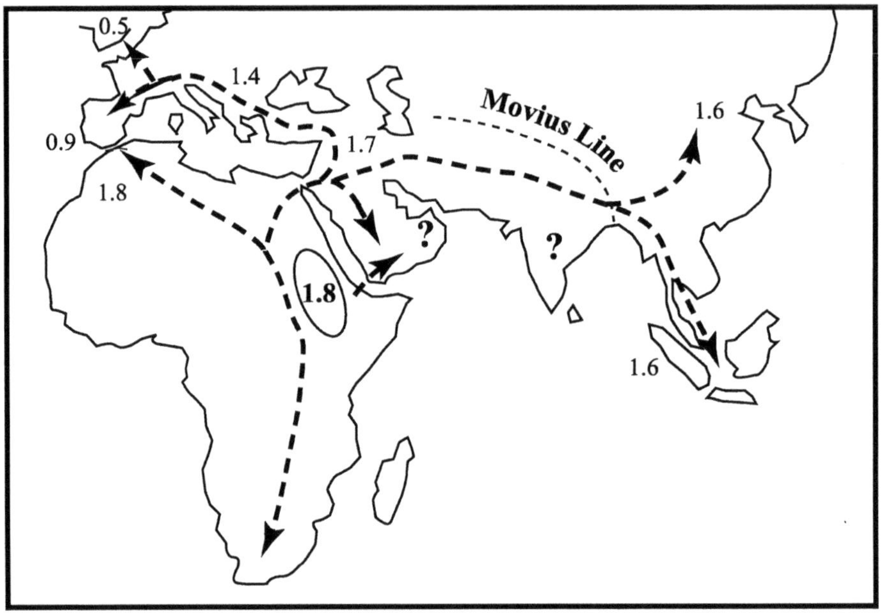

Fig. 2.1 The Out of Africa 1 model is based on the suppositions that *Homo erectus* i) originated in East Africa; ii) was the only hominin to leave Africa prior to *H. sapiens*; and iii) left (on current and very sparse evidence) ca. 1.8 Ma and rapidly dispersed as far east as Java and North China by 1.6 Ma, and somewhat later into Europe. The "Movius Line" indicates the basic Early Palaeolithic division in Asia between Southwest and South Asia, where Acheulean bifaces are common, and East Asia, where they are very scarce (see e.g. Movius 1948; Schick 1994). Numbers refer to the age in millions of years (Ma) for the principal points of observation.

This map attempts to indicate the probable level of precipitation across southern Asia during moist (i.e. interglacial) parts of the Early Pleistocene. It is impressionistic (i.e. there have been no computer simulations of the kind used for the PRISM 2 reconstruction of Late Pliocene climate 3.2 Ma), and is based on comparing the modern climate of Asia with that inferred from extensive reading of the Asian paleoclimatic evidence for the Late Pliocene, and the Early and Middle Pleistocene. (These sources are discussed in Dennell, in press). Arrows show the main rain-bearing winds: westerly ones in winter and spring from the East Mediterranean and Black Sea, and southwest and southeast ones in summer from the Indian and East Asian monsoon. The asterisks denote the Early Pleistocene lake systems in the Nahal Zihor (Israel) and Ain Nefud (Saudi Arabia), which are now desert.

The numbers in squares highlight current areas of uncertainty: 1) the timing and extent of the Akchagylian Transgression that linked the Black and Caspian Seas; 2) changes to the size of the Aral Sea and its possible connections to the Caspian; 3) the Early Pleistocene elevation of the Tien Shan; and 4) the height of the northern Tibetan Plateau.

In the Late Pliocene and Early Pleistocene, there were no ice-sheets over northern Europe, and therefore no northerly winds blowing southwards and blocking westerly winds from the East Mediterranean bringing rainfall eastwards across Southwest and much of Central Asia. Apart from small areas of Arabia and North China, few areas of Asia are likely to have received <100 mm. Loess was deposited in cold periods in both Central Asia and North China, but on a much smaller scale than during the Middle Pleistocene. The summer monsoon penetrated further inland than during much of the Middle Pleistocene, and thus the northern limit of the semi-arid zone (i.e. <600 mm) lay further north.

On this reconstruction, the desert barrier between the Sahara and Arabia was greatly reduced, and faunal movements (including hominins) would have been possible between Africa and Asia (and vice versa). The primary constraints to movement across most of continental Asia would have been topographic (notably the mountain ranges of the Himalayas, Karakorum, northern Tibet, and the Tien Shan etc.) rather than climatic.

The isohyets are derived from present-day ones, and adjusted so that effective precipitation in Southwest Asia is sufficient for the formation of substantial lakes in the deserts of the Negev and Saudi Arabia. They were also adjusted so that there was sufficient moisture in Central Asia, Northwest India and North China to prevent the spread of deserts, which were largely a Middle Pleistocene development. The base map for modern rainfall was derived from the Times World Atlas

(Source: the author)

both Africa and Asia (Wheeler 1991, 1992), as might its capacity for endurance running (Bramble and Lieberman 2004). Additionally, its large brain would have made it a more skillful competitor with other predators, even if its energetically expensive brain would also have required a larger proportion of meat in its diet (Aiello and Wheeler 1995). Their technological proficiency also enabled them to access and process a wide range of animal and plant resources. A further suggestion has been that *H. erectus* s.l. was a "weed taxon" that could flourish in the type of environments that were frequently disrupted by climatic or volcanic events, and was thus more resilient than its competitors (Cachel and Harris 1998). The unexpectedly early dates obtained by Swisher's team for the earliest Javan hominins implied that *H. erectus* s.l. had already left Africa before the development of the Acheulean handaxe and cleaver technologies c. 1.5 Ma, and did not need these items during its dispersal across southern Asia. The new dates from Java neatly explain why hand axes and cleavers are virtually absent east of the "Movius Line" (Fig. 2.1).

The addition of European evidence to the Out of Africa 1 model indicated that hominins had begun to enter southern Europe by 1.3–1.4 Ma (Carbonell and Rodríguez 2006) and northern Europe c. 0.6 Ma (Parfitt et al. 2005) as part of the same process of colonisation from a source that was directly or indirectly African. As such, the Out of Africa I model gave a satisfying coherence to Old World prehistory before 1.0 Ma in both Asia and Europe, and confirmed the primacy of Africa in human evolution (Dennell and Roebroeks 2005).

Problems with the "Out of Africa 1" Model

Although the "Out of Africa 1" model is widely accepted, there are several reasons why this "grand narrative" should be questioned (see e.g. Dennell 1998, 2004, 2009: 186–202; Dennell and Roebroeks 2005). One is why grasslands in Asia remained unoccupied by hominins before 2 Ma. Other reasons concern the origin of *H. erectus* s.l., whether it was uniquely adapted to living in grasslands in and especially outside Africa, and the Javan affinities of Olduvai hominin (OH) 9. An appropriate starting point is to consider the environmental changes associated with early *Homo* in East Africa in the Late Pliocene.

The Emergence and Consequences of Grasslands in Late Pliocene East Africa

Much attention has been paid in recent years to the development of grassland environments in East Africa in the Late Pliocene, and their possible consequences upon the evolution of hominins and other mammals (see, e.g. Vrba 1995, 1996; Behrensmeyer et al. 1997; Bobe and Behrensmeyer 2004; deMenocal 2004; Fernández and Vrba 2006). Vrba (1995, 1996) proposed that the developments relating to the emergence of *Paranthropus*, our own genus *Homo*, the making of stone tools, and the evidence for the consumption of meat and marrow resulted from increasingly more arid and seasonally distinct conditions after 2.5 Ma. She suggested that these climatic changes would have favored hominins that could access harder and coarser foods, and widen their resource base by acquiring animal protein. Hominins would also have faced more complex, difficult and unstable situations; complex because some resources (e.g. plants) would have become more seasonally distinct and resource scheduling thus became more complicated, difficult because as the climate became drier, resources would have been more dispersed, and thus harder to locate and obtain; and unstable because of numerous minor oscillations in temperature and precipitation. Grasslands would also have expanded, along with a range of dangerous predators, and those too would have made life more difficult for hominins. Those that could develop new strategies, such as using stone-tools, processing more information by acquiring larger brains, eating larger amounts of meat, or developing larger cheek-teeth that could process harder foods would also have been advantaged. Thus the reason why both *Paranthropus* and our own genus *Homo*, as well as stone tool-making and carnivory, all originated around or shortly after 2.5 Ma was primarily, in Vrba's view, the change in climate brought about by the onset of northern hemisphere glaciation. Furthermore, she argued that the hominin lineage was not the only one to experience rapid change at this time. Bovids also show an increase in hypsodonty (i.e. high-crowned teeth that are better adapted to grazing on tough, silicaceous grasses), and even rodents developed tougher teeth, and longer hind-legs that enabled those that hop to forage over greater distances (Vrba 1995, 1996).

Vrba's (1995, 1996) model was a powerful way of explaining a wide range of important changes across several lineages in terms of different responses to Late Pliocene climatic change in East Africa, and still retains much of its original validity. Conditions did become more arid in parts of East Africa after 2.5 Ma; for example, pollen data indicate that the vegetation of the Turkana Basin was closed and humid at 4 Ma and 3.4 Ma, but showed drier conditions at c. 2.35 Ma and maximum aridity at c. 1.8 Ma (Bonnefille 1995). By the Late Pliocene, semi-evergreen rain forest had been replaced by deciduous woodland, with some savannah grassland (Fernández and Vrba 2006), although the latter did not become dominant until well into the Pleistocene (Cerling 1992). At Olduvai, there was a trend towards greater aridity that peaked at 1.77 Ma, and some developments, including the use of stone tools, also seem to be linked to drier conditions (Bobe and Behrensmeyer 2004).

The main objections to Vrba's model have been either that the main evidence for grasslands in East Africa is not until c 1.8 Ma, rather than at 2.5 Ma, and/or that the faunal record, if rigorously scrutinized, does not support clear correlations with a change in climate. Cerling (1992: 244), for example, suggests from soil carbonate analyses that before 1.7 Ma, the main fossil hominin localities may have been "somewhat open [but] nothing like the grasslands or wooded grasslands of East Africa today". Various authors (e.g. Kappelman et al. 1997; Reed 1997; Spencer 1997) have shown from large mammal remains that both Australopithecines and *Homo habilis* lived in woodland environments, and Fernández-Jalvo et al. (1998) have shown from micro-mammalian evidence that Olduvai in Bed I times, c. 1.8 Ma, was more densely wooded than today.[4] Overall, there is little indication that East African hominins were habitually living in grasslands before 1.8 Ma. Nevertheless, they were living in environments that were more arid and open than previous ones, particularly between 1.6 and 1.8 Ma (deMenocal 2004: 18).

Other criticisms have focused on the timing of climatic changes, and the first and last appearance dates of key taxa. White (1995), for example, argued that there are no clear correlations between global climatic change and the evolution of pigs and hominins: most speciations and extinctions occurred between 1.6 and 2.0 Ma, during which time there were no clear global climatic changes. Kimbel (1995) also concluded that whilst hominin evolution may have been forced by climatic change, the fossil data available from Africa are inadequate to confirm or refute this suggestion. In a larger scale study, Behrensmeyer et al. (1997: 1593) argue that "late Pliocene evolution in East Africa was affected by the cumulative ecological consequences of cooler, drier, and more variable climatic conditions rather than by a sudden change toward more open habitats". As East Africa has the best-studied African sequence, this finding "weakens the case for rapid climatic forcing of continent-scale ecological change and faunal turnover" (Behrensmeyer et al. 1997: 1593).

Despite criticism of Vrba's argument, and counter-proposals by Behrensmeyer et al. (1997) and others, there is a consensual view that the development of grasslands and the emergence of *H. erectus* s.l. are closely linked, so to that extent, climatic change resulting in greater aridity and more open landscapes remains as a driving force behind an important phase in human evolution. Yet as DeMenocal (2004) points out, the increased climatic variability at this time may also have been an important influence on faunal evolution (including that of hominins).

In contrast with the amount of attention given to the emergence of grasslands in East African, and their likely importance to the evolution of our genus, almost no attention has been paid to similar developments in Asia. Four features of these developments are potentially significant. One is the great antiquity of grasslands in Asia compared to Africa; the second is that the emergence of grasslands in East Africa in the Late Pliocene was the final part of a westward expansion of grasslands across Asia that began in the Early Miocene; the third is the role that the rate and pattern of uplift of the Tibetan Plateau may have played in driving these environmental and climatic changes, rather than the onset of northern hemisphere glaciation; and the last is that, as in East Africa, these Asian grasslands experienced considerable climatic and local variability.

[4] This finding is consistent with anatomical evidence that *Homo habilis* retained arboreal capabilities as late as 1.8 Ma (Susman and Stern 1982).

The Development of "Savannahstan" – The Asian Grasslands

As noted above, the seasonal grasslands that were exploited by hominins (including early *H. erectus*) began to develop in East Africa in the Late Pliocene. In contrast, the grasslands of Asia are extremely ancient, and linked to the development of the Asian monsoonal system, with its pronounced seasonal contrasts between cool/cold, dry winters, and warm/hot moist summers driven by seasonally-reversed winds.[5] Because these Late Pliocene and Early Pleistocene Asian grasslands were so extensive, Dennell and Roebroeks (2005) suggested the term "Savannahstan" as a way of highlighting their significance (see also Kohn 2006). There are two likely causes for their formation. The first and earlier one is the retreat of the Paratethys Sea during the Miocene, which changed the climate of Central Asia and Eastern Europe from oceanic to continental. It thereby became drier, increased the amplitude of the seasonal cycle, and thus enhanced monsoon precipitation. Most of the increased monsoonal rains would have fallen on the southern side of the Himalayas and middle part of India, whereas central and northern Tibet became drier. Another consequence of the retreat of the Paratethys Sea was that Arabia and Northeast Africa also became more arid (Ramstein et al. 1997). The second and increasingly important one was the uplift of the Tibetan Plateau, particularly its northern and eastern parts.

The Tibetan Plateau and the Grasslands of North China

The Tibetan Plateau has been described as "the engine that drives the modern monsoon" (Dettman et al. 2001: 31). The climatic consequences of its uplift are well summarized by Guo et al. (2002: 161): "Uplift strengthens the summer monsoon and brings wetter climates to India and Southeast Asia, but this moisture cannot reach the Asian interior because uplifted Himalayan topography blocks flow from the south. As a result, central Asia becomes drier as uplift proceeds. Uplift also produces drier climates in central Asia in the winter season because dry winter monsoon winds blow out of the Asian interior. The combination of summer and winter drying produces year-round aridity and forms deserts. In addition, uplift strengthens the flow of winter monsoon winds from the northwest".Estimates of when the Tibetan

[5] The term "monsoon" is derived from the Arabic "mausim", meaning "a wind in South West Asia and the Indian Ocean blowing from the south-west from April to October, and from the north-east the rest of the year" (Baker 1949: 940).

Plateau was uplifted have ranged from the Late Eocene to the Upper Pleistocene. Recently, Rowley and Currie (2006) propose that the elevation of the central part of Tibet has been >4,000 m for the last 35 Ma. This is considerably earlier than Fort's (1996) assessment that the uplift of Tibet and the Himalayas started c. 25 Ma, and accelerated after 20 Ma, with a pulse of rapid uplift at c. 8 Ma. Other studies have concluded that the central and southern parts of the Plateau were at modern altitudes c. 15, 14 and 13.5 Ma, respectively (Coleman and Hodges 1995; Blisniuk et al. 2001; Spicer et al. 2003), as were parts of Nepal by c. 8 Ma (Garzione et al. 2000). Northern Tibet was uplifted much later. Zheng et al. (2000) concluded from studying c. 4.5 km of tilted fluvial and conglomeritic deposits on the southern edge of the Tarim basin, Northwest Tibet, that coarse debris flows (indicating the onset of uplift) replaced fluvial deposition c. 4.5 Ma. Further uplift of the entire section followed after 1.6 Ma. Li et al. (1997) also showed by dating the very thick Jishi Conglomerate adjacent to the Linxia Basin, North China, that rapid uplift in northern Tibet occurred c. 3.6 Ma. Recently, Zhou et al. (2006) have argued that a third of the crustal shortening and uplift of the Qaidam Basin of Northwest China occurred in only the last 2.8 Ma. As a further example of the scale on which geological processes have operated in this region in recent times, the Tian Shan Range, which rises to over 7,400 m, may be largely a product of the last 10 Ma (Abdrakhmatov et al. 1996), or even the last 5 Ma (Sun et al. 1999). Extreme aridity along the northern edge of Tibet would have resulted from such uplift as moisture would have been blocked from both the Indian and Pacific Oceans to the south and east respectively (Liu et al. 1996). The consequences of this uplift are detectable in oceanic records from both the North Pacific and South China Sea (see e.g. Rea et al. 1998; Tian et al. 2002), and may even have been felt as far away as the western Sahara (Ruddiman et al. 1989; see below).

A recent detailed review of paleobotanical data (Sun and Wang 2005) showed that the present-day contrasts in China between an arid north and northwest region, and a humid south and northeast one were established in the Early Miocene, and thus the East Asian monsoon must have a comparable antiquity. This has recently been confirmed by analysis of the pollen record of a 2,880 m sequence from north-central China that covers the last 20 Ma (Jiang and Ding 2008). A long sequence of brownish loesses and reddish soils with a basal age of 22 Ma from the western part of the Chinese Loess Plateau indicates that parts of Central Asia – the likely source of this dust – were already desert at this time (Guo et al. 2002). These researchers (2002: 161) suggest that "the southern margin of the Tibetan Plateau was sufficiently elevated by 22 Ma to cause year-round drying and desert formation in the Asian interior and to produce northwest winds strong enough to carry aeolian particles southeast into the Loess plateau". According to Guo et al. (2002), the winter monsoon was only moderately strong throughout the Miocene, and alternations of loess and paleosol formation were governed by cyclical variations in the earth's orbit. Dust accumulation rates increased in the Late Miocene c. 7–8 Ma and again after 3.6 Ma, when the climate became distinctly cooler (see, e.g. Ding et al. 1999). An (2000) noted that the accumulation rate of red clay increased from 2.8 to 6.6 cm/ka between 3.6 and 2.4 Ma, and the influx of dust into the North Pacific also increased at this time (Rea et al. 1998). Both the summer and winter monsoon appear to have intensified, as indicated by magnetic susceptibility, grain size and other indicators (An et al. 2001). This evidence is at variance with that from deep-sea cores, which indicate a shift to more glacial conditions at this time. Climate-model simulations of glacial climate suggest that the summer monsoon should have weakened under these conditions, rather than have strengthened, as indicated in the Red Clay records. They thus suggest the simultaneous strengthening of both the summer and winter monsoons was caused by additional uplift of the northern and eastern parts of the Tibetan Plateau. A similar conclusion was reached by Qiang et al. (2001), who note a substantial increase in sedimentation rates between 3.5 and 3.1 Ma, which suggests that the East Asian monsoon strengthened at this time, with a further intensification after 2.6 Ma, when loess instead of red clay was deposited (see below). They suggest that the strengthening of the East Asian winter monsoon c. 3.5 Ma coincided with an episode of rapid uplift of the Tibetan Plateau. Some scientists believe that these changes could have been the trigger for the onset of northern hemisphere glaciation. An et al. (2001), for example, suggest that strong winter winds and increased amounts of atmospheric dust (combined with Central Asian aridity) could have helped cool global climate and thereby influenced the development of northern hemisphere glaciation.

In the Late Pliocene c. 2.5 Ma, loess replaced red clay as the main aeolian component, and thereafter the loess-paleosol sequence of the Chinese Loess Plateau records 166 episodes of moist and arid climate (see e.g. Liu and Ding 1998). Liu et al. (1999) made a detailed comparison of the 156 m-thick loess section at Baoji with the record for global ice volume indicated in deep-sea core DSDP (Deep-sea Drilling Program) 607 (North Atlantic) and showed that the strength of the winter monsoon matched the record for global ice for the last 1.67 Ma. From 0.8 Ma to 1.67 Ma, the dominant periodicity was one of 41 ka (resulting from the obliquity of the earth's axis). However, the 41 ka cycle recorded in deep-sea cores such as DSDP 607 was not found in the loess profiles before 1.67 Ma, and they suggest that it was only after 1.7 Ma that northern hemisphere ice sheets may have reached the critical size necessary to influence the East Asian winter monsoon system. Once again, the height of the Tibetan Plateau is suggested as a factor affecting the East Asian winter monsoon.

Palynological studies indicate that the vegetation of North China in the Late Pliocene was predominantly steppe and grassland. Wang et al. (2006) have shown from a study of the Xifeng red clay profile on the Chinese Loess Plateau that the region was dominated by grassland between 6.2 and 2.4 Ma, and particularly after c. 3.7 Ma. Han et al. (1997) analyzed samples from the Guanzhong Basin of North China south of the Loess Plateau and showed that steppe vegetation predominated between 3.0 and 2.7 Ma, with non-arboreal taxa typically >90% of the total pollen count. *Artemisia* was the dominant plant genus (typically >50%), followed by Chenopods and Gramineae. *Pinus* was the commonest type of arboreal pollen (up to 4%) but probably derived from a long distance. The climate appears to have become more severe between 2.7 and 2.5 Ma; the total pollen count decreases sharply, and the plants are typically steppic. A thick loess layer in this part of the sequence is seen as consistent with the onset of loess deposition in the Loess Plateau c. 2.6 Ma (see below). Between 2.5 and 1.9 Ma, the climate appears to have alternated between cold and dry periods associated with an arid or semi-arid steppe flora, and warmer and wetter episodes, when forest grasslands developed. In these warmer phases, *Artemisia* was still the commonest plant but arboreal pollen reached frequencies of up to 46%. *Pinus* was the main constituent, followed by *Quercus, Juglans, Ulmus, Rhus,* and *Betula*. Analysis of stable carbon isotopes were consistent with the pollen data in showing a C_3 type of vegetation, which is entirely unsurprising as *Artemisia*, by far the dominant plant type, is a C_3 type.

The South Asian Grasslands and the Indian Monsoon

The Indian monsoonal system is more recent than the East Asian, and various analyses from Pakistan, India, Nepal and the Bay of Bengal indicate that it dates from the Late Miocene. Quade et al. (1989) sampled numerous paleosols between 16 Ma to 0.4 Ma from the Siwalik series in northern Pakistan, which lies on the western edge of the modern Indian monsoon. They showed that there was a pure or nearly pure C_3 biomass before 7.4 Ma. Between 7.4 and 7.0 Ma, there was an increase in C_4 grasses, indicating a mosaic of C_4 grassland and C_3 forest. From 5.0 Ma to 0.4 Ma ago, C_4 grasses formed 90% of the plant cover. They therefore suggested that the expansion of grassland after 7.4 Ma indicated an intensified Indian monsoon, with greater seasonal contrasts between dry winters and wet summers, and a warm growing season. This study was particularly persuasive because of the amount of supporting evidence. After 7.0 Ma, much less leaching is observed in soil horizons, and paleosols with organic A horizons become commoner. There were also major faunal changes: before 7.4 Ma ago, the larger herbivores were mostly browsers, whereas afterwards, they were mainly grazers. The rodent faunas also changed from forest to grassland types. Significantly, in terms of relevance to hominoid evolution, *Sivapithecus*, a hominoid related to forest-dwelling orangutans, also became extinct at this time (Barry et al. 2002; Nelson 2007).

Broadly similar conclusions to those reached by Quade et al. (1989) were also reached in a study of the $\partial^{18}O$ content of bivalves from Nepal and fossil teeth from Pakistan (Dettman et al. 2001). High ^{18}O values in the mollusks were interpreted as showing high rates of evaporation; the fossil teeth were sampled for annual variation encompassing both dry and wet seasons to assess the degree of seasonality. They concluded that the Indian monsoon was present by c. 10.7 Ma, and thus the Tibetan Plateau must have been sufficiently high by that time to drive the monsoon. However, the monsoon apparently intensified c. 7–8 Ma, roughly the time when C_4 plants were established in the area studied by Quade et al. (1989). Harrison et al. (1993) also identified a change from C_3 to C_4 vegetation at c. 7 Ma (as in Pakistan), and linked this to an intensification of the Indian monsoon resulting from Tibetan uplift. A similar vegetational shift c. 7 Ma was recorded in sediments in the Bay of Bengal (Derry and France-Lanord 1997).

Central and Southwest Asia

Central and Southwest Asia lie outside the area of the summer monsoon, and receive most of their precipitation in winter and spring, mainly from westerly winds driving inland from the Mediterranean and Black Sea. There are pronounced and long-standing vegetational differences between areas north or south of latitudes 35–40° N. As shown above, C_4 grasslands prevailed in South Asia after 7.0 Ma, and the same is likely of the Arabian Peninsula (see below). However, isotopic analysis of bovid teeth from the late Miocene site of Molyan (c. 6.6–7.5 Ma [Sen 1998]) in Afghanistan, which lies north of the Siwaliks and outside the monsoon region, shows an overwhelmingly C_3 vegetation at this time (Zazzo et al. 2002). This conclusion is similar to that reached in a study of dental microwear of Late Miocene herbivores in Afghanistan (Merceron et al. 2004), who concluded that C_3 grasses and evergreen bushes were probably the main types of plants because of the higher altitude and lower temperatures than in the area of Pakistan sampled by Quade's group. C_3 vegetation also prevailed (as it does today) in western Turkey and Greece during the Late Miocene and Early Pliocene (Quade et al. 1994; Bocherens and Sen 1998).

The most important climatic and paleoenvironmental data from Central Asia comes from the loess sections of Tajikistan, where loess deposition began c. 2.5 Ma (Dodonov and Baiguzina 1995; Dodonov 2002). In Tajikistan, loess accumulated to a thickness of 180–200 m, comparable to that seen in the Chinese Loess Plateau, and was first deposited c. 2.5 Ma. (Unlike in China, there is no equivalent of the Red Clays and there does not appear to have been any aeolian deposition prior to the loess). As in North China, numerous paleosols developed during interglacial periods,[6] when conditions were warmer and moister. At least 37 are recognized, of which 27 date from the Late Pliocene and Early Pleistocene (Dodonov 2002:183). Before 800 ka, each loess-paleosol cycle lasted c. 34–50 ka, with an average accumulation rate of 0.5 m/1,000 years, compared to c. 1.6 m/1,000 years in the Middle Pleistocene.

Faunal evidence from the Late Pliocene, c. 2.2–2.4 Ma from sites such as Kuruksay indicates an open grassland or steppe environment. Large carnivores include a large canid *Canis etruscus, Ursus* cf. *etruscus, Pliocrocuta, Lynx,* and the large cats *Acinonyx, Megantereon* and *Homotherium*, and herbivores such as *Archidiskodon, Dicerorhinus, Equus stenonis, Paracamelus, Sivatherium*, and a medium-sized bovid, *Damalops palaeindicus* (Sotnikova et al. 1997). There was also a Eurasian type of macacque, *Paradolichopithecus suskhini* (Nishimura et al. 2007) that was initially identified as a baboon, *Papio suschkini* (Maschenko 1994). Sotnikova et al. (1997) note that the fauna is overall similar to that of the Middle Villafranchian of Europe, but with some Oriental types such as camels, and the cervids and bovids *Sinomegaceros, Elaphurus, Axis* and *Antilospira*. Three taxa – *Damalops, Equus* and *Sivatherium* – are also recorded in both Pakistan and Africa at this time (Dennell et al. 2006). Broadly similar assemblages of the same age-range were recovered from the loess sequences at Karamaidan, Obigarm and Tutak. Animals represented at Karamaidan included *Equus* (which replaced *Hipparion* c. 2.5 Ma), *Gazella* and *Paracamelus* as well as *Dicerorhinus* and *Ursus* cf. *etruscus*. Similar evidence was found in the loess profile of Zil'fi, in a context dated to the Olduvai Subchron (1.77–1.95 Ma), and in the middle level of Koplay, where the main large mammals were *Canis etruscus, Pachycrocuta brevirostris, Homotherium, Equus stenonis, Dicerorhinus* and *Leptobos*, a large bovid.

The Late Pliocene and Early Pleistocene climatic and environmental record of Southwest Asia remains poorly documented, but it is likely to have been less arid than today during interglacial periods at least. In the late Pliocene, grasslands replaced woodland in Azerbaijan (Dowsett et al. 1994), and the Arabian Peninsula (now largely desert) experienced a long humid period from c. 3.5 to 1.2 Ma (Al-Sayari and Zötl 1978: 310), and its interior was drained by several large and now extinct river systems (see Petraglia 2003).

Early Pleistocene Lakes in Asia

Lakeside environments appear to have been major foci of activity for early hominins in both East Africa and Asia. 'Ubeidiya (Bar-Yosef and Goren-Inbar 1993) and Dmanisi (Gabunia et al. 2000a) are the two best known examples from Southwest Asia. Others are Dursunlu, Turkey (Güleç et al. 1999) and Kashafrud, Northeast Iran (Ariai and Thibault 1975/1977), as well as the Nihewan Basin, North China (e.g. Gao et al. 2005). These provided perennial water, flakable stone from in-flowing streams, shade and shelter in the woods and scrub along the lake margins, and attracted animals that would have grazed in the grasslands away from the lake perimeter. Large springs were also important for many of the same reasons (Por 2004). The major drawback to such locales was that they also attracted large predators.

Two recent examples of Early Pleistocene lakes come from the modern deserts of northern Saudi Arabia and the Negev, Israel. In the former, Thomas et al. (1998) investigated several that were probably not synchronous in the An Nefud desert of northern Saudi Arabia and which lay in good quality grassland in an area that is now desert with <50 mm rainfall. Three vertebrate fossil localities are reported. Fossils from locality 1 were collected from the surface of the lacustrine deposits from which they had been recently eroded. Those from localities 2 and 3 came from a thin siltstone between the basal aeolian sand and a 10 cm-thick layer of lacustrine carbonates.

The faunal assemblages share some features with Early Pleistocene ones from East Africa and 'Ubeidiya. They include a maxilla of a large fish, estimated to have been >1 m long, so the lake was clearly quite large, and a carapace fragment of *Geochelone sulcata*, the largest African land tortoise, and now found only in the Sahel. Carnivores are represented by specimens of *Crocuta crocuta*, *Panthera gombaszoegensis* (found also at 'Ubeidiya), and the fox, *Vulpes vulpes*. Herbivores include ones similar to those from Olduvai Upper Bed II, such as *Elephas* (possibly *E. recki*), *Pelorovis oldowayensis, and Equus*. The pygmy hippopotamus *Hexaprotodon* is also represented; this is found today only in West Africa, but in the Early Pleistocene it was also present in North Pakistan and Java; its presence here confirms that the lake was large, as they prefer standing water 2–5 m deep (Jablonski 2004). A camel, oryx and some kind of alcelaphine are also present. $\partial^{13}C$ measurements taken from tooth fragments of *Pelorovis, Elephas* and the alcelaphine indicate a C_4 grassland environment.

Ginat et al. (2003) have recently published a study of an Early Pleistocene lake system at Nahal (or Wadi) Zihor, in the Negev Desert of Israel. As at An Nefud, the climate today is

[6] A long-running debate amongst Russian scientists over whether loess formed in glacial or interglacial periods can be regarded as conclusively settled in favor of the former viewpoint: loess is glacial (see Dodonov and Baiguzina 1995: 708).

very arid, with only c. 50 mm of rainfall. Large parts of this area were covered in the Pliocene by fluvial-lacustrine deposits known as the Arava Formation. These were laid down by a large river system that drained the southern Negev, eastern Sinai and part of the eastern portions of the Dead Sea Rift, and flowed north in to the Dead Sea. As a consequence of tectonic movements, the deposits from this river system near the Dead Sea and Gulf of Aqaba are now buried below Late Pleistocene alluvial fans, but are still exposed in the central parts, including the Nahal Zihor (Ginat et al. 1998). Here, the Arava Formation is overlain unconformably by Early Pleistocene lake sediments known as the Zeheiha Formation, which is up to 15 m thick and covers c. 18 km^2. Pollen from this formation is ascribed to the Early Pleistocene (1.8–1.5 Ma), and reflects a wet Mediterranean climate (Horowitz 2001).

The ancient lake system at Nahal Zihor had three sedimentary cycles. Each lake is thought to have been 3–5 m deep; i.e. deep enough to maintain a viable fish population, and also extensive enough to prevent coarse stones being transported into the lake center. Some pedogenic features are present which indicate that the lake periodically dried out. The paleosols that developed when the lake was drying out indicate a semi-arid climate, with an annual rainfall of c. 150–200 mm. The incision of the present Wadi Zihor channel into the lake sediments indicates the onset of very arid conditions. Ginat et al. (2003) estimate that each lake at Nahal Zihor probably lasted between 45,000 and 150,000 years under a semi-arid climate, and suggest that the lake was maintained by rainfall as well as some ground water seepage. Further dating work is needed to establish how the three lake cycles correspond to the isotope stages in the marine records. Three concentrations of stone artefacts were found that might be linked to the ancient shoreline, and include bifaces and picks that are described as being similar to those from 'Ubeidiya and Latamne.

Lakes such as those mentioned above, as well as springs and perennial streams, would have provided the means by which Late Pliocene and Early Pleistocene hominins could have dispersed across much of continental Asia where summer drought is a major limiting factor on animal populations.

The Longevity and Importance of the Asian Grasslands

As noted above, the grasslands of Asia are immensely ancient compared to those in East Africa, and date from 22 Ma in North China and at least 7 Ma in South Asia. The best base-line for assessing their likely extent during the Late Pliocene is the PRISM2 reconstruction of northern hemisphere climate and vegetation in the late Pliocene, between 3.29 and 2.97 Ma (Dowsett et al. 1994, 1999; see Fig. 2.2).

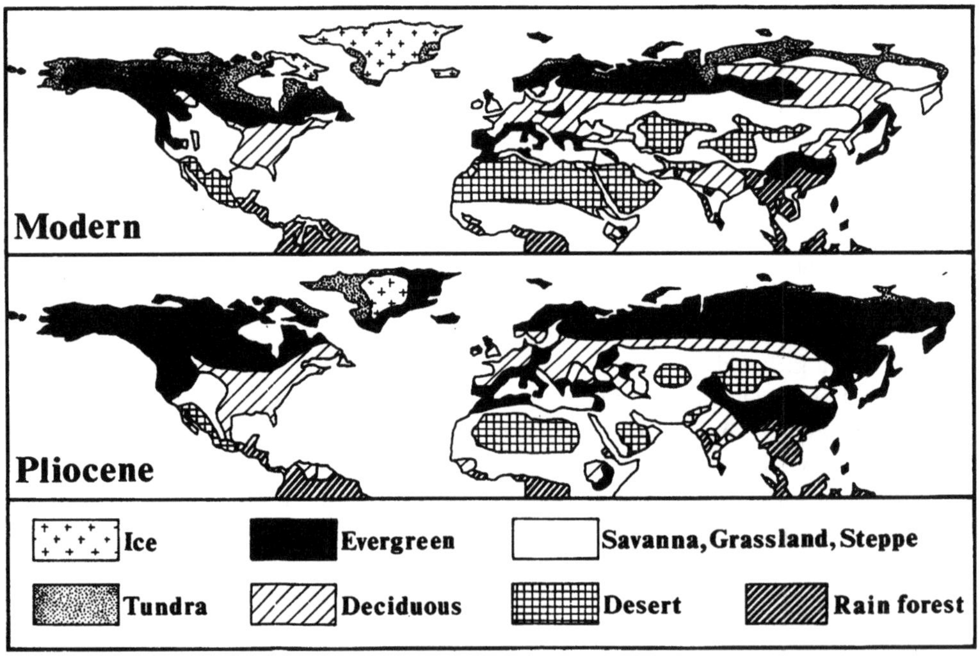

Fig. 2.2 The most striking contrasts between the Pliocene ca. 3 Ma and today are in the extent of grasslands and deserts. As shown, in the Late Pliocene, savannah or steppe grasslands extended from northern China to West Africa, and the present-day desert barrier between the Sahara and Arabia did not exist. Conditions for hominin dispersals out of (and perhaps into) Africa were thus more favourable then than in recent times.

Reproduced from Global and Planetary Change 9 Dowsett et al. "Joint investigations of the Middle Pliocene climate I: PRISM palaeoenvironmental reconstructions" 169–195, 1994, with permission from Elsevier. (Source: Dowsett et al. 1994, Figure 11)

At this time, the world was perhaps 3.5°C warmer than today, but CO_2 levels were much the same as now (Raymo et al. 1996). The primary data for this reconstruction are from deep-sea cores, with terrestrial data incorporated when sufficiently well-dated and environmentally-sensitive. The most interesting feature of this reconstruction is that grasslands c. 3 Ma were probably continuous from West Africa right across to northern China; in other words, the present-day desert barriers of the Sahara and southwest Asia did not then exist.

These grasslands probably persisted throughout the Early Pleistocene (Dennell 1998, 2004), particularly in Southwest, South and Central Asia as well as North China. Climatic conditions at this time were also not as severe as during the Middle Pleistocene, when cold dry periods (= glaciations) were much longer, and contrasts with warm, moist periods (= interglacials) much greater. Figure 2.3 shows estimated rainfall in the Early Pleistocene; the semi-arid zone (400–600 mm/p.a.) would have been largely grasslands, and there was no desert barrier between Africa and Southwest Asia. As noted above, lakes, springs and perennial streams would have been primary foci of hominin activity, especially in the summer dry season.

Sangiran: Were the Earliest Hominins Inhabiting a Swampy Estuary?

At first sight, the Javan evidence appears anomalous because the earliest indications of hominins at Sangiran is specimen 1996.04, below a tuff dated to 1.51 ± 0.08 Ma (Larick et al. 2001) at the top of the Sangiran (= Pucangan) Formation, in what would then have been a swampy estuary, and definitely not the open woodland or grassland habitat that was preferred elsewhere. Two points can be made. The first is a taphonomic one, namely that there is no reason to suppose that the earliest hominins at Sangiran were living or even dying in a swampy estuary. Rather, it is likely that they lived upstream in a different environment, and their remains floated downstream and were then incorporated into estuarine mud. As example, it has been demonstrated in the UK that the predominantly cranial

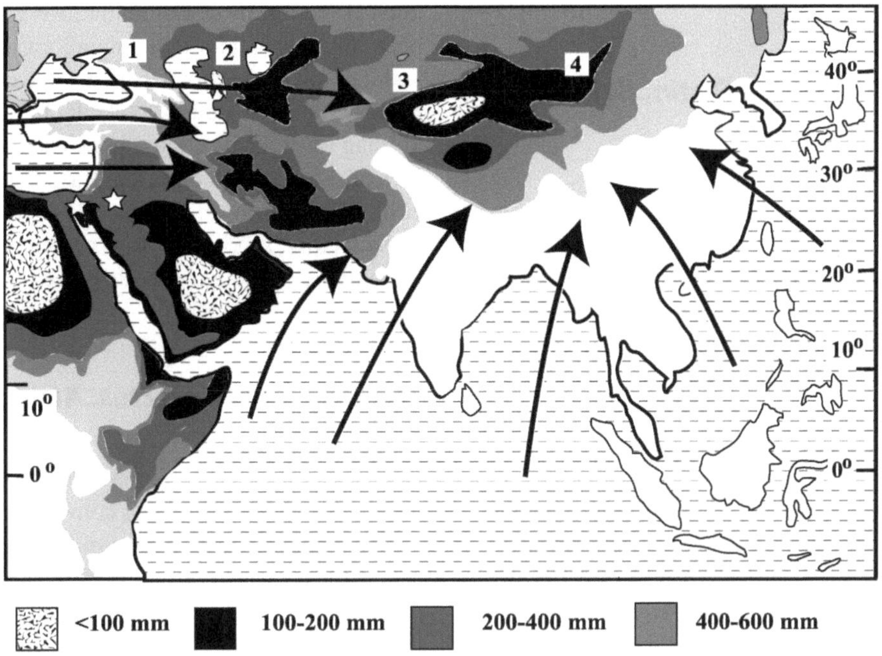

Fig. 2.3 This map attempts to indicate the probable level of precipitation across southern Asia during moist (i.e. interglacial) parts of the Early Pleistocene. The numbers in squares highlight current areas of uncertainty: 1) the timing and extent of the Akchagylian Transgression that linked the Black and Caspian Seas; 2) changes to the size of the Aral Sea and its possible connections to the Caspian; 3) the Early Pleistocene elevation of the Tien Shan; and 4) the height of the northern Tibetan Plateau.

In the Late Pliocene and Early Pleistocene, there were no ice-sheets over northern Europe, and therefore no northerly winds blowing southwards and blocking westerly winds from the East Mediterranean bringing rainfall eastwards across Southwest and much of Central Asia. Apart from small areas of Arabia and North China, few areas of Asia are likely to have received <100 mm. Loess was deposited in cold periods in both Central Asia and North China, but on a much smaller scale than during the Middle Pleistocene. The summer monsoon penetrated further inland than during much of the Middle Pleistocene, and thus the northern limit of the semi-arid zone (i.e. <600 mm) lay further north.

(Source: the author)

and mandibular human remains from Holocene deposits in the lower parts of the Thames and the Mersey estuaries were probably derived from corpses that had entered these rivers (e.g. through accidental deaths or by rivers exposing burials) several kilometres upstream (Knüsel and Carr 1995; Turner et al. 2002). When the crania (sometimes with the mandible attached) disarticulated from the rest of the body, they moved furthest downstream by floating, or by rolling along the stream bed, because of their greater transport potential. A similar argument has been made for the Ngandong (Solo) assemblage of hominin crania (Dennell 2005). The second point is that the bulk of the Javan hominin evidence comes from the Bapang (= Kabuh) Formation that overlies the Sangiran Formation (Larick et al. 2001). The Bapang Formation is predominantly fluvial rather than estuarine, and is associated with mixed woodland (Sémah 1984; Bettis et al. 2009) and the Trinil H.K. fauna that is classified as part of an open woodland community (van den Bergh et al. 2001). It is thus likely that the preferred habitat of the Javan hominins in both the Sangiran and Bapang Formations was open woodland with grasses and scrub, and that the hominins at the time of the Sangiran Formation were living (and probably dying) upstream of the swampy estuary.

The Uncertain Origins and Distinctiveness of Homo erectus

The origins of *H. erectus* remain unclear, despite intensive fieldwork in East Africa over the last 30 years. One reason is that the hominin fossil record remains poorly documented between 2.0 and 2.4 Ma (Kimbel 1995: 435). A second is that there are only three sets of finds (KNM-ER 803, 1808 and WT 15000) where post-cranial material has been found in direct association with cranio-dental specimens of early East African *H. erectus*; and a third is that one of its contemporaries, *H. rudolfensis,* is known only from cranial material, and thus isolated post-cranial specimens could belong to either. The earliest East African specimens attributed to *Homo erectus* s.l. are a pelvic specimen (KNM ER 3228), and femora and other leg bones (KNM ER 1472 and 1481) that are dated to c. 1.89 Ma. However, McHenry and Coffing (2000:128) suggested that these leg bones might have belonged to *H. rudolfensis*, the remains of which were found nearby. There is also a partial occipital bone (KNM ER 2598), also dated to c. 1.89 Ma, that have belong to *H. erectus* but which is not particularly diagnostic. A third reason why the origins of *H. erectus* remain unclear is that experts disagree whether the earliest cranial specimens form a coherent group; for example, an important cranial specimen (KNM ER 3733), dated to 1.78 Ma, may belong to *H. erectus* s.l., although some (e.g. Schwartz 2000; Schwartz and Tattersall 2003:133) feel that it does not share the same morphology as the cranium from the skeleton WT15000 or the isolated cranium KNM ER 3833, both of which are also attributed to *H. erectus*. The partial skeleton from Nariotokome (WT15000) is the most complete and best-known example of early *H. erectus* in Africa, and is dated to c. 1.53 Ma (Brown and McDougall 1993: 19).

Fourthly, there is currently some doubt over the dating of specimen KNM ER 3773. Recently, Gathogo and Brown (2006) have proposed that this specimen may be only 1.65 Ma, and the least ambiguous specimens of *Homo erectus* s.l. (i.e. those identified from cranial-dental specimens) in East Africa should also be re-dated to only c. 1.65–1.5 Ma. At the time of writing (September 2008), there is an on-going debate over this re-dating, and it may be some time before the dust settles over the age of this very important set of specimens.

The distinctiveness of early *H. erectus* also appears contentious, and it is no longer clear that it was uniquely adapted for life in savannah grasslands compared to its contemporaries. For example, Wood and Collard (1999) have claimed that there were fundamental differences between *H. ergaster* (i.e. early African *H. erectus* s.l.) and earlier types of hominins. However, its dental maturation does not appear to have been any faster than its contemporaries (Dean et al. 2001), and its limb proportions may also have been similar (see e.g. Haeusler and McHenry 2004). Its brain was not larger overall than those of its contemporaries, as small-brained examples of *H. erectus* persisted in East Africa after 1.6 Ma (Potts et al. 2004) alongside *H. habilis* (Spoor et al. 2007). Given these factors, it becomes hard to explain why *H. erectus* was the only hominin that ever dispersed out of Africa.

The Origin of H. erectus: Africa or Asia?

Despite the weight of opinion favoring an African origin of *H. erectus*, some have been less convinced.[7] Swisher et al. (1994: 1118) suggested that "*Homo erectus* may have evolved outside Africa", whilst White (1995: 383) suggested "It seems more likely that *Homo erectus* is an immigrant from Asia to the eastern African area than an anagenetic, in situ derivative from *Homo habilis*". Another example is provided by Asfaw et al. (2002: 319): "Uncertainties surrounding the taxon's appearance in Eurasia and southeast Asia make it impossible to establish accurately the time or place of origin of *H. erectus*. Available evidence is insufficient to detect the direction of its geographic dispersal. Given new perspectives afforded by the discoveries at Dmanisi in Eurasia, the assumption that the

[7] As an early example, see also "*Homo erectus* is an Asian lineage at least as old as *Homo habilis*" (Dennell et al. 1988: 105).

earliest *H. erectus* populations migrated from Africa to Eurasia, rather than invading Africa from Eurasia, is premature".

These doubts are particularly strengthened by the exceedingly archaic features of the Dmanisi cranial and mandibular specimens.

The Dmanisi Hominins

The Dmanisi hominins have proved difficult to assimilate within a simple model whereby *H. erectus* originated in Africa and later dispersed into Asia. The discoveries in 1999 of two crania (D2880 and D2882) that came from the same level and area as the mandible found in 1991 led to the conclusion that "these hominids may represent the species that initially dispersed from Africa and from which the Asian branch of *H. erectus* was derived" (Gabunia et al. 2000a: 1025). It was, in other words, a very early version of *H. ergaster*, as well as a putative ancestor of the East Asian *H. erectus.* Schwartz (2000), however, pointed out that the East African group of finds (KNM ER 992, 3773, 3883 and WT15000) that were classified as *H. ergaster* formed an ill-assorted heterogeneous group, and also suggested that the differences between D2880 and D2882 were such that two taxa were also likely to be represented. Given the depositional history of the site and the close proximity of the two skulls to each other, Gabunia et al. (2000b) thought that this was highly unlikely.

Following the discovery of the third cranium (D2700), a second mandible (D2735) and ten isolated teeth, Vekua et al. (2002: 85) suggested that if the new finds are added to the previous ones, "The Dmanisi specimens are the most primitive and small-brained fossils to be grouped with this species or any taxon linked unequivocally with genus *Homo* and also the ones most similar to the presumed *habilis*-like stem. We suggest that the ancestors of the Dmanisi population dispersed from Africa before the emergence of humans identified broadly with the *H. erectus* grade", and also noted that "it can be argued that this population is closely related to *Homo habilis* (sensu stricto)" (2002: 88).

Recently, following the discovery in 2000 of a mandible (D2600) that is dated to 1.8 Ma and was considerably more robust than the mandibular specimen D211, Gabounia[8] et al. (2002) proposed a new taxon, *H. georgicus*. According to them (2002: 245), "This species preserves several affinities with *Homo habilis* and *Homo rudolfensis*, and can be considered as one of their descendants, foretelling the emergence of *Homo ergaster.* It is close to the roots of the *Homo* branch and its presence indicates an early hominid diffusion from Africa towards Eurasia, between 2 and 1.8 Ma, by the Levantine corridor." They noted that the species appears to have been highly dimorphic sexually, and comprises a gracile (i.e. female) group (D211, D2280 and D2282) and a robust (i.e. male) one, represented by specimens D2600 and another mandible, D2735. The decision to classify the Dmanisi hominins as a new taxon is broadly consistent with the assessment of Schwartz and Tattersall (2003: 490) that "none of the Dmanisi fossils can be regarded as belonging to either Asian *Homo erectus* or to the species containing its supposed African relatives". However, the size differences between mandibles D211 and D2600 exceed the variation seen within fossil specimens of *Homo*, and may imply that they belonged to two species (Skinner et al. 2006; but see Rightmire et al. 2008). These researchers point out that the robust specimens D2600, D2700/D2735 were found c. 15 m distant from and in the layer below the gracile ones, D211, D2280 and D2282, and thus two populations might be represented.

The proposal that the Dmanisi hominins belong to a new taxon, *H. georgicus*, is at odds with the conclusions reached by Rightmire et al. (2006). They propose that the Dmanisi crania are most appropriately classified as *H. erectus* s.l., but are also the earliest types (in morphological terms) yet found, and suggest that the trinomen *H. erectus georgicus* is available to allow differentiation from the East African populations of *H. erectus* s.l. (or *H. erectus ergaster*) and the Javan ones of *H. erectus erectus* (= *H. erectus* s.s.; i.e. sensu stricto). The most provocative part of their analysis is their suggestion that the Dmanisi population may be ancestral to early *H. erectus* s.l. in East Africa; as Rightmire et al. (2006: 140) state "Dating does not presently rule out the possibility that *H. erectus* originated in Eurasia and that some groups then returned to Africa, where they evolved towards *H. erectus ergaster*".[9] This "Out of Asia" scenario is summarized in Fig. 2.4.

Debate will doubtless continue over whether the Dmanisi hominins should be classified as an incipient (in morphological terms) and highly sexually dimorphic form of *H. erectus*, as proposed by Antón (2002) and Rightmire et al. (2006); regarded as a very early form of *Homo* that is most appropriately placed in a new taxon, *H. georgicus* (Gabounia et al. 2002); or suspected of being a composite sample of two populations (Schwartz 2000; Skinner et al. 2006).

Were Hominin Migrations Always One-Way from Africa to Asia?

Although it is widely assumed that early hominin dispersals were always out of, rather than into Africa, this need not have been the case. Two-way faunal traffic between Africa and Asia is well-documented for non-hominin taxa: in the

[8]This is the French spelling of Gabunia, as is Dmanissi for Dmanisi.
[9]See also Dennell (1995: 24): "Another and more daring suggestion is that *Homo erectus* evolved in Asia and then migrated into Africa 1.5 million years ago, and was therefore not descended from *Homo habilis* and its African contemporaries".

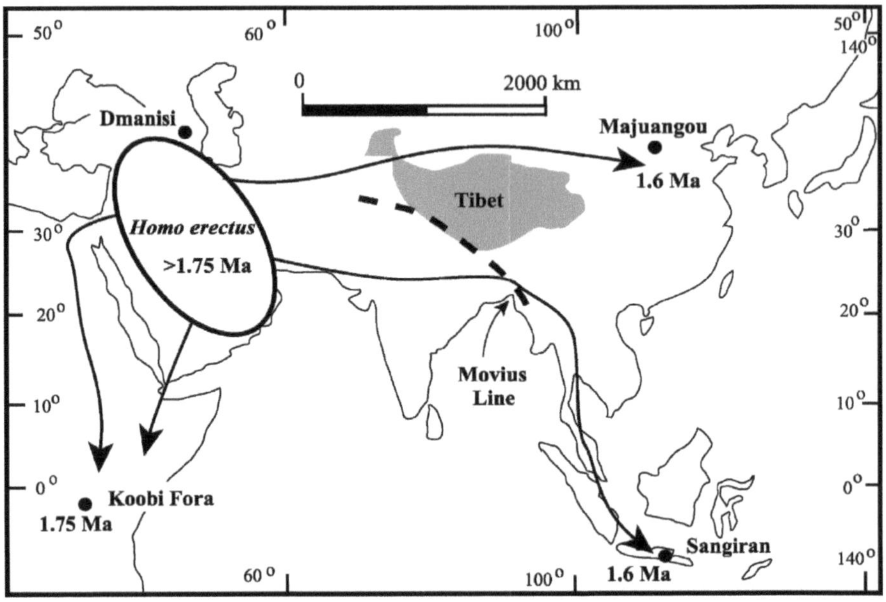

Fig. 2.4 This figure takes account of the evidence that the Dmanisi hominins are morphologically, and perhaps chronologically, the earliest forms of *H. erectus* yet found. It may therefore have originated in South West Asia, from which it may have dispersed eastwards across continental Asia, west to Europe and south into East Africa (Source: the author)

Late Pliocene, 13 types of bovids (as well as *Equus;* [Lindsay et al. 1980]) entered Africa from Asia, but only six types of bovids left Africa for Asia (Vrba 1995, Figure 27.8).

We should also remember than Asia and Africa are constructs of the classical and post-classical world, and are not necessarily appropriate geographical entities for the Late Pliocene. One example of a possible hominin migration from Asia into Africa is provided by Olduvai hominin (OH 9) from Upper Bed II, Olduvai Gorge. This specimen is dated to c. 1.4–1.5 Ma (Schwartz and Tattersall 2003: 194), and is considered to have close resemblances with cranial specimen Sangiran 17, Java, dated to c. 1.6 Ma.[10] Both Tattersall (1997) and Clarke (2000) have suggested it indicates a back-migration of *H. erectus* s.s. from Asia into Africa. As noted above, the Dmanisi hominins may also provide an example of an earlier migration of hominins from Asia into Africa.

Absence of Evidence and Evidence of Absence

The Asian component of the Out of Africa 1 model is based on the extraordinarily flimsy foundations of three unequivocal points of observation – (Dmanisi [1.75 Ma], the Nihewan Basin [1.66 Ma] and Sangiran [1.6 Ma]]) – several thousands of miles apart across Asia.[11] (There are also a few other claims that do not have widespread acceptance because of the lack of documentation, and/or doubts about the context, dating and identification of material as either hominin and/or stone artefacts. Ones that I would exclude (but opinions vary) are Erq el Ahmar and Yiron [Israel], Yuanmou, Jianshi, Longgupo and Renzidong [China], Ulalinka and Diring [Russia]: see Dennell 2009).[12] One unfortunate consequence of there being so few points of observation of the earliest hominins in Asia is that perceptions of how these observations might be interpreted is heavily contingent upon their dating, and this has often proved to be very volatile. As example, as recently as 1998, Donggutuo and Xiaochangliang were thought to be the oldest archaeological sites in North China, dated at c. 1 Ma (see Pope 1995, Figure 34.3) or 0.73–0.97 Ma (Pope and Keates 1994, Table 25.1), roughly the same age as Kuldara, Tajikistan, dated to c. 880–955 ka (Ranov 1995). These dates fitted an explanation that hominins did not expand

[10] This argument rests heavily upon the assumption that Sangiran 17 is older than OH 9. However, according to proponents of a "short chronology" for the Javan hominins, its age may be only ≤1.1 Ma (Hyodo et al. 1993; Itihara et al. 1994).

[11] 'Ubeidiya to Sangiran is c. 5,640 air-miles; London–Johannesburg is 5,617 air-miles, and Paris–San Francisco is 5,683 air-miles. 'Ubeidiya to Zhoukoudian is c. 4,455 air-miles; New York–Sarajevo is 4,477 air-miles. Nihewan to Sangiran is 3,300 air-miles, or about the same as Atapuerca, Spain, to the Omo Valley, Ethiopia (see Fitzpatrick and Modlin 1986).

[12] If a claim that hominins were present in a region is to be credible, it has to satisfy three criteria of context, dating and identification. That is to say, the material must be in a clear stratigraphic context, preferably shown by photographs and detailed section drawings; the dating has to be unequivocal; and the evidence itself has to be demonstrably hominin in origin (i.e. fossil specimens have to be unequivocally hominin, and claimed artefacts have to be clearly different from geofacts), and preferably published as both drawings and photographs. In the author's opinion, the sites mentioned in this group fail at least one of those criteria.

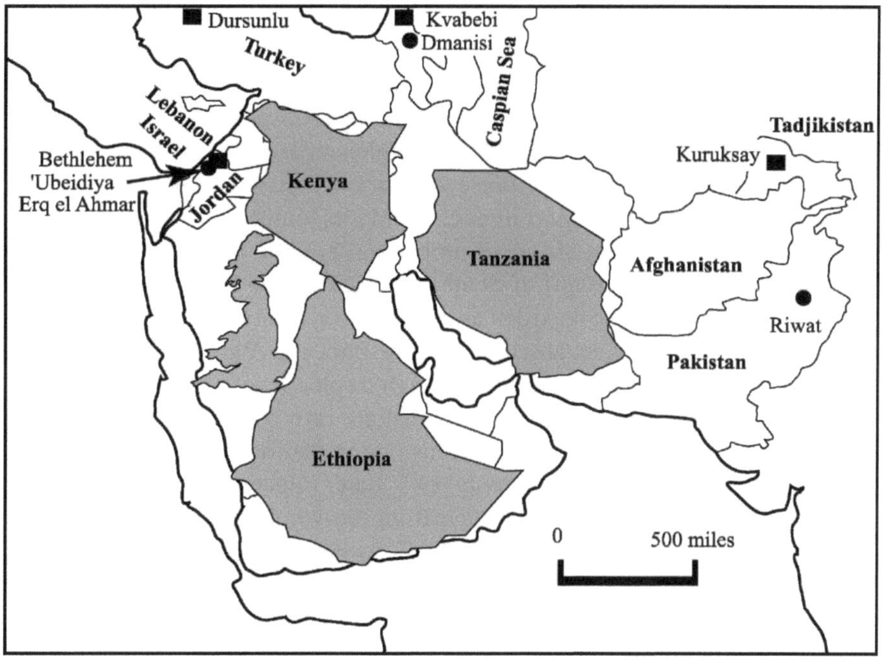

Fig. 2.5 Despite the impression gained from some map projections, Southwest Asia is considerably larger than the countries that normally define East Africa. Southwest Asia is normally defined as the countries within the area bounded by Iran, Turkey and the Arabian Peninsula, and thus excludes Afghanistan and Pakistan. The countries of the Caucasus (Armenia, Azerbaijan and Georgia) are variously included in Southwest Asia or Europe, but their combined area is only ca. 71,000 sq ml (roughly the size of Syria), and their exclusion from Southwest Asia does not significantly change the fact that this region is far larger than East Africa, and is almost unknown (unlike East Africa) regarding its Pliocene faunal record
(Source: the author)

into these continental grasslands until the late Early Pleistocene (see, e.g. Dennell 2004). However, in 2003, Xiaochangliang became the oldest site at c. 1.36 Ma in North China (Zhu et al. 2001), and then in 2004, Majuangou III, at 1.66 Ma (Zhu et al. 2004). The antiquity of hominins in North China has thus been increased by c. 66% in only 4 years, although it remains to be seen whether the oldest of these dates remain unchallenged as much depends upon estimates of sedimentation rates (see Gao et al. 2005). The age of Xiaochangliang, for example, has recently been revised downwards to ca. 1.26 Ma (Li et al. 2008). Similarly, the estimated age of Mojokerto has recently dropped from 1.8 Ma to ≤1.49 Ma (Morwood et al. 2003), leaving the earliest hominin at Sangiran dated to c. 1.6 Ma (Larick et al. 2001). When data sets are so small, as for the earliest hominin sites in Asia, it is difficult to establish whether an individual date is anomalous. Larger data sets (as in Europe) are less vulnerable to such uncertainties over the age and significance of individual sites.

The most important gap in current knowledge is Southwest Asia.

Southwest Asia: The Black Hole of Paleoanthropology

Southwest Asia is the critical area in any discussion of when hominins first appeared outside Africa, not only because it is its neighbour, but because this region acts as a cross-road into Europe, Central and South Asia. As shown in Fig. 2.5, Southwest Asia is considerably larger than East Africa. Southwest Asia covers 2.4 million square miles: the combined area of its three largest countries (Iran, Saudi Arabia and Turkey) is 1.77 million square miles, which is more than the 1.27 million square miles of East Africa,[13] or the 1.69 million square miles of all 27 member states of the European Union. Regarding its Early Pleistocene fossil hominin and archaeological records, Southwest Asia is also virtually empty apart from Dmanisi and 'Ubeidiya on its fringes. Its Pliocene faunal record comprises the Georgian localities of Dmanisi, two small and slightly earlier assemblages from Kočachuri and Calka, both dominated by large mammals, and Kvabebi (>2.6 Ma) (Vekua 1995), which has 26 mammalian taxa indicative of a riverine and marshy environment (Hemmer et al. 2004). There is also a small, poorly-preserved one from Bethlehem (c. 2.3 Ma), and an uninformative one from Çalta, western Turkey that had a small number of Pliocene taxa that include *Giraffa, Hipparion, Chasmaportetes* and *Nyctereutes;* isotopic analyses indicate an open steppic environment (Bocherens and Sen 1998). The Bethlehem assemblage comprised nine taxa, mainly of mammals >60 kg, that were found in a coarse gravel with clasts up to 0.5 m long in a clay matrix that mitigated against the preservation of small mammals or ones with fragile skeletons (Hooijer 1958). Most specimens were isolated and heavily weathered tooth fragments.

[13] Ethiopia = 435,521 square miles; Kenya = 224,081 square miles; Somalia = 246,201 square miles; and Tanzania = 364,899 square miles.

The taxa represented at Bethlehem were *Nyctereutes megamastoides*, *Homotherium* sp., *Archidiskodon* cf. *planifrons*, *Hipparion* sp., *Dicerorhinus etruscus*, *Sus* cf. *strozzii*, *Giraffa* cf. *camelopardalis*, *Leptobos* sp., and *Gazellospira torticornis* (Hooijer 1958: 289).[14] As the Bethlehem assemblage contained the three-toed horse *Hipparion* that was replaced by the one-toed *Equus* after c. 2.5 Ma in southern Asia (Lindsay et al. 1980), its age is probably ≤ 2.3 Ma. In no way can the faunal record of Southwest Asia (particularly from Bethlehem) be regarded as adequate for demonstrating that hominins were absent during the Pliocene. We should also remember that mammals at high trophic levels (such as carnivores and probably hominins) are rare compared to ones at lower levels, and are thus rare in the fossil record. As example, there are only two Late Pliocene records 3,000 miles apart of the puma from the whole of Asia (Hemmer et al. 2004), and even in the Early Pleistocene of much of Asia, the absence of large felids is very hard to demonstrate. In South Asia, for example, *Homotherium* is not recorded in the Early Pleistocene of India, Pakistan or Nepal, but was present in neighboring regions, at Bethlehem and Dmanisi; Longgupo (South China); and Kuruksay (Tajikistan) at that time (Dennell et al. 2007), and thus was probably also in South Asia. Large primates are also rare in the Asian fossil record: as example, the evidence for *Theropithecus* consists of only one specimen from 'Ubeidiya c. 1.4 Ma (Belmaker 2002), and a poorly provenanced one from Mirzapur, North India, c. 2,500 miles to the east and c. 1.0 – 0.1 Ma (Delson 1993).

The absence of a Pliocene faunal record from an area of Asia larger than East Africa raises the importance of discriminating between absence of evidence and evidence of absence.

Dispersal Events and the Importance of Absence of Evidence

The dangers of relying solely upon a few first appearance dates to demonstrate that a taxon such as *Homo erectus* s.l. dispersed into a new area, and the dangers of not knowing when it was last absent, are shown in Fig. 2.6a–d. In the first example (Fig. 2.6a), an initial set of observations suggests that a taxon evolved in one area, and then migrated into an adjacent territory some time after its first appearance – in much the same way that the timing of the dispersal of *Homo ergaster* was envisaged in the 1980s. In the second example (Fig. 2.6b), the timing of this dispersal appears to have happened at a much earlier date, shortly after the taxon first appeared in its core area: this of course is how the "Out of Africa I" model was envisaged after the re-dating of the earliest Javan hominins to 1.6–1.8 Ma. Further evidence may show a fundamentally different pattern (Fig. 2.6c), in which the taxon may have originated in the area in which it was thought to have colonized, and then later migrated into the area in which it was thought to have evolved.[15]

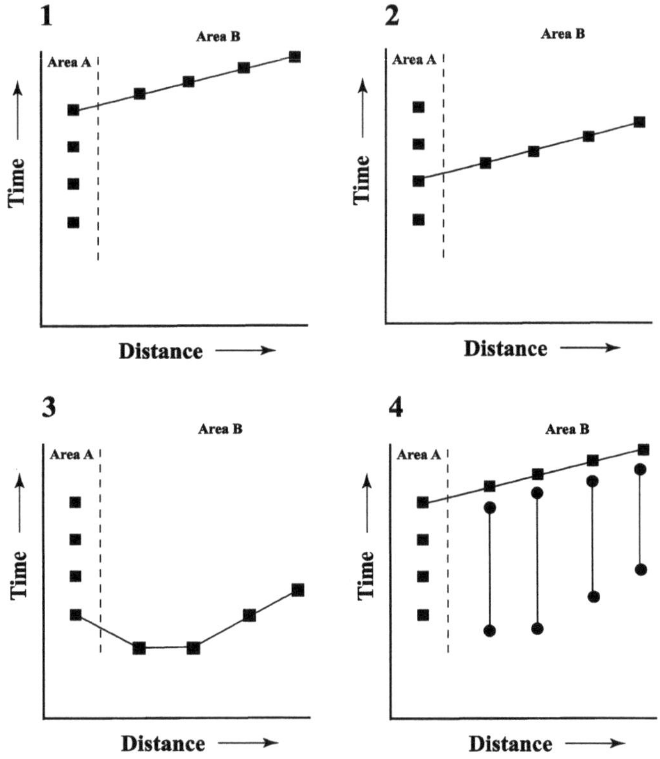

Fig. 2.6 First appearance dates, and problems of identifying core and peripheral areas of distribution (From The author)

In order to be confident that a set of observations about the first appearance of a taxon outside its apparent area of origin are correct about both the timing of its dispersal, and also its direction, we ideally need first appearance dates that can be matched by dates showing its last probable absence (Fig. 2.6d). At present, it is simply not possible to provide accurate dates of when hominins were last absent across South, and particularly Southwest Asia. Therefore, the absence of hominins in most of Asia during the Late Pliocene and Early Pleistocene cannot be demonstrated. Put more provocatively, we have no clear indication at present as to when hominins first appeared in Asia.

[14] Gardner and Bate (1937) also recognised *Testudo* (two types), *Hippopotamus* and *Stegodon*, but the last two of these were not confirmed by Hooijer (1958).

[15] As example, it was commonly believed in the 1920s and 1930s that hominins evolved in Central Asia and later migrated into Africa; not until the 1960s could it clearly be shown that the reverse pattern was more likely; see, e.g. Dennell (2001).

Some Alternative Perspectives

If we accept that current evidence from Dmanisi, the Nihewan Basin and Sangiran provides only a minimal estimate of when hominins first appeared in Asia, and that we cannot be in the slightest degree confident that they were previously absent (particularly in Southwest Asia), what alternatives might we consider? At least three have been proposed:

An Ultra-Long Chronology: Hominins Have Been in Asia as long as in Africa

Darwin (1871: 161) suggested "It is ... probable that Africa was formerly inhabited by extinct apes closely allied to the gorilla and chimpanzee; as these two species are now man's closest allies, it is somewhat more probable that our early progenitors lived on the African continent than elsewhere", but immediately qualified his remarks by stating "But it is useless to speculate on this subject: for ... since so remote a period the earth has certainly undergone many great revolutions, and there has been ample time for migration on the largest scale".

Although the canonical view is that the chimpanzee is our closest relative, and the fossil record shows unequivocally our African origins (as cautiously suggested by Darwin), a minority of researchers have proposed alternative perspectives. For example, Henneberg (2001) suggested that hominoids evolved everywhere as eurytopic species (i.e. ones adapted to a wide range of environments), and pointed out that they were present in Africa, Asia and Europe in the Miocene and Pleistocene. Their absence from Asia and Europe in the Pliocene, and the richness of the African Pliocene fossil hominin record could thus simply reflect the vagaries of fossil preservation and the massive investment of fieldwork in Africa. Consequently, he suggested, "At present, it is impossible to identify precisely which of the Miocene apes ... gave rise to the lineage leading to humans" and because the human lineage is eurotypic "It would be futile to ask for a precise small area of origin" (2001: 49). (This may seem an extreme viewpoint, but in a similar but less contentious line of reasoning, Cote 2004: 323 suggests that the absence of large hominoids in Africa in the Late Miocene reflects "small sample sizes, poor preservation, or inappropriate habitat sampling", and should not be taken as evidence of their absence.) Henneberg is not alone is seeing the quest for a place of origin for hominins as "futile": "If the origins of hominids occurred rapidly, followed by rapid range extension, as seems likely, it may be as futile to seek a specific and localized place of origin for hominids as it is for any other group" (Brunet et al. 1995: 274).

Darwin's (1871) suggestion that the gorilla and chimpanzee were our closest relatives was based primarily upon geographical proximity rather than comparative anatomy. Genetically, we are closer to these apes than the orangutan, and most researchers (and the public) are content to accept them as our closest living relatives. A minority view (e.g. Grehan 2006a; Schwartz 2005) holds that we are closer morphologically to the orangutan than to the chimpanzee, and it is therefore more credible as our closest living relative. In their view, the genetic evidence is deeply problematic, and there is no reason to prefer it over the findings of comparative morphology.

This unorthodox viewpoint need not imply that hominins evolved in Asia; indeed, Schwartz (2005: 23) believes that "human origins are to be sought in the African past". The modern distribution of orangutans in Southeast Asia is thus simply an example of "migration on the largest scale", as suggested by Darwin (see above). For others, however, our close morphological similarities to orangutans imply that hominins evolved over large areas of Asia and Africa over the last 12–14 Ma; i.e. subsequent to the divergence of the human and orangutan lineages. (Grehan 2006b even suggests that australopithecines are a sister group of orangutans and humans). A variant on this argument is presented by van Schaik et al. (2003): because chimpanzees and orangutans are equally adept at using tools, the origins of tool-use extend back to at least 14 Ma (when the common ancestor of orangutans and the African apes diverged), and thus evidence for such skills might be found in both Africa and Asia from the Miocene onwards.

The Implications of A. bahrelghazali (Chad): Could Hominins Have Dispersed out of Africa c. 3.0–3.5 Ma?

The discovery of a mandible from Koro Toro in Chad, c. 2,500 km west of the Rift Valley that was dated on faunal grounds to c. 3.0–3.5 Ma and attributed to a new taxon, *A. bahrelghazali* was interpreted as indicating that "hominids were distributed throughout the woodland and savannah belt from the Atlantic Ocean across the Sahel through eastern Africa to the Cape of Good Hope" (Brunet et al. 1995: 274). If so, there are no obvious reasons why hominins could not have dispersed the same distance northwards, and colonised similar grasslands in (and possibly beyond) the Arabian Peninsula (see Fig. 2.7; and also Dennell 1998; Gamble 2001: 7). As indicated by the sapropel evidence from the East Mediterranean (Fig. 2.8), there were numerous "windows of opportunity" for hominins to disperse into Asia after 3.2 Ma during warm, moist intervals, and some australopithecines might have done so. As argued recently by O'Regan et al. (2006: 311), "We must therefore consider at least the past 3 Ma as a possible time frame for actual and potential biotic contact and movements of early hominins between Africa and Eurasia". As a supporting argument, the

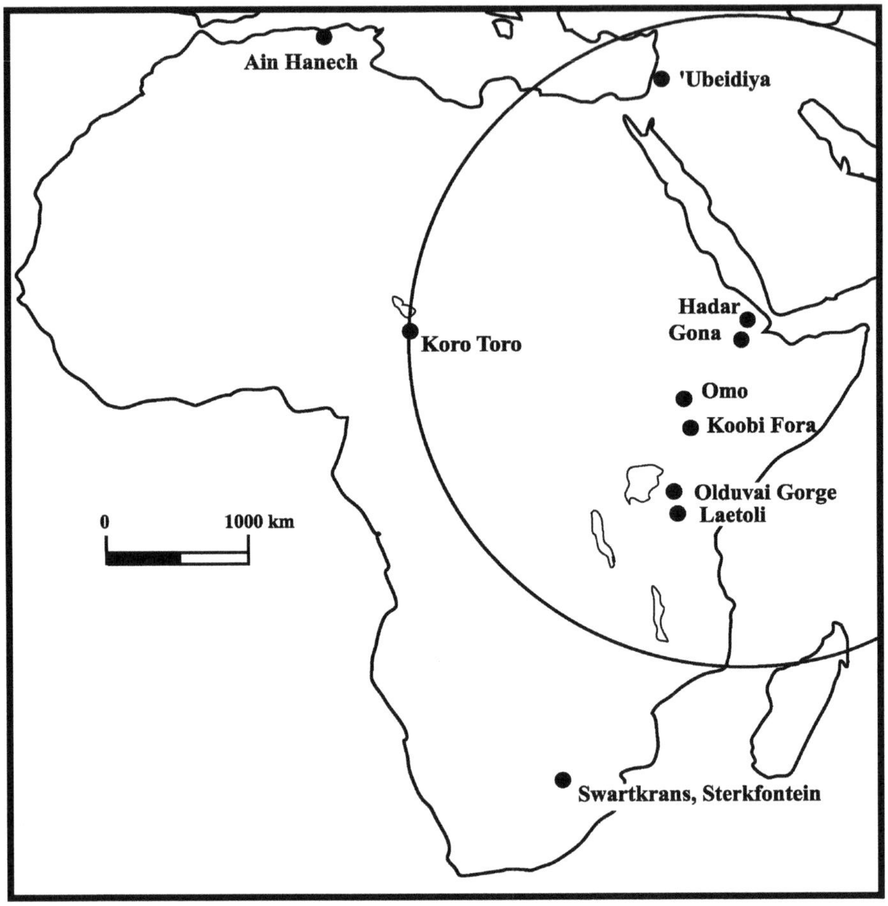

Fig. 2.7 The discovery of *Australopithecus bahrelghazali* ca. 3.0–3.5 Ma at Koro Toro, Chad, 2500 km west of the East African Rift Valley has prompted suggestions that australopithecines had colonised savannah grassland environments of Africa by this time. If they were able to disperse 2500 km westwards, there are no a priori reasons why they could not have dispersed the same distance northwards and colonised the grasslands of Southwest Asia, particularly those parts (such as lakes and springs) with sufficient year-round water
(Source: the author)

Dmanisi hominins clearly did not need large brains and modern body size to survive outside Africa as ancestors of *H. erectus* in both East Africa and Java. Once again, the virtual absence of a Pliocene fossil vertebrate record from Southwest Asia does not allow us to dismiss the possibility that some australopithecines ventured into those parts of this region that had similar habitats to those at Koro Toro.

The Implications of Kadar Gona (Ethiopia): Could Hominins Have Dispersed out of Africa c. 2.6 Ma?

A more modest suggestion is that the breakthrough that enabled hominins to leave Africa was the ability to flake stone tools and use them for butchering large carcasses, thus opening up a new source of food. On that scenario, hominins could have left Africa with those skills shortly after 2.6 Ma (Heinzelin et al. 1999; Semaw 2006), especially as there were no desert barriers at that time between Asia and Africa (see Fig. 2.2). Those that might have left could have included very early, or even ancestral, forms of *Homo*, and one of those could have given rise to the population of *H. erectus* at Dmanisi. Others might have included *A. garhi*, which is implicated in the tool-making at Kadar Gona; another might have been the ancestor of *H. floresiensis* (see Argue et al. 2006). Turner (1999) additionally suggested that because carnivore competition was less severe in the Late Pliocene than afterwards, it was easier for hominins to disperse out of Africa before than after 2 Ma.

As Ruddiman et al. (1989) suggested, one climatic consequence of the Late Pliocene uplift of the Tibetan Plateau was that summers became drier in East Africa (see Fig. 2.9). The expansion of grasslands in East Africa at this time can thus be seen as a consequence of Tibetan uplift and the strengthening of the Asian monsoon system in addition to the onset of northern hemisphere glaciation. If that was the case, Asian groups of *H. erectus* s.l. could then have entered Africa by following the advancing front of grasslands, as did several types of bovid and *Equus* in the Late Pliocene (see above).

Routes of Dispersal

Whatever the date(s) at which hominins left (and entered) Africa might turn out to have been, what were the most likely

Fig. 2.8 The sapropels indicate periods when the influx of fresh water (probably from the Nile) into the East Mediterranean increased, and can serve as a proxy indicator of increased rainfall. Such periods would have created opportunities for mammals (including hominins) to disperse out of (and possibly into) Africa

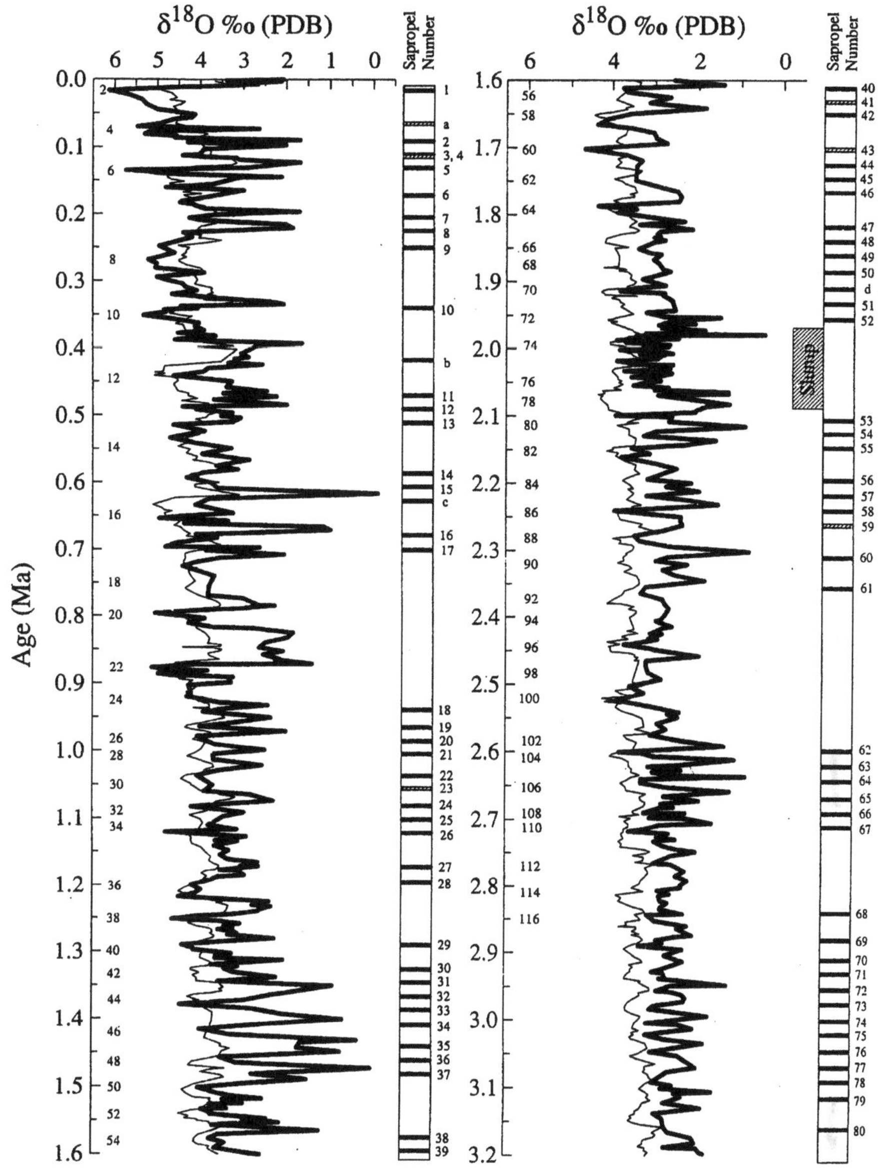

Fig. 2.9 The expansion of grasslands in East Africa in the Late Pliocene can be seen as a consequence of Tibetan uplift and the strengthening of the Asian monsoon system in addition to the onset of northern hemisphere glaciations after 2.5 Ma. Grasslands date from at least 7 Ma in South Asia, and 4 Ma in East Asia, and gradually expanded westwards, entering East Africa in the Late Pliocene after 2.5 Ma

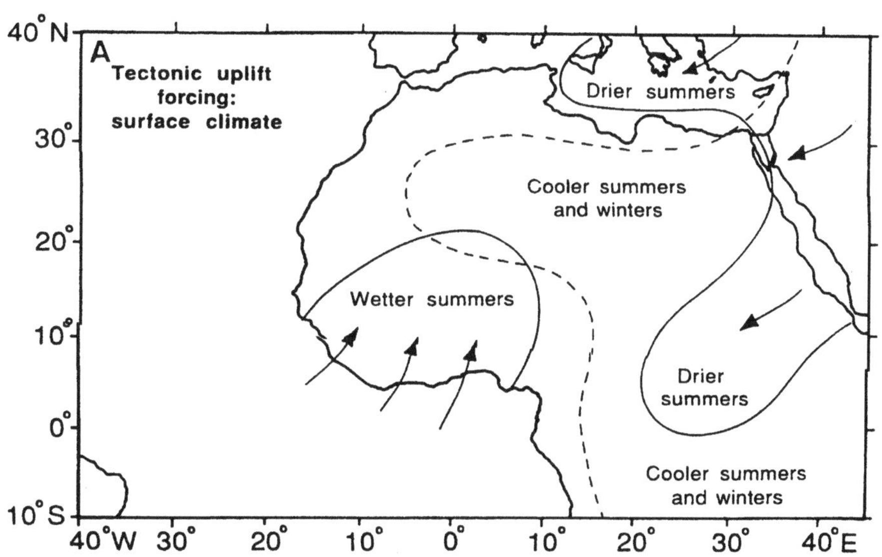

routes that they might have taken? The main corridors would have been either a northern route via the Red Sea and Sinai Desert into the Levant, or a southern one across the Bab al-Mandab Strait at the southern end of the Red Sea (Beyin 2006; Derricourt 2006). As its shallowest depth is 137 meters, it would always have been open to the Indian Ocean (Fernandes et al. 2006), although its width might have narrowed considerably from the present 37 km to c. 10 km. Once across, hominins might have taken either a coastal route towards India and Southeast Asia, or headed north or northeastwards and taken an inland route further east, targeting lakes and springs where water was available year-round, as at 'Ubeidiya, Dmanisi, and the Nihewan Basin. As the individual at Mojokerto, Java, inhabited a coastal deltaic environment (Huffman et al. 2006: 449), early coastal migrations by *H. erectus* are possible. The Levantine corridor leads northwards towards Anatolia and the Caucasus; both allow dispersal further east across Iran (either along the southern coastline of the Caspian Sea or the northern edge of the Iranian Plateau) and Central Asia towards North China, as has been suggested in a recent GIS simulation study (Holmes 2007). Major constraints to dispersal across this region would have been the patchy distribution of flakable stone (especially in loess and landscapes), and the need to target lakes and springs as there were few perennial rivers.

Dispersals or Colonisation?

Climatic changes across much of Asia during the Pliocene and Pleistocene were both considerable and in step with those recorded at higher latitudes (Lu et al. 1999; Vandenberghe et al. 2004 for North China; Kravchinsky et al. 2003; Williams et al. 1997 for Lake Baikal; Dodonov 2002; Dodonov and Baiguzina 1995 for Tajikistan), and closely linked to changes in the monsoonal weather system and variations in the earth's orbit. In cool, dry periods equivalent to high-latitude glaciations, the winter monsoon, dominated by cold, dry winds from Siberia and Central Asia would have significantly reduced the strength of the summer monsoon that delivers most of the rainfall over South, Southeast and East Asia. It would also have weakened the westerly winds that provide most of the rainfall over Southwest Asia (and vice versa in warm, moist intervals corresponding to high-latitude interglacials). It is therefore likely that there would have been significant vegetational shifts across southern Asia below latitude 40–45°N (currently the northernmost limit of hominins before 0.5 Ma) between desert and semi-desert, semi-desert and grassland, grassland and open woodland, and vice versa. It seems a reasonable expectation that animal populations (including hominins) would also have expanded and contracted in step with these changes in rainfall and temperature. In warm, moist interglacial periods, faunal and botanical communities would have expanded northwards and often longitudinally, but contracted southwards during cold, dry glacial periods. There would thus have been a constant ebbing and flowing within Asia of plant and animal communities (including hominins) throughout the Pleistocene, in much the same way as has been recorded for Europe (see e.g. Roebroeks 2001; Gamble 1999). It is also probable that Asian populations became fragmented under conditions of severely decreased precipitation. These changes make it likely that the hominin record across southern Asia during the Pleistocene is one of regional and chronological discontinuities (see Fig. 2.10), as well as "repeated, short-lived and modest dispersal events, rather than continuous residence" (Dennell 2003: 434). If so, "Out of Africa 1" was not an isolated, uni-directional continental-level colonization event (and was certainly not a "damburst" of hominins flooding out of Africa, as suggested by Lewin 1994), but a process of numerous, small-scale dispersal (and sometimes contraction) events within Asia both latitudinally and longitudinally, as well as between Southwest Asia, Northeast Africa, and Southeast Europe.

Summary

Whilst the fundamental aspects of the "Out of Africa 1" model are probably correct in that hominins (including the genus *Homo*) originated in Africa, a fundamental re-appraisal is required of the timing, pattern and scale of early hominin dispersals out of (and possibly into) Africa. The strength of the prevailing model rests heavily upon the abundance of evidence for *Homo* and earlier hominins in East Africa before 1.8 Ma, and the absence of evidence for the genus *Homo* outside Africa before this time. Nevertheless, estimates of when hominins first entered Asia are derived from only three very widely-spaced sets of observations from Georgia, North China and Java, and the present record from Asia is grossly insufficient to demonstrate when hominins were first resident in Asia.[16] In particular, the Late Pliocene faunal record from Southwest Asia is wholly inadequate for demonstrating the absence of hominins from this region prior to 1.8 Ma. An additional consideration is that the type of grassland environments associated with the emergence of the genus *Homo* in East Africa had been extensive in much of Asia during much of the Pliocene, and woodlands would also have been available around lakes, springs and many rivers. There is therefore no a priori reason why these should have remained

[16] No Africanist would suggest that three points of observation were sufficient to document the emergence of the genus *Homo*, or that the East African record is now so well known that any further information is redundant.

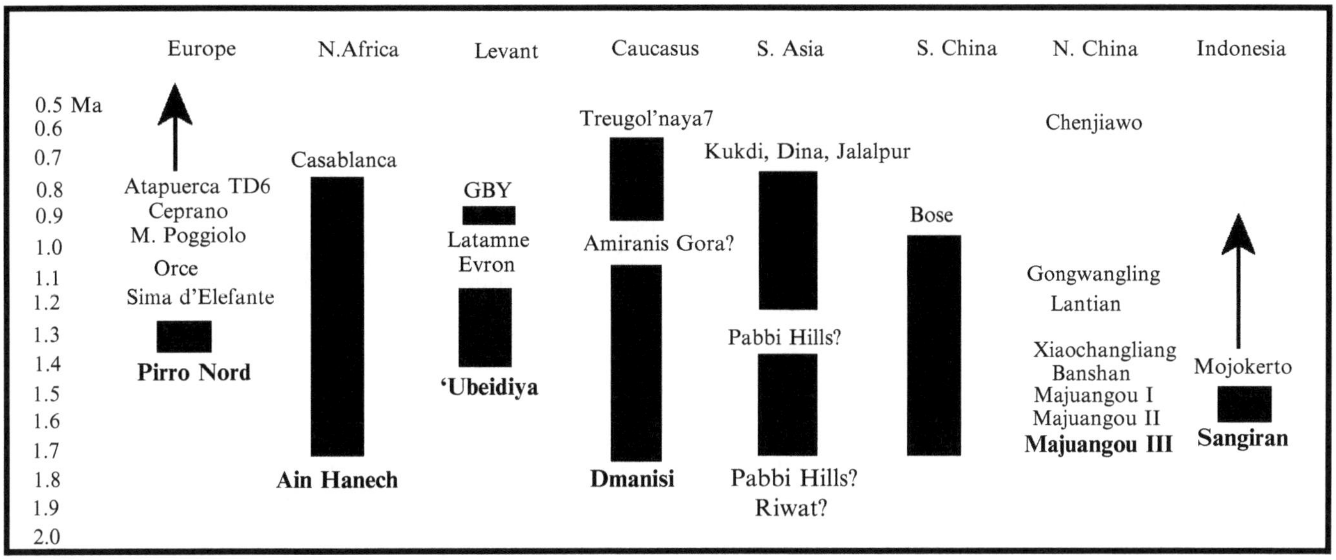

Fig. 2.10 Shaded areas denote periods for which there is no definite evidence that hominins were present. Controversial dates or finds are indicated by a "?". Although many regional discontinuities doubtless reflect a lack of fieldwork, Early Palaeolithic settlement across much of Asia is likely to have been discontinuous because of the consequences of climatic shifts corresponding to high-latitude glaciations. The following sites are considered too unreliable to include: Erq el Ahmar and Yiron (Israel), and Longgupo (Central South China), Renzidong (Central East China) and Yuanmou (South West China)

uninhabited until 1.8 Ma, when, it seems (on just three points of evidence), Asia south of latitude 45°N. was colonised rapidly by *H. erectus*.

Two hypotheses are currently available for testing, and neither can be rejected until the Asian Pliocene fossil record (particularly from Southwest Asia) is comparable to that from East Africa: (i) australopithecines c. 3.0–3.5 Ma colonised parts of the Asian grasslands as well as those in Africa and, (ii) hominins (early *Homo* and perhaps *Paranthropus*) left Africa c. 2.6 Ma shortly after they had learnt how to flake stone and butcher large mammals. The latter hypothesis is considerably strengthened by the evidence from Dmanisi for the (morphologically) earliest type of *H. erectus* yet found. One explanation of this evidence is that *H. erectus* evolved in Southwest Asia before 1.8 Ma and then dispersed back into Africa as well as eastwards to North China and Java (Rightmire et al. 2006). If so, an earlier resident population of hominins (including early *Homo*) remains a strong possibility, at least in Southwest Asia. OH9 may indicate a further instance of "Out of Asia" migration into East Africa c. 1.4–1.5 Ma.

Irrespective of when hominins first left Africa, they would probably have entered Asia via the Levant or across the southern end of the Red Sea, and then dispersed along the coast to Southeast Asia, as well as overland to North China through the better-watered parts of the grasslands of Southwest and Central Asia. The Early Pleistocene occupation of Asia is likely to have been episodically and spatially discontinuous as the climate fluctuated between cool and dry climates and warmer and moister ones, although on a more muted scale than during the Middle Pleistocene. The challenge now is to place the "Out of Africa 1" model on more solid foundations in Asia than a handful of widely-spaced observations across an area larger than Africa.

Acknowledgements I would like to thank Chris Norton for inviting me to the session that he co-organised at the AAPA meeting in Philadelphia in 2007, and the cogent criticism of an anonymous reviewer of an earlier version of this paper. Wil Roebroeks (Leiden) and Paul Pettitt (Sheffield) are thanked for their role in earlier formulations of some of the ideas expressed in this paper. The British Academy is also thanked for granting me a 3-year research sabbatical (2003–2006) in which to concentrate on researching early Asian prehistory.

References

Abdrakhmatov, K. Ye, Makarov, V. I., Molnar, P., Panasyuk, S. V., Prilepin, M. T., Reilinger, R. E., et al. (1996). Relatively recent construction of the Tien Shan inferred from GPS measurements of present-day crustal deformation rates. *Nature, 384*, 450–453.

Aiello, L. C., & Wheeler, P. (1995). The expensive tissue hypothesis. *Current Anthropology, 36*(2), 199–222.

Al-Sayari, S. S., & Zötl, J. G. (Eds.). (1978). *Quaternary period in Saudi Arabia*. Vienna/New York: Springer.

An, Z. (2000). The history and variability of the East Asian paleomonsoon climate. *Quaternary Science Reviews, 19*, 171–187.

An, Z., Kutzbach, J. E., Prell, W. L., & Porter, S. C. (2001). Evolution of Asian monsoons and phased uplift of the Himalaya-Tibetan plateau since Late Miocene times. *Nature, 411*, 62–66.

Antón, S. C. (2002). Evolutionary significance of cranial variation in Asian *Homo erectus*. *American Journal of Physical Anthropology, 118*, 301–323.

Argue, D., Donlon, D., Groves, C., & Wright, R. (2006). *Homo floresiensis*: Microcephalic, pygmoid, *Australopithecus*, or *Homo*? *Journal of Human Evolution, 51*, 360–374.

Ariai, A., Thibault, C. (1975/1977). Nouvelles précisions à propos de l'outillage paléolithique ancien sur galets du Khorassan (Iran). *Paléorient, 3*, 101–108.

Asfaw, B., Gilbert, W. H., Beyenne, Y., Hart, W. K., Renne, P. R., WoldeGabriel, W., et al. (2002). Remains of *Homo erectus* from Bouri, Middle Awash, Ethiopia. *Nature, 416*, 317–319.

Baker, E. A. (Ed.). (1949). *Cassell's New English Dictionary*. London: The Reprint Society.

Barry, J. C., Morgan, M. E., Flynn, L. J., Pilbeam, D., Behrensmeyer, A. K., Raza, S. M., Khan, I. A., Badgley, C., Hicks, J., Kelley, J. (2002). Faunal and environmental change in the late Miocene Siwaliks of northern Pakistan. *Paleobiology Memoirs, 3*(Supplement to Paleobiology 28(2)), 1–55.

Bar-Yosef, O., & Goren-Inbar, N. (1993). *The Lithic assemblages of 'Ubeidiya; a Lower Palaeolithic site in the Jordan Valley, Jerusalem*. Jerusalem: The Hebrew University of Jerusalem.

Behrensmeyer, A. K., Todd, N. E., Potts, R., & McBrinn, G. E. (1997). Late Pliocene faunal turnover in the Turkana Basin, Kenya and Ethiopia. *Science, 278*, 1589–1594.

Belmaker, M. (2002). First evidence of the presence of *Theropithecus* sp. in the southern Levant. *Israel Journal of Zoology, 48*(2), 165.

Bettis, E. A., Milius, A. K., Carpenter, S. J., Larick, R., Zaim, Y., Rizal, Y., et al. (2009). Way out of Africa: Early Pleistocene paleoenvironments inhabited by *Homo erectus* in Sangiran, Java. *Journal of Human Evolution, 56*, 11–24.

Beyin, A. (2006). The Bab al Mandab vs the Nile-Levant: An appraisal of the two dispersal routes for early modern humans out of Africa. *African Archaeological Review, 23*, 5–30.

Blisniuk, P. M., Hacker, B. R., Glodny, J., Ratschbacher, L., Siwen Bi, Wu, Z., McWilliams, M.O., Calvert, A. (2001). Normal faulting in central Tibet since at least 13.5 Myr ago. *Nature*, 412, 629–632.

Bobe, R., & Behrensmeyer, A. K. (2004). The expansion of grassland ecosystems in Africa in relation to mammalian evolution and the origin of the genus *Homo*. *Palaeogeography, Palaeoclimatology, Palaeoecology, 207*, 399–420.

Bocherens, H., & Sen, S. (1998). Pliocene vertebrate locality of Çalta, Ankara, Turkey. 11. Isotopic composition. *Geodiversitas 20*(3), 487–495.

Bonnefille, R. (1995). A re-assessment of the Plio-Pleistocene pollen record of East Africa. In E. S. Vrba, G. H. Denton, T. C. Partridge, & L. H. Burckle (Eds.), *Paleoclimate and human evolution* (pp. 299–310). New Haven: Yale University Press.

Bramble, D. M., & Lieberman, D. (2004). Endurance running and the evolution of *Homo*. *Nature, 432*, 345–352.

Brown, F. H., & McDougall, I. (1993). Geological setting and age. In A. Walker & R. Leakey (Eds.), *The Nariokotome Homo erectus skeleton* (pp. 9–20). Cambridge: Harvard University Press.

Brunet, M., Beauvillain, A., Coppens, Y., Heintz, E., Moutaye, A. H. E., & Pilbeam, D. (1995). The first australopithecine 2500 kilometres west of the Rift Valley (Chad). *Nature, 378*, 273–274.

Cachel, S., & Harris, J. W. K. (1998). The lifeways of *Homo erectus* inferred from archaeology and evolutionary ecology: A perspective from East Africa. In M. Petraglia & R. Korisettar (Eds.), *Early human behavior in global context: The rise and diversity of the Lower Palaeolithic record* (pp. 108–132). London: Routledge.

Carbonell, E., & Rodríguez, X. P. (2006). The first human settlement of Mediterranean Europe. *Comptes Rendues Palévolution, 5*, 291–298.

Cerling, T. E. (1992). Development of grasslands and savannahs in East Africa during the Neogene. *Palaeogeography, Palaeoclimatology, Palaeoecology, 97*, 241–247.

Clarke, R. J. (2000). Out of Africa and back again. *International Journal of Anthropology, 15*(3–4), 185–189.

Coleman, M., & Hodges, K. (1995). Evidence for Tibetan plateau uplift before 14 Myr ago from a new minimum age for east-west extension. *Nature, 374*, 49–52.

Cote, S. M. (2004). Origins of the African hominids: An assessment of the palaeobiogeographical evidence. *Comptes Rendues Palévolution, 3*, 323–340.

Darwin, C. (1871). *The descent of man*. New York: Prometheus Books (republished 1998).

Dean, C., Leakey, M. G., Reid, D., Schrenk, F., Schwartz, G. T., Stringer, C., & Walker, A. (2001). Growth processes in teeth distinguish modern humans from *Homo erectus* and earlier hominins. *Nature, 414*, 628–631.

Delson, E. (1993). *Theropithecus* fossils from Africa and India and the taxonomy of the genus. In N. G. Jablonski (Ed.), *Theropithecus, the rise and fall of a primate genus* (pp. 15–83). Cambridge: Cambridge University Press.

DeMenocal, P. B. (2004). African climate change and faunal evolution during the Pliocene-Pleistocene. *Earth and Planetary Science Letters, 220*, 3–24.

Dennell, R. W. (1995). Do human origins lie in Africa? New evidence from northern Pakistan. *Cranium, 12*(1), 21–24.

Dennell, R. W. (1998). Grasslands, tool-making and the earliest colonization of south Asia: A reconsideration. In M. Petraglia & R. Korisettar (Eds.), *Early human behavior in global context: The rise and diversity of the Lower Palaeolithic record* (pp. 280–303). London: Routledge.

Dennell, R. W. (2001). From Sangiran to Olduvai, 1937–1960: The quest for "centres" of hominid origins in Asia and Africa. In R. Corbey & W. Roebroeks (Eds.), *Studying human origins: Disciplinary history and epistemology* (pp. 45–66). Amsterdam: Amsterdam University Press.

Dennell, R. W. (2003). Dispersal and colonisation, long and short chronologies: How continuous is the Early Pleistocene record for hominids outside East Africa? *Journal of Human Evolution, 45*, 421–440.

Dennell, R. W. (2004). Hominid dispersals and Asian biogeography during the Lower and Early Middle Pleistocene, c.2.0–0.5 Mya. *Asian Perspectives, 43*(2), 205–226.

Dennell, R. W. (2005). The Solo (Ngandong) *Homo erectus* assemblage: A taphonomic assessment. *Archaeology in Oceania, 40*, 81–90.

Dennell, R. W. (2009). *The Palaeolithic Settlement of Asia*. Cambridge: Cambridge University Press.

Dennell, R. W., & Roebroeks, W. (2005). An Asian perspective on early human dispersal from Africa. *Nature, 438*, 1099–1104.

Dennell, R. W., Rendell, H. R., & Hailwood, E. (1988). Early toolmaking in Asia: Two-million-year-old artefacts in Pakistan. *Antiquity, 62*, 98–106.

Dennell, R. W., Coard, R., & Turner, A. (2006). The biostratigraphy and magnetic polarity zonation of the Pabbi Hills, northern Pakistan: An Upper Siwalik (Pinjor Stage) Plio-Pleistocene fluvial sequence. *Palaeogeography, Palaeoclimatology, Palaeoecology, 234*, 168–185.

Dennell, R. W., Coard, R., & Turner, A. (2007). Predators and scavengers in Early Pleistocene southern Asia. *Quaternary International, 192*, 78–88.

Derricourt, R. (2006). Getting "Out of Africa": Sea crossings, land crossings and culture in the hominin migrations. *Journal of World Prehistory, 19*(2), 119–132.

Derry, L. A., & France-Lanord, C. (1997). Himalayan weathering and erosion fluxes: climate and tectonic controls. In W. F. Ruddiman (Ed.), *Tectonic uplift and climate change* (pp. 289–312). New York: Plenum.

Dettman, D. L., Kohn, M., Quade, J., Ryerson, F. J., Ojha, T. P., & Hammidullah S. (2001). Seasonal stable isotope evidence for a strong Asian monsoon throughout the past 10.7 m.y. *Geology, 29*(1), 31–34.

Ding, Z. L., Xiong, S. F., Sun, J. M., Yang, S. L., Gu, Z. Y., & Liu, T. S. (1999). Pedostratigraphy and Paleomagnetism of a ~ 7.0 Maeolian loess-red clay sequence at Lingtai, Loess Plateau, north-central China and the implications for palaeomonsoon evolution. *Palaeogeography, Palaeoclimatology, Palaeoecology* 152(1–2), 49–66.

Dodonov, A. E. (2002). *Quaternary of Central Asia: Stratigraphy correlation, palaeogeography*. Moscow: Geos (in Russian).

Dodonov, A. E., & Baiguzina, L. L. (1995). Loess stratigraphy of Central Asia: Palaeoclimatic and palaeoenvionmental aspects. *Quaternary Science Reviews, 14*, 707–720.

Dowsett, H., Thompson, J., Barron, J., Cronin, T., Fleming, F., Ishman, S., et al. (1994). Joint investigations of the Middle Pliocene climate I: PRISM palaeoenvironmental reconstructions. *Global and Planetary Change, 9*, 169–195.

Dowsett, H. J., Barron, J. A., Poore, R. Z., Thompson, R. S., Cronin, T. M., Ishman, S. E., et al. (1999). Middle Pliocene paleoenvironmental reconstruction: PRISM2. *US Geological Survey Open File Report, 99–535*, 1–23.

Fernández, M. H., & Vrba, E. S. (2006). Plio-Pleistocene climatic change in the Turkana Basin (East Africa): Evidence from large mammal faunas. *Journal of Human Evolution, 50*, 595–626.

Fernandes, C. A., Rohling, E. J., & Siddall, M. (2006). Absence of post-Miocene Red Sea land bridges: Biogeographic implications. *Journal of Biogeography, 33*, 961–966.

Fernández-Jalvo, Y., Denys, C., Andrews, P., Williams, T., Dauphin, Y., & Humphreys, L. (1998). Taphonomy and palaeoecology of Olduvai Bed-I (Pleistocene, Tanzania). *Journal of Human Evolution, 34*, 137–172.

Fitzpatrick, G. L., & Modlin, D. L. (1986). *Direct-line distances*. Meteuchen: Scarecrow Press.

Fort, M. (1996). Late Cenozoic environmental changes and uplift on the northern side of the central Himalaya: A reappraisal from field data. *Palaeogeography, Palaeoclimatology, Palaeoecology, 120*, 123–145.

Gabounia, L., de Lumley, M. A., Vekua, A., Lordkipandize, D., & de Lumley, H. (2002). Découverte d'un nouvel hominide à Dmanissi (Transcaucasie, Georgie). *Comptes Rendues Palévolution, 1*, 243–253.

Gabunia, L., Vekua, A., Lordkipanidze, D., Swisher, C. C., Ferring, R., Justus, A., et al. (2000a). Earliest Pleistocene hominid cranial remains from Dmanisi, Republic of Georgia: Taxonomy, geological setting, and age. *Science, 288*, 1019–1025.

Gabunia, L., Vekua, A., & Lordkipanidze, D. (2000b). Response to Schwartz. *Science, 289*, 55–56.

Gamble, C. (1999). *The palaeolithic societies of Europe*. Cambridge: Cambridge University Press.

Gamble, C. (2001). Modes, movement and moderns. *Quaternary International, 75*(1), 5–10.

Gao, X., Wei, Q., Shen, C., & Keates, S. (2005). New light on the earliest hominid occupation in East Asia. *Current Anthropology, 46*(S5), 115–120.

Gardner, E. W., & Bate, M. A. (1937). The bone-bearing beds of Bethlehem: Their fauna and industry. *Nature, 140*, 431–433.

Garzione, C.N., Quade, J., DeCelles, P.G., & English, N.B. (2000). Predicting paleoelevation of Tibet and the Himalaya from $\partial^{18}0$ vs. altitude gradients in meteoric water across the Nepal Himalaya. *Earth and Planetary Science Letters, 183*(1–2), 215–229.

Gathogo, P. N., & Brown, F. H. (2006). Revised stratigraphy of Area 123, Koobi Fora, Kenya, and new age estimates of its fossil mammals, including hominins. *Journal of Human Evolution, 51*(5), 471–479.

Ginat, H., Enzel, Y., & Avni, Y. (1998). Translocated Plio-Pleistocene drainage systems along the Arava fault of the Dead Sea transform. *Tectonophysics, 284*, 151–160.

Ginat, H., Zilberman, E., & Saragusti, I. (2003). Early pleistocene lake deposits and Lower Palaeolithic finds in Nahal (wadi) Zihor, Southern Negev desert, Israel. *Quaternary Research, 59*, 445–458.

Grehan, J. R. (2006a). Mona Lisa smile: the morphological enigma of human and great ape evolution. *The Anatomical Record B: The New Anatomist, 289*(4), 139–157.

Grehan, J.R. (2006b). Ape biogeography. In H. J. Birx (Ed.), *Encyclopedia of anthropology* (pp. 202–207). Thousand Oaks: Sage.

Güleç, E., Howell, F. C., & White, T. D. (1999). Dursunlu – a new Lower Pleistocene faunal and artifact-bearing locality in southern Anatolia. In H. Ulrich (Ed.), *Hominid evolution: Lifestyles and survival strategies* (pp. 349–364). Edition Archaea: Gelsenkirchen.

Guo, Z. T., Ruddiman, W. F., Hao, Q. Z., Wu, H. B., Qiao, Y. S., Zhu, R. X., et al. (2002). Onset of Asian desertification by 22 Myr ago inferred from loess deposits in China. *Nature, 416*, 159–163.

Haeusler, M., & McHenry, H. M. (2004). Body proportions of *Homo habilis* reviewed. *Journal of Human Evolution, 46*, 433–465.

Han, J., Fyfe, W. S., Longstaffe, F. J., Palmer, H. C., Yan, F. H., & Mai, X. S. (1997). Pliocene-Pleistocene climatic change recorded in fluviolacustrine sediments in central China. *Palaeogeography, Palaeoclimatology, Palaeoecology, 135*(1–4), 27–39.

Harris, D. R. (Ed.). (1980). *Human ecology in Savannah environments*. London/New York: Academic.

Harrison, T. M., Copeland, P., Hall, S. A., Quade, J., Burner, S., Ojha, T. P., et al. (1993). Isotopic preservation of Himalyan/Tibetan uplift, denudation, and climatic histories of two molasse deposits. *Journal of Geology, 101*, 157–175.

Heinzelin, J.de, Clark, J. D., White, T., Hart, W., Renne, P., WoldeGabriel, Beyene, Y., & Vrba, E. (1999). Environment and behaviour of 2.5-million-year-old Bouri hominids. *Science, 284*, 625–629.

Hemmer, H., Kahlke, R.-D., & Vekua, A.K. (2004). The Old World puma – Puma pardoides (Owen, 1846) (Carnivora: Felidae) – in the Lower Villafranchian (Upper Pliocene) of Kvabebi (East Georgia, Transcaucasia) and its evolutionary and biogeographical significance. *Neues Jahrbuch Geologie und Paläontologie Abhandlungen*, 233, 197–231.

Henneberg, M. (2001). The gradual eurytopic evolution of humans: not from Africa alone. In E. Indriati (Ed.), *A scientific life: Papers in honor of Prof. Dr. T. Jacob* (pp. 41–51). Yogyakarta, Indonesia: BIGRAF.

Holmes, K. M. (2007). Using Pliocene palaeoclimatic data to postulate dispersal pathways of early hominins. *Palaeogeography, Palaeoclimatology, Palaeoecology, 248*(1–2), 96–108.

Hooijer, D.A. (1958). An early Pleistocene mammalian fauna from Bethlehem. *Bulletin of the British Museum (Natural History): Geology, 3*, 267–292.

Horowitz, A. (2001). *The Jordan Rift Valley*. Lisse: Balkema.

Huffman, O.F., Zaim, Y., Kappelman, J., Ruez, D.R. jr., de Vos, J., Rizal, Y., Aziz, F., & Hertler, C. (2006). Relocation of the 1936 Mojokerto skull discovery site near Perning, East Java. Journal of Human Evolution, *50*(4), 431–451.

Hyodo, M., Watanabe, N., Sunata, W., Susanto, E. E., & Wahyono, H. (1993). Magnetostratigraphy of hominid fossil bearing formations in Sangiran and Mojokerto, Java. *Anthropological Science, 101*(2), 157–186.

Itihara, M., Watanabe, N., Kadar, D., & Kumai, H. (1994). Quaternary stratigraphy of the hominid fossil bearing formations in the Sangiran area, Central Java. *Courier Forschungsinstitut Senckenberg, 171*, 123–128.

Jablonski, N. (2004). The hippo's tale: how the anatomy and physiology of Late Neogene *Hexaprotodon* shed light on Late Neogene environmental change. *Quaternary International, 117*, 119–123.

Jiang, H., & Ding, Z. (2008). A 20 Ma pollen record of East-Asian summer monsoon evolution from Guyuan, Ningxia, China. *Palaeogeography, Palaeoclimatology, Palaeoecology, 265*, 30–38.

Kappelman, J., Plummer, T., Bishop, L., Duncan, A., & Appleton, S. (1997). Bovids as indicators of Plio-Pleistocene paleoenvironments in East Africa. *Journal of Human Evolution, 32*, 229–256.

Kimbel, W. H. (1995). Hominid speciation and Pliocene climatic change. In E. S. Vrba, G. H. Denton, T. C. Partridge, & L. H. Burckle (Eds.), *Paleoclimate and human evolution* (pp. 425–437). New Haven/London: Yale University Press.

Knüsel, C. J., & Carr, G. C. (1995). On the significance of the crania from the River Thames and its tributaries. *Antiquity, 69*, 162–169.

Kohn, M. (2006). Made in Savannahstan. *New Scientist, 191*, 34–39.

Kravchinsky, V. A., Krainov, M. A., Evans, M. E., Peck, J. A., King, J. W., Kuzmin, M. I., et al. (2003). Magnetic record of Lake Baikal sediments: chronological and paleoclimatic implication for the last 6.7 Myr. *Palaeogeography, Palaeoclimatology, Palaeoecology, 195*, 281–298.

Kroon, D., Alexander, I., Little, M., Lourens, L. J., Matthewson, A., Robertson, A. H. F., et al. (1998). Oxygen isotope and sapropel stratigraphy in the eastern Mediterranean during the last 3.2 million years. *Proceedings of the Ocean Drilling Program, Scientific Results, 160*, 181–189.

Larick, R., Ciochon, R. L., Zaim, Y., Sudijono, S., Rizal, Y., Aziz, F., et al. (2001). Early Pleistocene 40Ar/39Ar ages for Bapang Formation hominins, Central Jawa, Indonesia. *Proceedings of the National Academy of Sciences of the United States of America, 98*(9), 4866–4871.

Lewin, R. (1994). Damburst of humans flooded from Africa. *New Scientist, 141*, 14.

Li, J., Fang, X., Van der Voo, R., Zhu, J., Niocaill, C. M., Cao, J., et al. (1997). Late Cenozoic magnetostratigraphy (11–0 Ma) of the Donghanding and Wangjiashan sections in the Longzhong Basin, western China. *Geologie en Mijnbouw, 76*, 121–134.

Li, H., Yang, X., Heller, F. & Li, H. (2008). High resolution magnetostratigraphy and deposition cycles in the Nihewan Basin (North China) and their significance for stone artifact dating. *Quaternary Research,* 69, 250–262.

Lindsay, E. H., Opdyke, N. D., & Johnson, N. M. (1980). Pliocene dispersal of the horse *Equus* and late Cenozoic mammalian dispersal events. *Nature, 287*, 135–138.

Liu, T., & Ding, Z. (1998). Chinese loess and the paleomonsoon. *Annual Review of Earth and Planetary Sciences, 26*, 111–145.

Liu, T., Ding, M., & Derbyshire, E. (1996). Gravel deposits on the margins of the Qinghai-Xizang Plateau, and their environmental significance. *Palaeogeography, Palaeoclimatology, Palaeoecology, 120*, 159–170.

Liu, T., Ding, Z., & Rutter, N. (1999). Comparison of Milankovitch periods between continental loess and deep sea records over the last 2.5 Ma. *Quaternary Science Reviews, 18*(10–11), 1205–1212.

Lu, H., Liu, X., Zhang, F., An, Z., & Dodson, J. (1999). Astronomical calibration of loess-paleosol deposits at Luochan, central Chinese Loess Plateau. *Palaeogeography, Palaeoclimatology, Palaeoecology, 154*, 237–246.

Maschenko, E. N. (1994). *Papio (Paradolichopithecus) suschkini*: A revision of systematics, morphofunctional peculiarities of the skull and mandible (in Russian). *Paleoteriologiya* 15–57.

McDougall, I., & Brown, F. H. (2006). Precise $^{40}Ar/^{39}Ar$ geochronology for the upper Koobi Fora Formation, Turkana Basin, northern Kenya. *Journal of the Geological Society, London, 173*, 205–220.

McHenry, H. M., & Coffing, K. (2000). *Australopithecus* to *Homo*: Transformations in body and mind. *Annual Review of Anthropology, 29*, 125–146.

Merceron, G., Blondel, C., Brunet, M., Sen, S., Solounias, N., Viriot, L., et al. (2004). The Late Miocene paleoenvironment of Afghanistan as inferred from dental microwear in artiodactyls. *Palaeogeography, Palaeoclimatology, Palaeoecology, 207*, 143–163.

Morwood, M. J., O'Sullivan, P., Susanto, E. E., & Aziz, F. (2003). Revised age for Mojokerto 1, an early *Homo erectus* cranium from East Java, Indonesia. *Australian Archaeology, 57*, 1–4.

Movius, H. L. (1948). The lower Palaeolithic cultures of southern and eastern Asia. *Transactions of the American Philosophical Society, 38*(4), 329–420.

Nelson, C. V. (2007). Isotopic reconstructions of habitat change surrounding the extinction of *Sivapithecus*, a Miocene hominoid, in the Siwalik Group of Pakistan. *Palaeogeography, Palaeoclimatology, Palaeoecology, 243*, 204–22.

Nishimura, T. D., Takai, M., & Maschenko, E. N. (2007). The maxillary sinus of *Paradolichopithecus sushkini* (late Pliocene, southern Tajikistan) and its phylectic implications. *Journal of Human Evolution, 52*(6), 637–646.

O'Regan, H., Bishop, L., Elton, S., Lamb, A., & Turner, A. (2006). Afro-Eurasian mammalian dispersal routes of the Late Pliocene and Early Pleistocene, and their bearing on earliest hominin movements. *Courier Forschungsinstitut Senckenberg, 256*, 305–314.

Parfitt, S. A., Barendregt, R. W., Breda, M., Candy, I., Collins, M. J., Coope, G. R., et al. (2005). The earliest record of human activity in northern Europe. *Nature, 438*, 1008–1012.

Petraglia, M. D. (2003). The lower palaeolithic of the Arabian Peninsula: Occupations, adaptations, and dispersals. *Journal of World Prehistory, 17*(2), 141–179.

Pope, G.G. (1995). The influence of climate and geography on the biocultural evolution of the Far Eastern hominids. In E.S. Vrba, G.H. Denton, T.C. Partridge, & L.H. Burckle (Eds.), *Paleoclimate and evolution with emphasis on human origins* (pp. 493–506). New Haven/London: Yale University Press.

Pope, G. G., & Keates, S. G. (1994). The evolution of human cognition and cultural capacity. In R. Corruchini, & S. Ciochon (Eds.), *Integrative paths to the past: Paleoanthropological advances in honor of F. Clark Howell* (pp. 531–567). New Jersey: Prentice Hall.

Por, D. (2004). The Levantine waterway, riparian archaeology, paleolimnology, and conservation. In N. Goren-Inbar & J. D. Speth (Eds.), *Human paleoecology in the Levantine corridor* (pp. 5–20). Oxford: Oxbow Books.

Potts, R., Behrensmeyer, A. K., Deino, A., Ditchfield, P., & Clark, J. (2004). Small mid-Pleistocene hominin associated with East African Acheulean technology. *Science, 305*, 75–78.

Qiang, X. K., Li, Z. X., Powell, C. Mc A, & Zheng, H. B. (2001). Magnetostratigraphic record of the Late Miocene onset of the East Asian monsoon, and Pliocene uplift of northern Tibet. *Earth and Planetary Science Letters, 187*(1–2), 83–93.

Quade, J., Cerling, T. E., & Bowman, J. R. (1989). Development of Asian monsoon revealed by marked ecological shift during the latest Miocene in northern Pakistan. *Nature, 342*, 163–166.

Quade, J., Solounias, N., & Cerling, T. E. (1994). Stable isotopic evidence from paleosol carbonates and fossil teeth in Greece for forest or woodlands over the past 11 Ma. *Palaeogeography, Palaeoclimatology, Palaeoecology, 108*, 41–53.

Ramstein, G., Fluteau, F., Besse, J., & Joussaume, S. (1997). Effect of orogeny, plate motion and land-sea distribution on Eurasian climate change over the past 30 million years. *Nature, 386*, 788–795.

Ranov, V. A. (1995). The 'Loessic Palaeolithic' in South Tadjikistan, Central Asia: Its industries, chronology and correlation. *Quaternary Science Reviews, 14*(7–8), 731–745.

Raymo, M. E., Grant, B., Horowitz, M., & Rau, G. H. (1996). Mid-Pliocene warmth: Stronger greenhouse and stronger conveyor. *Marine Micropalaeontology, 27*, 313–326.

Rea, D. K., Snoeckx, H., & Joseph, L. H. (1998). Late Cenozoic eolian deposition in the North Pacific: Asian drying, Tibetan uplift, and cooling of the northern hemisphere. *Paleoceanography, 13*(3), 215–224.

Reed, K. E. (1997). Early hominid evolution and ecological change through the African Plio-Pleistocene. *Journal of Human Evolution, 32*, 289–322.

Rightmire, G. P., Lordkipanidze, D., & Vekua, A. (2006). Anatomical descriptions, comparative studies and evolutionary significance of the hominin skulls from Dmanisi, Republic of Georgia. *Journal of Human Evolution, 50*(2), 115–141.

Rightmire, G. P., van Arnsdale, A. P., & Lordkipanidze, D. (2008). Variation in the mandibles from Dmanisi, Georgia. *Journal of Human Evolution, 54*, 904–908.

Roebroeks, W. (2001). Hominid behaviour and the earliest occupation of Europe: An exploration. *Journal of Human Evolution, 4*, 437–461.

Rowley, D. B., & Currie, B. S. (2006). Palaeo-altimetry of the late Eocene to Miocene Lunpola basin, central Tibet. *Nature, 439*, 677–681.

Ruddiman, W. F., Sarnthein, M., Backman, J. G., Baldauf, J. G., Curry, W., Dupont, L. M., et al. (1989). Late Miocene to Pleistocene evolution of climate in Africa and the low-latitude Atlantic: overview to Leg 108 results. *Proceedings of the Ocean Drilling Program, Scientific Results, 108*, 463–484.

Ruff, C., & Walker, A. (1993). Body size and shape. In A. Walker & R. Leakey (Eds.), *The Nariokotome Homo erectus Skeleton* (pp. 234–265). Cambridge, MA: Harvard University Press.

Schick, K. D. (1994). The Movius Line reconsidered: perspectives on the Earlier Paleolithic of Eastern Asia. In R. Corruchini, & S. Ciochon (Eds.), *Integrative paths to the past: paleoanthropological advances in honor of F. Clark Howell* (pp. 569–594). New Jersey: Prentice Hall.

Schwartz, J. H. (2000). Taxonomy of the Dmanisi cranium. *Science, 289*, 5776.

Schwartz, J. H. (2005). *The Red Ape: orangutans and human origins*. Cambridge, MA: Westview Press.

Schwartz, J. H., & Tattersall, I. (2003). The human fossil record. Volume Two: Craniodental morphology of genus *Homo* (Africa and Asia). New York: Wiley-Liss.

Sémah, A.-M. (1984). Palynology and the Javanese *Pithecanthropus* palaeoenvironment. *Courier Forschungsinstitut Senckenberg, 69*, 237–252.

Semaw, S. (2006). The oldest stone artifacts from Gona (2.6–2.5 Ma), Afar, Ethiopia: Implications for understanding the earliest stages of stone knapping. In N. Toth & K. D. Schick (Eds.), *The oldowan: Case studies into the Earliest Stone Age* (pp. 43–76). Gosport: Stone Age Institute Press.

Sen, S. (1998). The age of the Molayan mammal locality, Afghanistan. *Geobios, 31*(3), 385–391.

Skinner, M. M., Gordon, A. D., & Collard, N. J. (2006). Mandibular size and shape variation in the hominins at Dmanisi, Republic of Georgia. *Journal of Human Evolution, 51*(1), 36–49.

Sotnikova, M. V., Dodonov, A. E., & Pen'kov, A. V. (1997). Upper Cenozoic bio-magnetic stratigraphy of Central Asian mammalian localities. *Palaeogeography, Palaeoclimatology, Palaeoecology, 133*(3–4), 243–258.

Spencer, L. M. (1997). Dietary adaptations of Plio-Pleistocene Bovidae: Implications for hominid habitat use. *Journal of Human Evolution, 32*, 201–228.

Spicer, R. A., Harris, N. B. W., Widdowson, M., Herman, A. B., Guo, S., Valdes, P. J., et al. (2003). Constant elevation of southern Tibet over the past 15 million years. *Nature, 421*, 622–625.

Spoor, F., Leakey, M. G., Gathogo, P. N., Brown, F. H., Anton, S. C., McDougall, I., et al. (2007). Implications of new early *Homo* fossils from Ileret, east of Lake Turkana, Kenya. *Nature, 448*, 688–691.

Sun, X., & Wang, P. (2005). How old is the Asian monsoon system? – Palaeobotanical records from China. *Palaeogeography, Palaeoclimatology, Palaeoecology, 222*, 181–222.

Sun, Z., Feng, X., Li, D., Yang, F., Qu, Y., & Wang, H. (1999). Cenozoic ostracoda and palaeoenvironments of the northeastern Tarim Basin, western China. *Palaeogeography, Palaeoclimatology, Palaeoecology, 148*, 37–50.

Susman, R. L., & Stern, J. T. (1982). Functional morphology of *Homo habilis*. *Science, 217*, 931–934.

Swisher, C. C. III, Curtis, G. H., Jacob T., Getty, A. G., Suprijo A., & Widiasmoro (1994). Age of the earliest known hominids in Java, Indonesia. *Science, 263*, 1118–1121.

Tattersall, I. (1997). Out of Africa again ... and again? *Scientific American, 276*(4), 46–53.

Thomas, H., Geraads, D., Janjou, D., Vaslet, D., Memseh, A., Billiou, D., Bocherens, H., Dobigny, G., Eisenmann, V., Gayet, M., Lapparent de Broin, F. de, Petter, G., & Halawani, M. (1998). Découverte des premières faunes pléistocenes de la péninsule Arabique dans le désert du Nafoud (Arabie Saoudite). *Compte Rendu de l'Academie des Sciences*, Paris, *326*, 145–152.

Tian, J., Wang, P., Cheng, X., & Li, Q. (2002). Astronomically tuned Plio-Pleistocene benthic $\partial^{18}0$ record from the South China Sea and Atlantic-Pacific comparison. *Earth and Planetary Science Letters, 203*, 1015–1029.

Turner, A. (1999). Assessing earliest human settlement of Eurasia: Late Pliocene dispersions from Africa. *Antiquity, 73*, 563–570.

Turner, A., Gonzalez, S., & Ohman, J. C. (2002). Prehistoric human and ungulate remains from Preston Docks, Lancashire, U.K: Problems of river finds. *Journal of Archaeological Science, 29*, 423–433.

van den Bergh, G. D., de Vos, J., & Sondaar, P. Y. (2001). The Late Quaternary palaeogeography of mammal evolution in the Indonesian Archipelago. *Palaeogeography, Palaeoclimatology, Palaeoecology, 171*, 385–408.

van Schaik, C. P., Ancrenaz, M., Borgen, G., Galdikas, B., Knott, C. D., Singleton, I., et al. (2003). Orangutan cultures and the evolution of material culture. *Science, 299*, 102–105.

Vandenberghe, J., Lu, H., Sun, D., van Huissteden, J., & van Konert, M. (2004). The late Miocene and Pliocene climate in East Asia as recorded by grain size and magnetic susceptibility of the Red Clay deposits (Chinese Loess Plateau). *Palaeogeography, Palaeoclimatology, Palaeoecology, 204*, 239–255.

Vekua, A. (1995). Die Wirbeltierfauna des Villafranchium vom Dmanisi und ihre biostratigraphische Bedeutung. *Jahrbuch des Römisch-Germanischen Zentralmuseums, Mainz, 42*, 77–180.

Vekua, A., Lordkipanidze, D., Rightmire, G. P., Agusti, J., Ferring, R., Maisuradze, G., et al. (2002). A new skull of early *Homo* from Dmanisi, Georgia. *Science, 297*, 85–89.

Vrba, E. S. (1995). The fossil record of African antelopes. In E. S. Vrba, G. H. Denton, T. C. Partridge, & L. H. Burckle (Eds.), *Paleoclimate and Human Evolution* (pp. 385–424). New Haven/London: Yale University Press.

Vrba, E. S. (1996). Climate, heterochrony, and human evolution. *Journal of Anthropological Research, 52*, 1–28.

Wang, L., Lü, H. Y., Wu, N. Q., Pei, Y. P., Tong, G. B., & Peng, S. Z. (2006). Palynological evidence for Late Miocene-Pliocene vegetation evolution in the red clay sequence of the central Chinese Loess Plateau and implication for palaeoenvironmental change. *Palaeogeography, Palaeoclimatology, Palaeoecology, 241*, 118–128.

Wheeler, P. E. (1991). The thermoregulatory advantages of hominid bipedalism in open equatorial environments: the contribution of increased convective heat loss and cutaneous evaporative cooling. *Journal of Human Evolution, 21*, 107–115.

Wheeler, P. E. (1992). The thermoregulatory advantages of large body size for hominids foraging in savannah environments. *Journal of Human Evolution, 23*, 351–362.

White, T. D. (1995). African omnivores: Global climatic change and Plio-Pleistocene hominids and suids. In E. S. Vrba, G. H. Denton, T. C. Partridge, & L. H. Burckle (Eds.), *Paleoclimate and human evolution* (pp. 369–384). New Haven/London: Yale University Press.

Williams, D. F., Peck, J., Karabanov, E. B., Prokopenko, A. A., Kravchinsky, V., King, J., et al. (1997). Lake Baikal record of continental climate response to orbital insolation during the past 5 million years. *Science, 278*, 1114–1117.

Wood, B., & Collard, M. (1999). The human genus. *Science*, 284, 65–71.

Zazzo, A., Mariotti, A., Lécuyer, C., & Heintz, E. (2002). Intra-tooth isotope variations in late Miocene bovid enamel from Afghanistan: paleobiological, taphonomic and climate implications. *Palaeogeography, Palaeoclimatology, Palaeoecology, 186*, 145–161.

Zheng, H., Powell, C.M., An, Z., Zhou, J., & Dong, G. (2000). Pliocene uplift of the northern Tibetan Plateau. *Geology, 28*, 715–718.

Zhou, J., Xu, F., Wang, T., Cao, A., & Yin, C. (2006). Cenozoic deformation history of the Qiadam Basin, NW China: Results from cross-section restoration and implications for Qinghai-Tibet Plateau tectonics. *Earth and Planetary Science Letters, 243*, 195–210.

Zhu, R. X., Hoffman, K. A., Potts, R., Deng, C. L., Pan, Y. X., Guo, B., et al. (2001). Earliest presence of humans in northeast Asia. *Nature, 413*, 413–417.

Zhu, R. X., Potts, R., Xie, F., Hoffman, K. A., Deng, C. L., Shi, C. D., et al. (2004). New evidence on the earliest human presence at high northern latitudes in northeast Asia. *Nature, 431*, 559–562.

Chapter 3
On the Road to China: The Environmental Landscape of the Early Pleistocene in Western Eurasia and Its Implication for the Dispersal of *Homo*

Miriam Belmaker

Abstract Many adaptations of the genus *Homo* have been attributed to the expansion savanna environments in East Africa during the past 5 million years. It has been hypothesized that this expansion of savanna habitats northward into the Levant as well as eastward into India and China in the Late Pliocene/Early Pleistocene is the impetus for *Homo*, as well as other African taxa, to disperse out of Africa.

Paleoecological reconstructions based on the identification of grazing taxa and ecological diversity analyses have supported models that favor the role of extrinsic forces in the dispersal of early hominins. However, critical analysis of Indicator Species and Ecological Diversity Analysis methods of paleoecological reconstructions suggest that their application to higher latitude sites may not provide robust results. Community-wide taxonomic analysis for western Eurasian sites suggests that the 'Ubeidiya, Dmanisi and the Orce basin sites are most similar to Mediterranean and temperate woodland and differ from African savanna. While Mediterranean and temperate biomes include a mixture of both open and closed habitats, there is no evidence for the presence of subtropical savannas of African origin.

Key ecological factors such as temperature, precipitation and seasonality differ between the northern latitudinal region and East Africa. This suggests that grassland habitats present in northern latitudes may have provided a novel environment for dispersing hominin populations. This supports the intrinsic model of 'Out of Africa I', and the variability selection hypothesis in particular.

Keywords 'Out of Africa I' • Ecological diversity • Grassland • Paleoecology • Savanna

M. Belmaker (✉)
Department of Anthropology, Harvard University, Peabody Museum, 11 Divinity Ave, Cambridge, MA, USA 02138
e-mail: Belmaker@fas.harvard.edu

Introduction

Sometime during the Early Pleistocene hominins dispersed from Africa into Eurasia (Schick and Dong 1993; Gabunia and Vekua 1995; Larick and Ciochon 1996; Bar-Yosef 1998; Arribas and Palmqvist 1999; Bar-Yosef and Belfer-Cohen 2001; Antón and Swisher 2004; Langbroek 2004). The large biogeographic range, from the Iberian Peninsula to China, reached by hominins less than 0.2 Ma after the initial dispersal c. 1.8–1.9 Ma (Bar-Yosef and Belfer-Cohen 2001) attests to hominins' ecological success. Paleoanthropological sites in Eurasia which predate 1.0 Ma are rare and include Dmanisi (Georgia) dated to c. 1.7–1.9 Ma (Dzaparidze et al. 1992; Dean and Delson 1995; Gabunia and Vekua 1995), Fuente Nueva-3 and Barranco León (Orce, Granada), dated to c. 1.3 Ma (Gibert and Palmqvist 1995; Martínez-Navarro and Palmqvist 1995; Palmqvist et al. 1996; Arribas and Palmqvist 1998; Gibert et al. 1998; Arribas and Palmqvist 1999; Martínez-Navarro and Palmqvist 1999), Modjokerto and Sangiran (Java) (Swisher et al. 1994; Huffman 2001; Larick et al. 2001) dated as early as c. 1.8–1.6 Ma, sites in the Nihewan and Yuanmou basins in China dated to c. 1.71–1.66 Ma (Zhu et al. 2001, 2003, 2004), and 'Ubeidiya in Israel which has been dated c. 1.6–1.2 Ma (Tchernov 1987; Sagi et al. 2005; Sagi 2005) (Fig. 3.1).

Models that aim to explain the dispersal of hominins from Africa into Eurasia during the early Pleistocene can be broadly divided into two main groups. Intrinsic explanations focus on hominin adaptability to a wide range of habitats and variable climates in Africa. Intrinsic characteristics include morphological and behavioral traits such as the capacity for long distance walking (Steudel 1994), endurance running (Bramble and Lieberman 2004), heat adaptation (Walker and Leakey 1993), greater brain capacity (Aiello 1993; Aiello and Wheeler 1995), social structure (Tappen 2009; Kroll 1994) and the ecological niche occupied by hominins within the carnivore guild (Walker 1981; Turner 1992; Brantingham 1998). The cultural adaptation that facilitated hominin long-range dispersal was probably not the advent of stone tool technology since it predates the earliest dispersal event by

C.J. Norton and D.R. Braun (eds.), *Asian Paleoanthropology: From Africa to China and Beyond*, Vertebrate Paleobiology and Paleoanthropology, DOI 10.1007/978-90-481-9094-2_3,

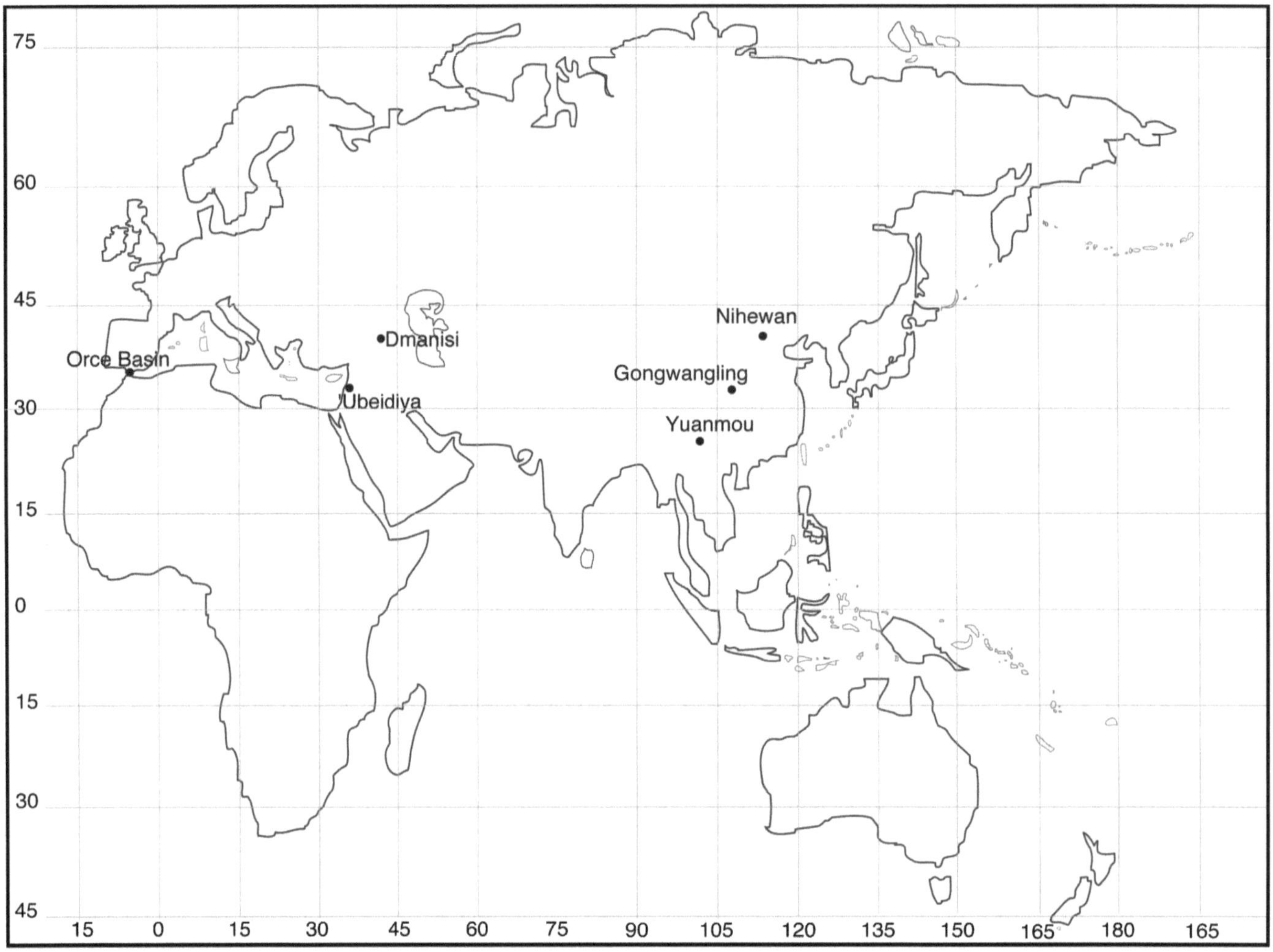

Fig. 3.1 Eurasian Plio-Pleistocene sites mentioned in the text

more than 0.5 Ma (Semaw et al. 2003; Bar-Yosef and Belfer-Cohen 2001). However, hominin adaptability may have provided a pre-adaptation to higher latitudinal environments (Potts 1998a, 2002).

Extrinsic explanations focus on the role of climate change in the 'Out of Africa I' hominin dispersal event. It has been suggested that the increase in the range of savanna grasslands observed prior to the Olduvai Subchron (1.98–1.79 Ma) may have promoted the dispersal of hominins and their subsequent ecological success by extending their preferred habitats (Dennell 2003, 2004; Dennell and Roebroeks 2005).

This study compares published paleoecological reconstructions for higher latitude Plio-Pleistocene sites that predate 1.0 Ma obtained from three methods: Indicator Species (IS), Ecological Diversity Analysis (EDA) and community-wide taxonomic diversity. Western Eurasian sites include Dmanisi (Georgia), 'Ubeidiya (Israel) and Venta Micena, as a representative of the Orce Basin sites Fuente Nueva-3, and Barranco León-3 in Spain. Eastern Asian sites include Yuanmou in southern China, Gongwangling in central China, and Xiaochangliang and Majuangou in northern China.

The Use of Large Mammals to Identify Grasslands in Paleoecological Analysis

Paleoecological analyses may be derived from several lines of evidence including stable isotope analysis, micro and macrofossils and aeolian dust (Williams et al. 1996). Unfortunately, due to preservation bias not all methods can be applied to various sites (Behrensmeyer 1991). Large mammal remains are often the best-preserved materials recovered from Plio-Pleistocene paleontological and paleoanthropological sites. Studies of macromammal fossils provide insight into paleohabitats, paleoenvironments and paleoclimates (Andrews et al. 1979; Andrews 1992, 1996; Montuire 1995, 1996, 1999; Reed 1996, 1998; Aguilar et al. 1998) as well as documenting changes in community structure and environments through time (Bobe and Eck 2001; Bobe et al. 2002; Alemseged 2003).

Each species has a unique niche and life history strategy (Ricklefs and Miller 2000). A niche is comprised of external environmental factors such as maximum or minimum temperatures, precipitation, shelter and food supplies as well as biological factors such as activity time, growth rate and habitat utilizations. Conditions that are harsh for one species are often optimal for another (Ricklefs and Miller 2000). By approximating the environmental niche requirements of a fossil species, it is possible to infer the habitat in which it lived. Several methods have been developed for paleoecological reconstructions of Plio-Pleistocene environments.

Indicator Species

The method of identifying certain taxa to document changes in the paleoenvironment is the Indicator Species (IS) method (Williams et al. 1996 and references therein). An example is the presence of reindeer, found today in higher latitudes, in the lower latitudes of Europe which indicates the farthest reach of glaciations during the Pleistocene (Delpech and Heintz 1976). This method can be applied using both presence/absence or abundance quantification methods. The IS method relies on the identification of fossils to species level and the association of the species to preferred habitat. The presumed habitat associated with extinct species is usually based on morphological and phylogenetic similarities to related extant species. Since many species have very narrow niche requirements dependent on temperature, precipitation, seasonality and other habitat variables, this method has the potential to provide a high-resolution paleoecological reconstruction (Andrews et al. 1979).

Ecological Diversity Analysis

Ecological Diversity Analysis (EDA) considers the relative abundance of the ecomorphological adaptations (e.g., size, diet, and locomotion) of the species present in the assemblage (Andrews et al. 1979). For example, *Tragelaphus scriptus* (kudu) is a bovid while *Dama dama* (fallow deer) is a cervid. However, both have a similar dietary regime and can be classified as browsers (<30% graze in diet) and both are large terrestrial quadrupeds. This allows for ecological comparison of sites that have no species in common due to spatial and temporal distances (Andrews et al. 1979; Nesbit Evans et al. 1981; Andrews 1989, 1992, 1995a, b; 1996; Cerling et al. 1992; Gunnell 1995; Reed 1996, 1998; Gagnon 1997).

Community-Wide Taxonomic Diversity

A paleoecological method for comparing a wide range of biomes and habitats is the community-wide taxonomic diversity method. Multidimensional analysis of the taxonomic diversity of the entire community can be based on different ordination methods (Legendre and Legendre 1998), most commonly Correspondence Analysis (CA). For example, CA was applied to census data abundance of bovid tribes to distinguish between habitats in several modern African regions (Greenacre and Vrba 1984; Vrba 1980). Two clusters of sites were identified based on the abundance of bovid tribes. The first included sites such as Serengeti Woodland, Manyara and Kruger Park and the second group included sites such as the Ngorongoro, Savanna grassland and Nairobi Park. The two groups differed in average rainfall and vegetation pattern.

Paleoecological Reconstructions of Plio-Pleistocene Higher Latitude Sites

The paleoenvironment of several Eurasian paleoanthropological sites has been reconstructed using different paleoecological methods. The section below details published reconstructions of sites using different methods and discusses their discordance.

Paleoecological Reconstructions of Plio-Pleistocene Higher Latitude Sites Using the Indicator Species Method

The high percentage of African grazers at ‘Ubeidiya (*Oryx* sp., *Kolpochoerus olduvaiensis*, and *Pelorovis oldowayensis*), led several researchers to suggest that the paleoecological reconstruction of ‘Ubeidiya was open, semi-arid, and included savanna habitats (Dennell 2004; Martínez-Navarro 2004, 2010). An IS analysis of the Venta Micena fauna suggested that southern Iberia was similar to Africa not only in terms of the climatic regime but also in terms of the exact species of animals that existed in both of these regions (Arribas and Palmqvist 1999). An IS analysis of Dmanisi reconstructed the habitat around this site as semi-dry to warm (Gabunia et al. 2000), including savanna-like grasslands with open to closed woodland (Dolukhanov 2000).

The IS method has been used often to reconstruct the paleoecology of East Asian sites (Pope 1995). Specifically, forest adapted species such as the *Ailuropoda meanoleuca fovealis* and *Stegadon orientalis* (also known as the *Aliuropoda - Stegodon* complex), occurred in warm interglacial periods. Open-area adapted species such as *Equus,* were indicative of cold glacial periods (Keates 2003). The site of Majuangou, in the Nihewan Basin (1.66 Ma) in northern China included a high proportion of cervids, equids and rhinos, indicative of a temperate woodland and grassland fauna. A similar list of species can be found at the near-by site of

Xiaochangliang (1.67–0.97 Ma) (Peterson et al. 2003; Zhu et al. 2004) and Yuanmou in southern China (1.67 Ma) (Qian and Zhou 1991). However, the younger site of Gongwangling (1.25–1.15 Ma) in central China includes fauna indicative of a subtropical humid forest (*Ailuropoda meanoleuca fovealis*, *Stegadon orientalis* and *Tapirus sinensis*) (Keates 2003).

Paleoecological Reconstructions of Plio-Pleistocene Higher Latitude Sites Using the Ecological Diversity Method

Based on EDA, the paleoenvironment of Dmanisi was interpreted by Palmqvist (2002) as "African savanna with tall grasses, shrubs and low bush/tree cover." Similar analysis of Venta Micena has suggested that "Orce at Venta Micena was very similar to that represented in modern African savannas with tall grass and low bush/tree cover, suggesting that the countryside in the Guadix-Baza basin was relatively unforested during early Pleistocene times, as happens today" (Palmqvist et al. 2003:46). However, a different EDA of Venta Micena suggested a temperate woodland forest (Mendoza et al. 2005). An EDA of the faunal assemblage of 'Ubeidiya suggested a paleoenvironment that was different from any of the modern African comparative sites (Belmaker 2002). EDA of the faunal assemblages from Yuanmou (1.67 Ma) in southern China suggests that the species are indicators of grassland, bushland and forest (Teague and Potts 2007). The EDA of Majuangou in the Nihewan basin (1.66 Ma) is suggestive of woodland, open country and steppe, and Gongwangling (1.25–1.15 Ma) can be assigned to montane forest (Teague and Potts 2007).

Paleoecological Reconstructions of Plio-Pleistocene Higher Latitude Sites Using the Community-Wide Taxonomic Diversity Method

Belmaker (2010) applied a community-wide taxonomic diversity analysis using Correspondence Analysis (CA) (Legendre and Legendre 1998) to both abundance and presence/absence data of mesoherbivores from the sites of Dmanisi, 'Ubeidiya, and Venta Micena, and to data from modern and other Plio-Pleistocene assemblages. The similar CA values of western European sites, and specifically that of 'Ubeidiya, Dmanisi, and Venta Micena and those of African woodland sites, suggests a paleoecological reconstruction for the western Eurasian sites as closed forest habitats rather than open grassland within the Mediterranean and temperate regions (Belmaker 2006, 2010).

Are the Methods Discordant?

A comparison between results obtained using the three methods presented here exemplifies a discordance between the methods (Table 3.1).

Using the IS method it is suggested that Dmanisi, 'Ubeidiya and Venta Micena are situated within a dry and savanna environment while community wide taxonomic diversity analysis indicates woodland Mediterranean or temperate habitats. Results for EDA are ambiguous. Dmanisi is interpreted as grassland, similar to results obtained using IS, while the two EDA for Venta Micena resulted in two different reconstructions: savanna grasslands and temperate woodland. The EDA for 'Ubeidiya could not be assigned to any of the known African comparative biomes. Paleoecological reconstructions of the East Asian sites revealed greater similarity between methods. IS and EDA produced similar results and suggested the presence of temperate woodlands and grassland in Yuanmou and Nihewan. The exception is Gongwangling, which was interpreted as subtropical forests using IS and as montane forest using EDA.

Can These Paleoecological Reconstructions Be Reconciled?

I suggest that a possible reconciliation of the discordant results may be obtained by critical evaluation of the methods applied and comparison with other paleoecological proxies. Reconstructions based on IS alone may be problematic. The IS method often focuses on a single species rather than the entire community thereby reducing the overall fidelity of the reconstruction. Specifically, IS may be less robust in detecting ecological shifts across large regions. Indicator Species reconstructions of higher latitude Plio-Pleistocene localities are often based on the presence of African taxa in these sites and may not account for the local adaptation of these taxa in their new habitats. The behavioral flexibility of these species may be confounding paleoecological interpretation.

There are several examples of this phenomenon. An ecomorphological analysis of limb bone measurements of two subspecies of *E. stenonsis* has indicated that they occupied different habitats (Eisenmann and Guérin 1984). The *E. stenonsis* from Saint-Vallier were inferred to occupy more closed, humid and soft ground habitats while *E. stenonsis* from la Puebla de Valverde occupied open, dry and hard ground habitats (Eisenmann and Guérin 1984). Moreover, although the family equidae are generally classified as grazers based on gross dental morphology, mesowear analysis has suggested that *E. capensis* from South Africa was a mixed feeder rather than a grazer (Kaiser and Franz-Odendaal

Table 3.1 Comparison of paleoecological reconstructions using three alternative methods for Eurasian Plio-Pleistocene sites which pre date 1.0 Ma

Site	Date	Indicator species	Ecological diversity analysis	Community wide taxonomic diversity
Dmanisi, Georgia	1.7–1.8 Ma	Semi Dry to warm Mediterranean (Gabunia et al. 2000); savanna-like grasslands woodland (Dolukhanov 2000)	Savanna grassland (Palmqvist 2002)	Mediterranean and temperate woodland (Belmaker 2006, 2010)
ʻUbeidiya, Israel	1.6–1.4 Ma	Arid and Semi-Arid (Dennell 2004);Savanna (Martínez-Navarro 2004)	Non African (Belmaker 2002)	Mediterranean and temperate woodland (Belmaker 2006, 2010)
Venta Micena (Orce Basin), Spain	1-6–1.3 Ma	"...faunally and climatically a part of Africa..."(Arribas and Palmqvist 1999)	Savanna grassland (Palmqvist et al. 2003); temperate woodland (Mendoza et al. 2005)	Mediterranean and temperate woodland (Belmaker 2006, 2010)
Yuanmou, S. China	1.67 Ma	Temperate woodland (Qian and Zhou 1991)	Grassland, bushland and forest (Teague and Potts 2007)	NA
Gongwangling, central China	1.25–1.15 Ma	Sub tropics (Keates 2003)	Montane forest (Teague and Potts 2007)	NA
Nihewan basin, N. China	1.66–1 Ma	Temperate woodland and grassland (Peterson et al. 2003)	Woodland, open country and steppe (Teague and Potts 2007)	NA

2004). This suggests that species-specific analysis is needed to work out the ecomorphological characteristics in order to discern the preferred habitat of local populations within a single species. Another case in point is the different ecological adaptation of the woolly rhinoceros in different regions. *Coelodonta antiquitatis* (woolly rhinoceros) is considered an indicator for glacial climates in Europe. However, it has also been found in both northern and southern localities in China where it is believed to indicate a more temperate climate as the population is smaller and lighter than their European and Siberian conspecifics (Keates 2003).

This is further confounded by the fact that IS relies heavily on precise taxonomic and biogeographic identification of fossil specimens. Disagreement among scholars may radically alter paleoecological reconstruction using this method. For instance, while Martínez-Navarro (2004, 2010) suggests the presence of several African taxa in Dmanisi, Tappen (2009) does not see such paleontological evidence with the exclusion of the genus *Homo*. Similarly, a revision of the faunal list of 'Ubeidiya between 1966 (Haas 1966), 1986 (Tchernov 1986) and 2004 (Martínez-Navarro 2004) has suggested a higher relative proportion of Africa taxa than previously reported, in accordance with the development of the 'Out of Africa I' extrinsic models.

One of the main criticisms of EDA is that it is applied only to general vegetation structure (e.g., lowland forest, montane forest, floodplain and woodland/bushland) and cannot incorporate variables like climate or habitat (e.g., Mediterranean woodland, tropical forest) into the model. For the study of the expansion of Plio-Pleistocene grasslands into Eurasia, aspects of habitat and climate rather than vegetation structure need to be specifically addressed. In EDA, identification of the relative abundance of grazers vs. browsers in a mammalian community has been used as a proxy for the extent of grassland in the environment. If the number of browsers to grazers in an African tropical forest and a non-African temperate forest are compared, both will have a high ratio of browsers to grazers, indicative of a closed environment and low proportion of grasslands. However, the former biome will be comprised of bovids and the latter by cervids. Thus, results from EDA will indicate similar vegetation structure reconstructions but will reflect forest types from very different habitats, climates and seasonality patterns. While EDA can provide robust paleoecological reconstruction of general structural forms of the vegetation, they tell us little about the climate (temperature, precipitation or seasonality) and habitat. This will be demonstrated by comparing the East African woodland-savanna with the Mediterranean park-forest.

African savannas contain widely scattered trees or shrubs. These form mosaic landscapes in which groves of woody plants are dispersed throughout a grassy matrix. Thus, savanna landscapes denote areas where savanna vegetation is dominant but may be interspersed with riparian or gallery forest, or patches of woodland, swamps, or marshes (Scholes and Archer 1997). Similarly, sclerophyllous evergreen shrubs and trees dominate Mediterranean woodlands. Within this generalized biome, different communities are recognized based on gradients of moisture and temperature, which result in a range of ground coverage of various heights and density. Thus, woodland are regions with dense forest coverage (>80%) with decreasing levels of ground coverage as well as decreasing density and height of vegetation (Allen 2001). Mediterranean woodland and African savanna-woodland exhibit a mosaic environment including both woodland and grassland. However, African savanna woodlands and Mediterranean woodlands differ in precipitation regime and seasonality, as well as in the geographic distribution of sub-habitats. The East African savanna is largely dominated by two wet periods and two dry periods (Delany and Happold 1979). In the Mediterranean region, there is only one dry season annually, which is long and is associated with the hottest months of the year (Blondel and Aronson 1999). Therefore, EDA applied to higher latitude Plio-Pleistocene sites, can identify general vegetation types but cannot account for the difference between African and Mediterranean woodland or between an African savanna and grassland.

Indicator Species is based on the adaptation of specific species and therefore is sensitive to differences in climate and habitat. Since analyses are at the species level, broader regional comparisons are often not possible. Ecological Diversity Analysis incorporates a community-wide approach and phylogeny free method. However, EDA does not have the higher resolution of habitat and climate as IS. Unlike the previous methods, Community-wide taxonomic diversity accounts for the taxonomic specific adaptation and allows for geographically broad comparisons. Community-wide taxonomic diversity (either abundance or presence/absence) provides a robust paleoecological reconstruction by using the concept of the entire community derived from ecological diversity, but focusing on select species with narrow niche requirements. Moreover, by investigating taxonomic distinctions above the species level, broader range comparisons are possible.

Community-wide taxonomic diversity is also the only paleoecological reconstruction that is consistent with other lines of evidence obtained for Early Pleistocene higher latitude paleoanthropological sites. Tchernov's (1980) analysis of the avifauna of 'Ubeidiya indicates that the Palaearctic groups dominate the assemblages and only a few are tropical (Oriental or Ethiopic). The development of the Mediterranean elements from the Asian species took place shortly after the Messinian Event (5 Ma) but increased during the Pliocene and Early Pleistocene around the Mediterranean basin and resulted in a high proportion of endemic species (Tchernov 1980). Further support for this reconstruction can be obtained from pollen spectra obtained from Eurasian and Levantine Plio-Pleistocene sites. Pollen analysis from the site of Dmanisi suggested a Mediterranean type climate analogous to recent Mediterranean woodlands (Gabunia et al. 2000). Macrofloral remains of fossilized leaves retrieved from 'Ubeidiya have been identified as *Pistacia lentiscus*, *Rhus tripratita* and *Myriophyllum* (Lorch 1966). A pollen spectrum analysis of a single stratum in 'Ubeidiya (III 12) indicated 82% arboreal pollen of which the overwhelming majority can be attributed to *Quercus* sp. followed by *Juniperus* sp. and *Olea* sp. Non-arboreal families include Gramineae, Cruciferae and Compositae. Cyperaceae comprise 8.4% of the pollen and attest to the water habitat present at the site. This pollen composition was interpreted as indicator of a pluvial environment (Bar-Yosef and Tchernov 1972). The paleoecological reconstruction obtained for western Eurasian sites by community-wide taxonomic diversity provides a robust reconstruction indicative of Mediterranean or temperate woodland habitat. The mosaic nature of this habitat does not preclude the presence of open regions or grasslands within the Mediterranean woodland biome, but they are not consistent with a subtropical African savanna.

Discussion and Conclusion

Paleoecological reconstructions based on the large mammalian assemblages from higher latitude Plio-Pleistocene sites provide inconclusive results regarding the expansion of grassland habitats during the Early Pleistocene. Critical evaluation of IS and EDA suggests they may not be applicable to higher latitude sites. Using community-wide taxonomic diversity, the paleoecological reconstruction for the western Eurasian Plio-Pleistocene sites of 'Ubeidiya, Dmanisi and Venta Micena is that of Mediterranean and temperate woodland and not an African savanna as previously suggested (Martínez-Navarro 2004, 2010; Dennell 2004; Dennell and Roebroeks 2005).

The ongoing debate between the intrinsic and extrinsic models has focused on finding evidence for the expansion of grasslands into northern latitudes during the Late Pliocene and Early Pleistocene or evidence for unique hominin adaptations in early Eurasian sites. The main line of argument used to promote the extrinsic models of 'Out of Africa I' is the Late Pliocene expansion of savannas called "Savannahstan" by Dennell (2004; 2003). The term savanna represents grasslands in the tropics and subtropics with a distinct habitat, climate, and environment. While other terms used in the literature such as "savanna" (using quotation marks) or grasslands refers to a large-scale vegetation type describing scattered trees or shrubs over continuous ground layer of herbs and grasses (Eiten 1992). The "Savannahstan" model (Dennell 2004) can be labeled the extrinsic model *sensu strictu* that suggests late Pliocene expansion of savannas into Eurasia. We could also hypothesize a climatic change that caused the expansion of grasslands, which are not African in origin; this can be labeled the extrinsic model *sensu latu*. Proponents of alternative intrinsic models, which suggest that hominins dispersed into novel habitats, do not argue against the evidence for the presence of grasslands in Eurasia but only of savannas of subtropical African origin. The presence of grasslands as an increasing part of the habitats in central and eastern Asia is well supported by other independent lines of evidence. Evidence from pollen in the Chinese loess plateau indicates a shift toward more steppe vegetation between 2.6 and 1.5 Ma (Wu et al. 2007 and references therein). The shift from C_3 to C_4 vegetation has been attributed to the intensification of the monsoon system resulting from the uplift of the Himalayan mountain range (Cerling et al. 1993).

While the general paleoevironmental reconstructions of higher latitude western Eurasian sites during the Plio-Pleistocene are suggestive of Mediterranean or temperate woodland, the presence of African grazer taxa in these regions supports paleoecological reconstructions that may include some grasslands habitats within a mosaic environment. However, there is no evidence to support the expansion of tropical or subtropical African savannas into higher latitudes during this time interval and the extrinsic model *sensu strictu* cannot be supported.

What are the implications for 'Out of Africa I'? If it can be shown that early *Homo* was adapted to savanna environments, then the ecological success of *Homo* in western Eurasian grasslands can be viewed as support of the intrinsic models. However, if it can be shown that early *Homo* was adapted to grasslands habitats, then the dispersal into Eurasia supports the extrinsic models *sensu latu*. Many of the theories that have dealt with the evolution of *Homo* in East Africa (e.g., Rightmire 1981, 1990, 2004; Wolpoff 1984; Day 1986; Cachel and Harris 1998, 1999; Aiello and Wells 2002; O'Connell et al. 2002; Antón 2003; Coqueugniot et al. 2004; Plummer 2004;), have not directly addressed this question. The research on the *Australopithecus* – *Homo* transitions suggest adaptations to open habitats and endurance running (Bramble and Lieberman 2004). However, none of these theories have addressed the issue of savanna vs. grassland as part of the unique *Homo* adaptive package. Bramble and Lieberman (2004) suggest that endurance running evolved in part to allow for hunting large mammals and that persistence running is only effective in open and hot areas and not in higher latitudes (Lieberman et al. 2007; Pickering and Bunn 2007). This would suggest that this biological adaptation might be related to the unique climate of savannas rather than grasslands.

While more specific research is needed to address the question, it would appear that the habitats in western Eurasia, both woodland and grassland, presented a novel environment to the dispersing hominin populations. Several key ecological factors differ between the northern latitude and African grasslands such as temperature, precipitation, and seasonality. It is important to stress that much of the focus of Plio-Pleistocene paleoecological and evolutionary research has been done in East Africa. The recent advance in the study of the early Pleistocene dispersal of hominins out of Africa raises the question of the extent to which we can transfer taphonomic and paleoecological actualistic models developed for one region (e.g., African savanna) to others (e.g., Mediterranean and temperate regions) and illustrates the need to study region specific models such as detailed here. This point has been made even more pertinent by the discoveries of early Pleistocene hominin sites in China (Zhu et al. 2003), which due to large geographic distances, may display even more marked differences in the past ecologies between the different regions.

Campbell (1972) suggested that hominin dispersal followed two broad categories, which were followed by biological and cultural adaptation: tropical to temperate dispersal and a much later temperate to arctic dispersal. These two phases represent an increased ability to cope with and exploit harsh environments (Turner 1984; Dennell 2004). While the conditions in the Mediterranean and temperate zone are not as harsh as in the arctic, they are more seasonal compared to African habitats, and thus represent adaptive "stepping stones." Such adaptation may include the controlled use of fire and changes in subsistence patterns. The success of dispersing hominins in northern latitudes suggests that they were able to adapt to novel environments. This ability would have required a necessary pre-adaptation to variability (Potts 1998a, b, 2002), which facilitated their successful dispersal into novel environments.

Acknowledgments I would like to thank the editors for the invitation to participate in this volume. I am grateful to Ofer Bar-Yosef for his guidance, encouragement and continuing assistance. My thanks to D. Zohary, H. Leschner and R. Flad for insightful discussions during the research. I would like to thank three anonymous reviewers for their very useful comments on earlier drafts of this manuscript. This research was funded by generous grants from the L.S.B. Leakey Foundation and Irene Levy Sala CARE Foundation. Miriam Belmaker is an American School of Prehistoric Research (ASPR) postdoctoral research fellow, Peabody Museum Harvard University (2006–2007) and a Wenner Gren Richard Carley Hunt Postdoctoral Fellow (2007–2008).

References

Aguilar, J.-P., Legendre, S., Michaux, J., & Montuire, S. (1998). Pliocene mammals and climatic reconstruction in the Western Mediterranean area. In J. H. Wrenn, J. P. Suc, & S. A. G. Leroy (Eds.), *The Pliocene: time of change* (pp. 109–120). Utah: American Association of Stratigraphic Palynologists Foundation.

Aiello, L. C. (1993). Neocortex size, group size, and the evolution of language. *Current Anthropology, 34*, 184–193.

Aiello, L. C., & Wheeler, P. (1995). Expensive-tissue hypothesis: the brain and digestive system in human and primate evolution. *Current Anthropology, 36*, 199–221.

Aiello, L. C., & Wells, J. C. K. (2002). Energetics and the evolution of the genus *Homo*. *Annual Review of Anthropology, 31*, 323–338.

Alemseged, Z. (2003). An integrated approach to taphonomy and faunal change in the Shungura Formation (Ethiopia) and its implication for hominid evolution. *Journal of Human Evolution, 44*, 451–478.

Allen, H. D. (2001). *Mediterranean ecogeography*. London: Prentice Hall.

Andrews, P. (1989). Palaeoecology of Laetoli. *Journal of Human Evolution, 18*, 173–181.

Andrews, P. (1992). Community evolution in forest habitats. *Journal of Human Evolution, 22*, 423–438.

Andrews, P. (1995a). Time resolution of the Miocene fauna from Paşalar. *Journal of Human Evolution, 28*, 343–358.

Andrews, P. (1995b). Mammals as paleoecological indicators. *Acta Zoologica Cracoviensa, 38*, 59–72.

Andrews, P. (1996). Palaeoecology and hominoid paleoenvironments. *Biological Reviews, 71*, 257–300.

Andrews, P., Lord, J. M., & Nesbit-Evans, E. M. (1979). Patterns of ecological diversity in fossil and modern mammalian faunas. *Biological Journal of the Linnaean Society, 11*, 177–205.

Antón, S. C. (2003). Natural history of *Homo erectus*. *Yearbook of Physical Anthropology, 46*, 126–1170.

Antón, S. C., & Swisher, C. C. (2004). Early dispersals of *Homo* from Africa. *Annual Review of Anthropology, 33*, 271–296.

Arribas, A., & Palmqvist, P. (1998). Taphonomy and paleoecology of an assemblage of large mammals: hyaenid activity in the Early Pleistocene site at Venta Micena (Orce, Guadix – Baza Basin, Granada, Spain). *Geobios, 31*, 3–47.

Arribas, A., & Palmqvist, P. (1999). On the ecological connection between saber-tooths and hominids: faunal dispersal events in the Early Pleistocene and a review of the evidence of the first human arrival in Europe. *Journal of Archaeological Science, 26*, 571–585.

Bar-Yosef, O. (1998). Early colonizations and cultural continuities in the Lower Palaeolithic of Western Asia. In M. D. Petraglia & R. Korisettar (Eds.), *Early human behaviour in global context: the rise and diversity of the Lower Palaeolithic record* (pp. 221–279). London: Routledge.

Bar-Yosef, O., & Tchernov, E. (1972). *On the palaeo-ecological history of the site of 'Ubeidiya*. Jerusalem: Israel Academy of Sciences and Humanities.

Bar-Yosef, O., & Belfer-Cohen, A. (2001). From Africa to Eurasia – early dispersals. *Quaternary International, 75*, 19–28.

Behrensmeyer, A. K. (1991). Terrestrial vertebrate accumulations. In D. E. G. Briggs (Ed.), *Taphonomy: releasing the data locked in the fossil record* (pp. 291–235). New York: Plenum.

Belmaker, M. (2002). Community structure changes through time – 'Ubeidiya as a case study. In H. Buitenhuis, A. M. Choyke, M. Mashkour, & A. H. Al-Shiyab (Eds.), *Proceedings of The 5th international conference of archaeozoology of the southwest Asia and adjacent areas (62)* (pp. 9–21). Center for Archeological Research and Consultancy, Rijksuniversiteit, Groningen, The Netherlands: ARC publications.

Belmaker, M. (2006). Large mammalian community structure through time: 'Ubeidiya as a case study. Ph.D. Dissertation. The Hebrew University of Jerusalem.

Belmaker, M. (2010). Early Pleistocene faunal connections between Africa and Eurasia: An ecological perspective. In: J. G. Fleagle, J. J., Shea, F. E., Grine, A. L., Baden, R. E., Leakey (Eds.), *Out of Africa I: The first human colonization of Eurasia*. Dordrecht: Springer.

Blondel, J., & Aronson, J. (1999). *Biology and wildlife of the mediterranean region*. Oxford: Oxford University Press.

Bobe, R., & Eck, G. G. (2001). Responses of African bovids to Pliocene climatic change. *Paleobiology Memoirs 2*(Supplement to 27), 1–47.
Bobe, R., Behrensmeyer, A. K., & Chapman, R. E. (2002). Faunal change, environment variability and late Pliocene Hominin evolution. *Journal of Human Evolution, 42*, 475–497.
Bramble, D. M., & Lieberman, D. E. (2004). Endurance running and the evolution of *Homo. Nature, 432*, 345–352.
Brantingham, J. P. (1998). Hominid-carnivore coevolution and invasion of the predatory guild. *Journal of Anthropological Archaeology, 17*, 327–353.
Cachel, S., & Harris, J. W. K. (1998). The lifeways of *Homo erectus* inferred from archaeology and evolutionary ecology: a perspective from East Africa. In M. D. Petraglia & R. Korisettar (Eds.), *Early human behaviour in global context* (pp. 108–132). London: Routledge.
Cachel, S., & Harris, J. W. K. (1999). The adaptive zone of *Homo erectus* from an African perspective. In H. Ullrich (Ed.), *Hominid evolution: lifestyles and survival strategies* (Edition Archaeath ed., pp. 128–137). Germany: Gelsenkirchen.
Campbell, B. (1972). Man of all seasons. In B. Cambell (Ed.), *Sexual selection and the decent of man (1871–1971)* (pp. 40–58). Chicago: Aldine.
Cerling, T. E., Kappelman, J., Quade, J., Ambrose, S., Sikes, N., & Andrews, P. (1992). Paleoenvironment of *Kenyapithecus* at Fort Ternan. *Journal of Human Evolution, 23*, 371–377.
Cerling, T. E., Wang, Y., Quage, J. (1993). Expansion of C_4 ecosystems as an indicator of global ecological change in the Late Miocene. *Nature*, 344–345.
Coqueugniot, H., Hublin, J.-J., Veillon, F., Houët, F., & Jacob, T. (2004). Early brain growth in *Homo erectus* and implications for cognitive ability. *Nature, 431*, 299–302.
Day, M. H. (1986). Bipedalism: Pressures, origins and modes. In B. Wood, L. Martin, & P. Andrews (Eds.), *Major topics in human evolution* (pp. 188–202). Cambridge: Cambridge University Press.
Dean, D., & Delson, E. (1995). *Homo* at the gates of Europe. *Nature, 373*, 472–473.
Delany, M. J., & Happold, D. C. D. (1979). *Ecology of African mammals*. London: Longman.
Delpech, F., & Heintz, E. (1976). Les faunas de Paleolithique dans le sud-ouset de la France. In H. de Lumley (Ed.), *Le Prehistoire Francaise I* (pp. 386–394). Bordeaux: Editions de Centre National de la Rechercher Scientifique.
Dennell, R. W. (2003). Dispersal and colonisation, long and short chronologies: how continuous is the Early Pleistocene record for hominids outside East Africa? *Journal of Human Evolution, 45*, 421–440.
Dennell, R. W. (2004). Hominid dispersals and Asian biogeography during the Lower and Early Middle Pleistocene, c. 2.0–0.5 Mya. *Asian Perspectives, 43*, 205–226.
Dennell, R. W., & Roebroeks, W. (2005). An Asian perspective on early human dispersal from Africa. *Nature, 138*, 1099–1104.
Dolukhanov, P. M. (2000). Microenvironments and the initial hominid settlement in Western Asia. In: D. Lordkipanidze, O. Bar-Yosef, M. Otte (Eds.), *Early humans at the gates of Europe* (pp. 93–98). Liège: Etudes et Recherches Archéologiques de l'Université de Liège.
Dzaparidze, V., Bosinski, G., Bugianisvili, T., Gabunia, L., Justus, A., Klopotovskaja, N., Kvavadze, E., Lordkipanidze, D., Majsuradze, G., Mgeladze, N., Nioradze, M., Pavlenisvili, E., Schminke, H.-U., Sologasvili, D., Tusabramsvili, D., Tvalcrelidze, M., & Vekua, A. (1992). Der altpaläolithische Fundplatz Dmanisi in Georgien (Kaukasus). *Jahrbuch Römisch-Germanischen Zentralmuseums Mainz, 36*, 67–116.
Eisenmann, V., & Guérin, C. (1984). Morphologie functionnelle et environnement chez les Périssodactyles. Geobios, Mém. *Spécial, 8*, 69–74.
Eiten, G. (1992). How names are used for vegetation. *Journal of Vegetation Science, 3*, 419–424.
Gabunia, L., & Vekua, A. (1995). A Plio-Pleistocene hominid from Dmanisi, East Georgia, Caucasus. *Nature, 373*, 509–512.
Gabunia, L., Vekua, A., & Lordkipanidze, D. (2000). The environmental context of early human occupation of Georgia (Transcaucasia). *Journal of Human Evolution, 38*, 785–802.
Gagnon, M. (1997). Ecological diversity and community ecology in the Fayum sequence (Egypt). *Journal of Human Evolution, 32*, 133–160.
Gibert, J., & Palmqvist, P. (1995). Fractal analysis of the Ocre skull sutures. *Journal of Human Evolution, 28*, 561–575.
Gibert, J., Campillo, D., Arqués, J. M., Borja, C., & Lowenstein, J. (1998). Hominid status of the Orce cranial fragment reasserted. *Journal of Human Evolution, 34*, 203–217.
Greenacre, M. J., & Vrba, E. S. (1984). A correspondence analysis of biological census data. *Ecology, 65*, 984–997.
Gunnell, G. (1995). Comparative paleoecology of Paleogene and Neogene mammalian faunas: trophic structure and composition. *Palaeogeography, Palaeoclimatology, Palaeoecology, 115*, 265–286.
Haas, G. (1966). *On the Vertebrate Fauna of the Early Pleistocene Site of 'Ubeidiya'*. Jerusalem: Israel Academy of Sciences and Humanities.
Huffman, F. O. (2001). Geological context and age of Perning/Mojokerto *Homo erectus*, East Java. *Journal of Human Evolution, 40*, 353–363.
Kaiser, T. M., & Franz-Odendaal, T. A. (2004). A mixed-feeding *Equus* species from the Middle Pleistocene of South Africa. *Quaternary Research, 62*, 316–323.
Keates, S. K. (2003). Biostratigraphy, taphonomy, paleoenvironment and hominid diet in the Middle and Late Pleistocene of China. In C. Shen, S.K. Keates (Eds.), *Current research in Chinese Pleistocene rchaeology* (pp. 37–56). Oxford: Bar International Series 1179.
Kroll, E. M. (1994). Behavioral implications of Plio-Pleistocene archaeological site structure. *Journal of Human Evolution, 27*, 1–3.
Langbroek, M. (2004). *'Out of Africa'. An investigation into the earliest occupation of the old world.* Oxford: Bar International Series 1244.
Larick, R., & Ciochon, R. L. (1996). The African emergence and early Asian dispersal of the genus *Homo. American Scientist, 84*, 538–551.
Larick, R., Ciochon, R. L., Zaim, Y., Suminto, S., Rizal, Y., Aziz, F., et al. (2001). Early Pleistocene 40Ar/ 39Ar ages for Bapang Formation hominins, Central Java, Indonesia. *Proceedings of the National Academy of Sciences, 98*, 4866–4871.
Legendre, P., & Legendre, L. (1998). *Numerical Ecology*. Amsterdam: Elsevier.
Lieberman, D. E., Bramble, D. M., Reichlen, D. A., & Shea, J. J. (2007). The evolution of endurance running and the tyranny of ethnography: a reply to Pickering and Bunn (2007). *Journal of Human Evolution, 53*, 434–437.
Lorch, Y. (1966). A Pleistocene florule from the Jordan Valley. *Israel Journal of Botany, 15*, 31–34.
Martínez-Navarro, B. (2004). Hippos, pigs, bovids, saber-toothed tigers, monkeys, and hominids: dispersals through the Levantine Corridor during Late Pliocene and Early Pleistocene times. In N. Goren-Inbar & J. D. Speth (Eds.), *Human Paleoecology in the Levantine Corridor* (pp. 37–52). Oxford: Oxbow Books.
Martínez-Navarro, B. (2010). Early Pleistocene faunas of Eurasia and hominid dispersals. In: J. G. Fleagle, J. J., Shea, F. E. Grine, A. L., Baden, R. E. Leakey (Eds.), *Out of Africa I: The first human colonization of Eurasia*. Dordrecht: Springer.
Martínez-Navarro, B., & Palmqvist, P. (1995). Presence of the African Machairdont *Megantereon whitei* (Broom, 1937) (Felidae, Carnivora, Mammalia) in the Early Pleistocene site of Venta Micene (Orce, Granada, Spain), with some considerations on the origin, evolution and dispersal of the genus. *Journal of Archaeological Science, 22*, 569–582.
Martínez-Navarro, B., & Palmqvist, P. (1999). Venta Micena (Orce, Granada, Spain): Human activity in a hyaena den during the Early Pleistocene. In G. Bosinski (Ed.), *The role of early humans in the accumulation of European Lower and Middle Palaeolithic bone assemblages* (pp. 57–71). Mainz: Römisch-Germanisches Zentralmuseum.
Mendoza, M., Janis, C., & Palmqvist, P. (2005). Ecological patterns in the trophic-size structure of large mammal communities: a 'taxon-free' characterization. *Evolutionary Ecology Research, 7*, 505–530.

Montuire, S. (1995). Evolution climatique et diversité chez les memmiféres en Europe Centrale depuis le Pliocene. *Geobios, 8*, 313–327.
Montuire, S. (1996). Rodents and climate: quantitative climatic estimates for Plio-Pleistocene faunas from Central Europe. *Acta Zoologica Cracoviensa, 39*, 373–379.
Montuire, S. (1999). Mammalian faunas as indicators of environmental and climatic changes in Spain during the Pliocene – quaternary transition. *Quaternary Research, 52*, 129–137.
Nesbit Evans, E. M., Van Covering, J. A. H., & Andrews, P. (1981). Paleoecology of Miocene sites in Western Kenya. *Journal of Human Evolution, 10*, 99–116.
O'Connell, J. F., Hawkes, K., Lupo, K. D., & Blurton Jones, N. G. (2002). Male strategies and Plio-Pleistocene archaeology. *Journal of Human Evolution, 43*, 831–872.
Palmqvist, P. (2002). On the community structure of the large mammals assemblage from Dmanisi (East Georgia, Caucasus). In: M. De Renzi, M. Alonso, M. Belinchòn, E. Peñalver, P. Montoya, A. Màrquez-Aliaga (Eds.), *Current topics on taphonomy and fossilization* (pp. 151–160). Ajuntament de Valencia.
Palmqvist, P., Pérez-Claros, A., Gibert, J., & Santamaria, J. L. (1996). Comparative morphometric study of the human phalanx from the Early Pleistocene site at Cueva Victoria (Murcia, Spain), by means of Fourier analysis, shape coordinates of landmarks, principle and relative warps. *Journal of Archaeological Science, 23*, 95–107.
Palmqvist, P., Groöcke, D. R., Arribas, A., & Fariña, R. A. (2003). Paleoecological reconstruction of a Early Pleistocene large mammal community using biogeochemical (δ13C, δ15N, δ18O Sr:Zn) and ecomorphological approaches. *Paleobiology, 29*, 205–229.
Peterson, E. C., Shen, C., Chen, C., Chen, W., Tank, Y. (2003). Taphonomy of an Early Pleistocene archaeofauna from Xiaochangliang, Nihewan Basin, North China. (Eds.), *Current research in Chinese Pleistocene archaeology* (pp. 83–98). Oxford: Bar International Series 1179.
Pickering, T. R., & Bunn, H. T. (2007). The endurance running hypothesis and hunting and scavenging in savanna woodland. *Journal of Human Evolution, 53*, 438–442.
Plummer, T. W. (2004). Flaked stones and old bones: biological and cultural evolution at the dawn of technology. *Yearbook of Physical Anthropology, 47*, 118–164.
Pope, G. G. (1995). The influence of climate and geography on the biocultural evolution of the Far Eastern Hominids. In E. Vrba, G. Denton, T. Partridge, & L. Burckle (Eds.), *Paleoclimate and evolution, with emphasis on human origins* (pp. 493–506). New Haven: Yale University Press.
Potts, R. (1998a). Environmental hypotheses of hominin evolution. *Yearbook of Physical Anthropology, 41*, 93–136.
Potts, R. (1998b). Variability selection in hominid evolution. *Evolutionary Anthropology, 7*, 81–96.
Potts, R. (2002). Complexity and adaptability in human evolution. In M. Goodman & A. S. Moffat (Eds.), *Probing human origins* (pp. 33–58). Cambridge, MA: American Academy of Arts and Sciences.
Qian, F., & Zhou, G. (1991). *Quaternary Geology and Paleoanthropology of Yuanmou, Yunnan, China*. Beijing: Science Press.
Reed, K. E. (1996). The Paleoecology of Makapansgat and other African Plio-Pleistocene hominid localities. Ph.D. Dissertation. SUNY at Stony Brook.
Reed, K. E. (1998). Using large mammals communities to examine ecological and taxonomical structure and predict vegetation in extant and extinct assemblages. *Paleobiology, 24*, 384–408.
Ricklefs, R. E., & Miller, G. L. (2000). *Ecology*. New York: W.H. Freeman.
Rightmire, G. P. (1981). Patterns of evolution in *Homo erectus*. *Paleobiology, 7*, 241–246.
Rightmire, G. P. (1990). *The evolution of Homo erectus*. Cambridge: Cambridge University Press.
Rightmire, G. P. (2004). Brain size and encephalization in early to mid-pleistocene homo. *American Journal of Physical Anthropology, 124*, 109–123.
Sagi, A. (2005). Magnetostratigraphy of 'Ubeidiya formation, Northern Dead Sea transform, Israel. M.Sc. Dissertation. The Hebrew University.
Sagi, A., Belmaker, M., Ron, H., Enzel, Y., Agnon, A., Bar Yosef, O. (2005). Paleomagnetic dating of 'Ubeidiya Formation. In S. Abramovich (Ed.), *Abstracts of the Israel geological society annual meeting 2005* (pp. 101). Israel Geological Society.
Schick, K. D., & Dong, Z. (1993). Early Paleolithic of China and eastern Asia. *Evolutionary Anthropology, 2*, 22–35.
Scholes, R. J., & Archer, S. R. (1997). Tree-grass interactions in savannas. *Annual Review of Ecological Systems, 28*, 517–544.
Semaw, S., Rogers, M. J., Quade, J., Renne, P. R., Butler, R. F., Dominguez-Rodrigo, M., Stout, D., Hart, W. S., Pickering, T., & Simpson, S. W. (2003). 2.6-Million-year-old stone tools and associate bones from OGS-6 and OGS-7, Gona, Afar, Ethiopia. *Journal of Human Evolution, 45*, 169–177.
Steudel, K. L. (1994). Locomotor energetics and hominid evolution. *Evolutionary Anthropology, 3*, 42–48.
Swisher, C. C., Curtis, G., Jacob, T., Getty, A. G., & Suprijo, A. (1994). Age of the earliest known hominids in Java, Indonesia. *Science, 263*, 1180–1121.
Tappen, M. (2009). The wisdom of the aged and out of Africa I. In J.J. Shea, D.E. Lieberman(Eds.), *Transitions in prehistory: essays in honor of Ofer Bar Yosef* (pp. 24–41). American School of Prehistoric Research.
Tchernov, E. (1980). The Pleistocene birds of 'Ubeidiya, Jordan Valley. Jerusalem: The Israel Academy of Science and Humanities.
Tchernov, E. (1986). *Les Mammifères Du Pléistocene Inférieur De La Vallée Du Jordain a Oubeidiyeh*. Paris: Association Paléorient.
Tchernov, E. (1987). The age of the 'Ubeidiya Formation, an early hominid site in the Jordan Valley. *Israel Journal of Earth Sciences, 36*, 3–36.
Teague, R.L., & Potts, R. (2007). Faunal and behavioral context of the earliest Pleistocene *Homo* in East Asia. *American Journal Physical Anthropology Supplement* 44, 230.
Turner, A. (1984). Hominids and fellow travelers: human migration into high latitudes as part of a large mammal community. In R. A. Foley (Ed.), *Hominid evolution and community ecology* (pp. 193–217). London: Academic Press.
Turner, A. (1992). Large carnivores and earliest European hominids: changing determinants of resource availability during the Lower and Middle Pleistocene. *Journal of Human Evolution, 22*, 109–126.
Vrba, E. S. (1980). The significance of bovid remains as indicators of environment and predation patterns. In A. K. Behrensmeyer & A. P. Hill (Eds.), *Fossils in the making* (pp. 247–271). Chicago: University of Chicago Press.
Walker, A. (1981). Dietary hypotheses and human evolution. *Philosophical Transactions of the Royal Society of London, 292*, 57–64.
Walker, A., & Leakey, R. (Eds.). (1993). *The Nariokotome Homo Erectus Skeleton*. Cambridge: Harvard University Press.
Williams, M. A. J., Dunkerley, D. L., De Deckker, P., Kershaw, A. P., & Stokes, T. (1996). *Quaternary Environments*. London: Arnold.
Wolpoff, M. H. (1984). Evolution in *Homo erectus*: the question of stasis. *Paleobiology, 10*, 389–406.
Wu, F., Fang, X., Ma, Y., Herrmann, M., Mosbrugger, V., An, Z., & Miao, Y. (2007). Plio-Quaternary stepwise drying of Asia: evidence from a 3-Ma pollen record from the Chinese Loess Plateau. *Earth and Planetary Science Letters, 257*, 160–169.
Zhu, R. X., Hoffman, R., Potts, R., Deng, C. L., Pan, Y. X., Guo, B., Shi, C. D., Guo, G. T., Yuan, B. Y., M., H. Y., Huang, W. W. (2001). Earliest presence of humans in northeast Asia. *Nature, 413*, 413–417.
Zhu, R. X., An, Z., Potts, R., & Hoffman, K. A. (2003). Magnetostratigraphic dating of early humans in China. *Earth Science Reviews, 61*, 341–359.
Zhu, R. X., Potts, R., Xie, F., Hoffman, K. A., Deng, C. L., Shi, C. D., Pan, Y. X., Wang, H. Q., Shi, R. P., Wang, Y. C., Shi, G. H., & Wu, N. Q. (2004). New evidence regarding the earliest human presence at high northern latitudes in northeast Asia. *Nature, 431*, 559–562.

Chapter 4
Africa and Asia: Comparisons of the Earliest Archaeological Evidence

David R. Braun, Christopher J. Norton, and John W.K. Harris

Abstract The earliest archaeological sites in East Asia suggest that making and using stone artifacts was a consistent part of the subsistence strategy of these earliest immigrants to East Asia. Although there are many differences between the earliest archaeological record in Africa and Asia, a few aspects of these industries allow formal comparisons. Here we review aspects of the African Oldowan archaeological record and compare it to the large and well-studied archaeological record from the Nihewan Basin. We suggest that the technological strategies shown in these East Asian Early Paleolithic assemblages are consistent with a subsistence pattern where stone tool mediated resources played a very different role than that found in East Africa. We suggest that the poor quality of available materials were not conducive to the maintenance of complex toolkits. Early Pleistocene hominins in East Asia may have exploited a series of diverse resources that had distinct technological requirements. In this sense the ecology of these hominins may have been very different from their African counterparts.

Keywords Early Stone Age • Lower Paleolithic • Oldowan • Koobi Fora • Olduvai • Nihewan • Kanjera

Introduction

The appearance of stone tool assemblages in Asia has important implications for the place of technology in human evolution. As new research uncovers older archaeological sites outside of Africa (Zhu et al. 2004) the role of stone tools in the dispersals of hominins from Africa needs to be addressed. Although the archaeological record outside of East Africa is unfortunately relatively meager, the overwhelming evidence is that the earliest stone tool use evolved in Africa (Semaw et al. 1997). However, in order to address the role of stone tool technology in the dispersal of hominins into Asia it is vital to understand the similarities and differences between hominin toolkits between these two regions. Here we review what is currently known about the use and manufacture of stone tools in East Africa. We then develop hypotheses about what the earliest assemblages outside of East Africa may look like based on a brief review of the ecology of latitudinal changes, as well as the local contextual information known from the earliest sites outside of Africa. Some assemblages from Asia are particularly well-studied, and therefore provide the information necessary for comparisons with the African record. Finally, we formulate some hypotheses that may explain some of the patterning seen in the artifact assemblages of East Asia.

The Oldowan of Africa

There is an emerging view about the complexity of the earliest Oldowan stone tool assemblages (Roche 2000; Semaw 2000; de la Torre 2004; Delagnes and Roche 2005). Although initial discussion suggested that hominins before 2 Ma lacked the full capacity to produce long reduction sequences (Kibunjia et al. 1992; Kibunjia 1994), it is now clear that there is great variability in the ability of hominin toolmakers (Roche 2000). There is a growing body of evidence from the sites in the Kada Gona region that early hominins were able to select stones based on the flaking quality (Stout et al. 2005) and that these hominins were adept tool makers (Semaw 2000). Further evidence from the Nachukui Formation suggests that at least some hominins followed a specific set of rules that governed tool production (Delagnes and Roche 2005). These rules allowed hominins to extend the sequence of removals in particular core forms. Although there were apparently some technological obstacles that these hominins could not overcome, many Pliocene industries show a remarkable degree of technological flexibility (de la Torre 2004; Delagnes and Roche 2005).

The Oldowan is often characterized as merely simple core and flake tools (Foley and Lahr 2003); however the techniques and systems employed in these earliest assemblages

D.R. Braun (✉)
Archaeology Department, University of Cape Town, Rondebosch 7701, South Africa
e-mail: david.braun@uct.ac.za

C.J. Norton and D.R. Braun (eds.), *Asian Paleoanthropology: From Africa to China and Beyond*, Vertebrate Paleobiology and Paleoanthropology, DOI 10.1007/978-90-481-9094-2_4, © Springer Science+Business Media B.V. 2010

appear to be more complex than the simplest solution to producing a sharp edge (Isaac and Harris 1997). Indeed, many studies now suggest that stone tool use in the Oldowan represented a complex system of resource acquisition (Braun and Harris 2003; Goldman et al. 2006; Harmand 2006). Although there is still some debate as to whether or not Pliocene stone tool industries represent an advanced technological system, major changes in the Early Pleistocene suggest a dramatic increase in the dependence of hominins on stone tool technology. Various Developed Oldowan industries show increases in the complexity of tool reduction (de la Torre et al. 2003), landscape use (Rogers et al. 1994), and tool transport (Kimura 2002), as well as the intensity of stone tool resource utilization (Braun and Harris 2003). This is probably most dramatically displayed in simple comparisons of mean maximum flake size and mean maximum core size in 20 Pliocene and Pleistocene archaeological assemblages from East Africa. These sites show that before 1.7 Ma hominins were not emphasizing the size of flakes being produced from cores. When hominins had access to larger cores they appear to have made similarly sized flakes irrespective of initial cobble size. After 1.7 Ma this pattern changes and the size of flakes tracks the size of cores. It is possible this trend is the result of increasing need by hominins in the Pleistocene to increase the distance between raw material procurement and activity loci on the ancient landscape by producing the largest flake possible from a given core (Braun and Harris 2001). This change that represents a shift to Developed Oldowan Industries represents a shift to standardized reduction systems like the single platform core (Ludwig and Harris 1998) and the centripetal hierarchical bifacial core structure (de la Torre et al. 2003).

These technological developments through the course of the Oldowan are likely associated with an increase in the utilization of stone artifact mediated resources: as seen in the increase in the utilization of carcasses on the landscape at sites at Olduvai (Monahan 1996) and Koobi Fora (Harris et al. 2002). In other words, the need for stone artifacts increased extensively and as a consequence of natural selection favoring technological systems that increased the efficiency (higher yield per unit of cost: (Christenson 1982) of stone artifact manufacture and use (Jeske 1989; Torrence 1989b). At many East African localities, the appearance of a more organized and standardized technological system, is often associated with the habitation of more arid and grassland environments. Certainly for sites like Gadeb (Williams et al. 1979) and Melka Kontoure (Isaac 1971) which are situated at extreme elevations, these xeric conditions would have imparted new constraints on hominin behavior/ecology – some of which may have necessitated shifts in lithic technologies. At Koobi Fora, the Karari Industry is associated with a shift to more xeric environments as documented by a recent analysis of paleosols (Wynn 2004). Developed Oldowan Industries at Olduvai may also be associated with a shift to more xeric conditions (Leakey 1971; Monahan 1996).

An Asian Perspective on the Oldowan

Previous efforts to characterize stone tool use outside of Africa have been hindered by small collections excavated from contexts that are not fully understood. This was largely the result of reduced infrastructure for prehistoric studies relative to the long history of research in Africa (Dennell 1998, 2001). Although there is clearly more research to be done on Early Stone Age archaeology outside of Africa, a clearer picture of the earliest Asian industries is beginning to appear (Keates 2004). Yet it is clear that, at least superficially, there are major differences between the African and Asian Early Stone Age records (Norton et al. 2006; Lycett and Norton 2010). As yet there are no sites in Asia that possess similar densities of lithic materials found in localities such as Koobi Fora, Hadar, Gona and Olduvai. While assemblage size is likely the result of a number of factors, including but not limited to: raw material availability, raw material quality, excavation strategy, and/or ancient hominin group size (Ammerman and Feldman 1974; Potts 1991; McHenry and Coffing 2000). Unfortunately, many of these factors cannot be controlled for or easily modeled. Yet a few certainties can be deduced in the comparisons of Asian and Oldowan industries. First, hominins in East Asia used stone tools to procure resources, some of which appear to be animal resources as seen by the percussion notched bone at Majuangou (Shen and Chen 2000; Zhu et al. 2004). Second, hominins that produced stone tools in East Asia had an understanding of fracture mechanics that was similar, at a minimum, to that expressed in the earliest Oldowan industries (Schick et al. 1991). Any further comparison between Asian and African lithic industries requires a more detailed discussion of the context of the Asian localities.

The Context of the Asian Early Paleolithic

Recent reviews of the dispersals of hominins out of Africa have remarked on the apparent lack of technological development associated with this range expansion (Anton and Swisher 2004). However, there may be some reason to believe that stone artifacts performed a different role in the adaptation of hominins in Asia than Africa. The work of Robin Dennell (1998, 2003) is particularly relevant to this explanation. Dennell (2003) has reviewed the environmental context of many of the earliest sites in Asia and his models of dispersal and colonization provide a framework that can be used to investigate the nature of the earliest tool-use outside of Africa. Dennell's (2003) assertion that early hominin occupation of Eurasia probably represents several failed colonization attempts following the expansion of grasslands into western Eurasia seems very plausible.

Many of the changes associated with the expansion into Asian habitats can be associated with variability in latitude. If hominins continued to gain some portion of their diet from large mammal remains, they would need to expand their annual ranges as opportunities for carcass acquisition would likely have been more widely dispersed. Increased distances from the equator are correlated with decreases in lower net production as well as total above ground productivity (Rosenzweig 1968; Binford 2001). Species diversity has an inverse relationship with latitude and therefore the availability of alternative resources decreases (Torrence 2001). The result is that the risk of hunter–gatherers failing to meet dietary requirements becomes more intense with movement toward the poles. These ecological differences, combined with a more diverse carnivore paleoguild in Asia in the Early Pleistocene and the increase in carnivore size with latitude (Klein 1986), would have meant that carcass acquisition would have been far more difficult in East Asia than in parts of East Africa (Turner 1992). High frequencies of carnivore modification on several East Asian Paleolithic localities attests to the high competition for large mammal resources in Asian contexts (Bakken 1997), though by the latter part of the Early Paleolithic hominins appear to have become dominant members of the carnivore guild (Norton and Gao 2008). Other differences associated with increases in latitude would relate to the slower rates of putrification of carcasses in higher latitudes associated with the interplay of temperature, humidity, and exposure to sunlight (Haynes 1982, 1988). Therefore, the reduced availability of flesh on carcasses may have been countered by an increased time frame for accessing marrow in carcasses. The implications for stone tool assemblages would be an increased reliance in pounding technologies associated with marrow processing, as opposed to cutting technologies. Another major change would have been the habitation of environments that have not been previously recorded in African archaeological sites. These habitats such as tropical rainforests (Pope 1995) or steppe and woodland environments (Aigner 1981; Belmaker 2006; Teague and Potts 2007) would have had a different resource structure than African savannas. These environments have lower percentages of their biomass in leafy material available for grazing and browsing animals to eat (Binford 2001). As a consequence these ecosystems would have had much lower frequencies of secondary biomass (large ungulates).

The broad reconstruction of Pliocene habitats by Dennell (2003) and Holmes (2007) using data from the PRISM project (Dowsett et al. 1994) shows the expansion of grasslands into much of the Asian continent but also the persistence of many forested habitats in much of Asia. In many of these habitats the majority of resources available to Pleistocene hunter–gatherers would have had low activity failure rates (i.e. plant tissues) (Jeske 1989). Thus, we can expect that Early Pleistocene toolmakers in East Asia would have invested less energy into the procurement, and maintenance of their toolkits (Jeske 1989; Torrence 1989a; Bamforth and Bleed 1997). The toolkits of East Africa are associated with the energy costs of transport and selection of high quality raw materials (Stout et al. 2005; Braun al. 2008a), as well as the time costs associated with the consistent production of numerous sharp edges (Delagnes and Roche 2005). These costs are offset by the high activity failure rates of large mammal butchery that was clearly a part of the East African Oldowan subsistence pattern (de Heinzelin et al. 1999; Harris et al. 2002; Dominguez-Rodrigo et al. 2005). However, if the East Asian Pleistocene toolmakers were focused on plant tissues as a higher percentage of their tool-assisted resource base, then we may expect a limited investment of energy into the production of large cutting tools. Subsequently, we may expect shorter tool use-lives and higher discard rates associated with these East Asian toolkits (Jeske 1989; Bousman 1993). We would predict that toolkits from East Asian contexts would not have required the long systematic reduction sequences found in East Africa, because the large mammal carcasses that require numerous small sharp edges may not have been a regular aspect of the resource base of East Asian hominins. If Early Pleistocene hominins in East Asia did shift their resource base away from large mammal tissue it may have been associated with a major shift in diet. Current understanding of *Homo erectus* sensu lato physiology suggests that this species required access to high quality diets (Wood and Collard 1999; McHenry and Coffing 2000). However, in some instances, these dietary requirements were met with the assistance of technologies that did not focus exclusively on the sharp edges required for acquisition of large mammal tissues (e.g. Goren-Inbar et al. 2002; Mora and De la Torre 2005). A focus on percussion technology or expedient flake and core technology to access high quality, predictable resources such as nuts, fruits or underground storage organs may have become a major focus of East Asian hominin behavior.

The Asian Early Paleolithic: Predictions and Current Data

The Early Paleolithic artifacts from East Asia differ from some the African Oldowan industries. Four main factors distinguish Asian and African Early Pleistocene assemblages.

First, in many Asian sites, it appears that there is a low density of lithic materials (Clark and Schick 1988; Shen and Qi 2004). Second, cores at Asian localities appear to have relatively short reduction sequences (Gao et al. 2005). Third, there is a high incidence of pounded pieces in some Asian assemblages (Pope 1988). Some assemblages show consistent evidence of core tools that also have evidence of pitting or bruising (Pope and Keates 1994). Fourth, the artifacts from East Asian assemblages appear to be demonstratively smaller than African Oldowan assemblages (Clark 1998). This is by no means an exhaustive review of East Asian Early Pleistocene localities. These observations are over-generalizations of a diverse group of assemblages and there are particular instances that will refute each one of these observations. However, these generalizations allow a framework that facilitates comparisons with East African artifacts. Unfortunately, very few of the East Asian localities in this time frame have enough artifacts to investigate patterns of tool manufacture and use (e.g. Xihoudu: 30 artifacts). Subsequently, we restrict our discussion to the best known Early Pleistocene sites, those located in the Nihewan Basin, northern China.

Nihewan Basin

In contrast to other regions in Asia, the large size and extensive exposures make assemblages from the Nihewan basin more comparable to Oldowan localities in Africa. Further, paleoenvironmental reconstruction of the sites in the Nihewan Basin suggests that these sites were found in habitats that are more similar to East African sites relative to other East Asian sites (Zhu et al. 2004). The tremendous preservation of vast deposits of the Nihewan formation makes it possible to examine hominin artifact transport and selection of stone raw materials. It is likely the Nihewan hominins were collecting raw materials from local deposits (Schick et al. 1991). Schick and colleagues (1991) suggest that the high frequency of localities on the eastern side of the Cenjiawan Platform is because of the proximity to these exposed Precambrian quartzites and cherts. Keates' (2000) analysis of the Xiaochangliang and a subset of the Donggutou collections indicate that this chert is the material most often selected. Interestingly, basalt represents a small portion of the Xiaochangliang assemblage (Keates 2000). Although there are exposed volcanic materials near Xiaochangliang the only real exposures of basalts are some 100 km to the east (Schick and Dong 1993). These raw materials may have been collected from secondary deposits much closer to the site. Extensive sampling of conglomeratic deposits would be necessary to fully understand these behaviors in detail.

Like many Early Pleistocene sites in Africa, the Nihewan sites display a pattern where hominins appear to have collected and used artifacts near to where they were eventually deposited. Local sources are a defining feature of many Oldowan assemblages (Stout et al. 2005; Goldman et al. 2006; Harmand 2006). Although chert was locally abundant, it was not a high quality material for artifact production. Schick and colleagues (1991) note that, within one chert nodule, one half may be completely homogenous while the other half may be riddled with impurities that cause the piece to fracture in unpredictable ways. This likely explains why Schick and colleagues' (1991) analysis of the Donggutou material show an extremely high incidence of angular fragments and very few cores. Similar patterns are found in a recent analysis of the Xiaochangliang materials (Shen and Chen 2003). Schick and colleagues analysis of the core forms from Donggutou note that none of the cores could be described as formal core forms in Leakey's (1971) typology. This pattern may be an adaptation to the particular constraints of the available raw materials. In the absence of raw material that fractures in predictable ways, the development of standardized flake removal system offers few advantages (Brantingham et al. 2000). On a similarly intractable raw material at the Oldowan site of Kanjera South in western Kenya, hominins employed a haphazard reduction strategy (Braun 2006). This highlights some basic parallels between African and Asian industries. We predict that the lack of high quality material in the Nihewan Basin and the potentially decreased reliance on high activity failure rate resources (Jeske 1989) mediated by sharp edged stone tools in these high latitude ecosystems (Binford 2001; Torrence 2001), resulted in a technology that does not involve the utilization of high quality raw materials intensively.

When compared to the patterns that characterize the Developed Oldowan, the Nihewan assemblages do not show similar patterns of flake production (Fig. 4.1). Artifacts from the Nihewan Basin seem to fall within the pattern of flake production similar to that expressed in Pliocene African assemblages, despite the fact that the Nihewan sites are in fact younger than the Developed Oldowan sites of East Africa (Zhu et al. 2004). We believe these patterns are to be expected. The available raw materials in the Nihewan Basin did not allow the systematic continuous production of large flakes that is seen in the Developed Oldowan (Ludwig and Harris 1998; Braun and Harris 2001; de la Torre et al. 2003). The parallels between the Nihewan and Pliocene Oldowan assemblages we have outlined here does not suggest some type of cultural stagnation on the part of East Asian Pleistocene toolmakers. Non-standardized core and flake industries are very well adapted to a number of different ecological scenarios. Modern humans have employed a simple core and flake technology with great success (Hayden 1979, 1989; Gould 1980; Shott and Sillitoe

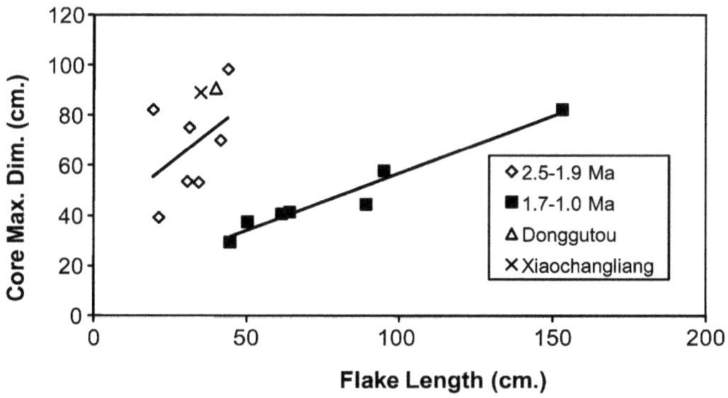

Fig. 4.1 Mean flake length and mean maximum core size for several East African Oldowan assemblages and two assemblages from the Nihewan basin. Date from Ludwig 1998 and Keates 2004

2001). It is possible that the resources that were being acquired with the stone tools in the Pleistocene of the Nihewan Basin did not require the consistent production of large sharp edged flakes. It is interesting to note that very few modified bones have been recorded from the Nihewan assemblages despite excellent fossilization of bones (>75% in Behrenesmeyer's (Behrensmeyer 1978) weathering stage 1 or 2 at Xiaochangliang) found in association with artifacts (Peterson et al. 2003). Interestingly, several examples of percussion fractured bones have been recovered (Peterson et al. 2003; Zhu et al. 2004). It is possible that the incidence of percussion-fractured bones in the absence of cut-marks reflects the slower rate of carcass decay found in higher latitude ecosystems. If hominins in the Nihewan were not dependent on a sharp edged stone tool technology and instead practiced a more diverse foraging strategy that included a heavier reliance on plant foods relative to their African counterparts this may be reflected in these Asian industries. Unfortunately, pounding tools are not well known from the archaeological record of the Nihewan Basin (Shen and Chen 2003). However, other East Asian localities do show some evidence of a percussive technology. The site of Chenjiawo in the Lantian region displays examples of very large (>16 cm) cobbles with pitting and bruising (Keates 2000). Anvils and hammerstones have also been identified at the Middle Pleistocene Zhoukoudian Locality 1 site (Chiu et al. 1973; Zhang 1985).

Another possible scenario that has been forwarded to explain the differences between East African Oldowan assemblages and the Early Pleistocene assemblages of East Asia is that the stone tool technology found in East Asian localities represents the basis of a more extensive but archaeologically invisible technology. Actualistic experiments show that the technology found in the Nihewan Basin could be used to manufacture a broad suite of bamboo implements (Clark 1998). Comparisons between African and Asian assemblages show much promise for teasing apart the impetus behind the differences and similarities in these assemblages.

Discussion

It might be tempting to use the superficially rudimentary aspects of the earliest evidence of stone tools in Asia to suggest a series of cultural waves out of Africa associated with different technological systems. It is possible to then associate these different movements with different cognitive capacities (e.g. Pre-Oldowan, Oldowan, Developed Oldowan) (Rolland 1998; Carbonell et al. 1999). This may be an oversimplification of the Early Paleolithic archaeological record. In relation to the African Oldowan record, the East Asian record represents a similar behavioral system applied to a unique context. Although it appears that many Asian sites were found in grasslands (Teague and Potts 2007) that may have been similar to East African grasslands, the differences in latitude associated with changes in secondary biomass (Binford 2001; Torrence 2001) and the persistence of diverse carnivore forms throughout the Pleistocene (Turner 1992) would have made for vastly different ecologies in these two areas. One aspect of the Chinese Early Paleolithic that seems to contrast the African record is the long periods of what appears to be static technological change (Gao and Norton 2002). This may be because many of the environments in Asia are best exploited using a simple core and flake technology.

There is good reason to believe that the African archaeological record represents a consistent increase in technological efficiency (Christenson 1982) associated with adaptations to increasingly variable habitats in Africa (Potts 1994, 1996; Rogers et al. 1994; Braun and Harris 2001; Braun et al. 2008b). It is possible that the habitats seen in some Asian localities represented unique circumstances that did not always require the flaked stone toolkit that environments in East Africa necessitated. Subsequently, the archaeological record of Asia may represent the application of a similar technological system to a new context. The hypotheses we have developed here show some promise for understanding the similarities and differences between the East African Oldowan and the East Asian Early Paleolithic; however, two attributes of the East Asian record do not support this hypothesis. The first is the general lack of pounding tools currently described at East Asian Early Paleolithic industries. Current reviews of the Early and Middle Pleistocene archaeological record of East Asia (Keates 2000) describe small quantities of lightly flaked diminutive quartz pebbles at many of these localities. The second piece of data that contradicts the hypothesis forwarded here derives from microwear evidence from the Nihewan Basin (Shen and Chen 2003). This analysis suggests that at least some of the tools from Xiaochangliang were used in the processing of animal material (Shen and Chen 2003).

Conclusion

The differences between the African record and the Early Paleolithic record in the Nihewan Basin represent the variability of a generalized Oldowan behavioral system. The variability seen in the core and flake industries of East Asia need not imply the presence of a relict population of Mode I toolmakers in the Early Pleistocene (Rolland 1998; Carbonell et al. 1999). It is possible that the hominins that first dispersed into East Asia arrived with a simple toolkit that was well adapted to the environments these hominins encountered. The presence of Mode I industries in central and south Asia and Europe well into the Middle Pleistocene attest to the success of these industries (Dennell 1998, 2003; Rolland 1998; Lycett 2007). We caution the attribution of different archaeological assemblages to specific waves of hominins moving out of Africa with specific toolkits. Regardless of cognitive or technical associations with Mode I technologies (Inizan et al. 1992; Stout and Chaminade 2007), the presence of simple core and flake toolkits over wide temporal and geographic ranges suggests they were extremely adaptive for particular ecological contexts.

Acknowledgements The authors would like to thank Professors Gao Xing, Wei Qi, Jin Changzhu, and other members of the Institute of Vertebrate Paleontology and Paleoanthropology staff that provided us with an enlightening tour of the archaeological sites in China in the summer of 2006. We thank Alison Brooks for imparting some of her expansive understanding of the East Asian archaeological record. We also thank the helpful suggestions of three anonymous reviewers. David R. Braun would like to thank the support of the University of Cape Town Emerging Researchers Program and constant careful assistance of Kathryn Underwood.

References

Aigner, J. S. (1981). *Archaeological remains in Pleistocene China.* Munchen: C.H. Beck Press.

Ammerman, A. J., & Feldman, M. (1974). On the 'Making' of an assemblage of stone tools. *American Antiquity, 39*, 610–616.

Anton, S. C., & Swisher, C. C. (2004). Early dispersals of *Homo* from Africa. *Annual Review of Anthropology, 33*, 271–296.

Bakken, D. A. (1997). Taphonomic parameters of Pleistocene hominid sites in China. *Bulletin of the Indo-Pacific Prehistory Association, 16*, 13–26.

Bamforth, D. B., & Bleed, P. (1997). Technology, flaked stone technology, and risk. In C. M. Barton & G. A. Clark (Eds.), *Rediscovering Darwin: evolutionary theory and archeological explanation* (7th ed., pp. 109–139). Washington, DC: American Anthropological Association.

Behrensmeyer, A. K. (1978). Taphonomic and ecologic information from bone weathering. *Paleobiology, 4*, 150–162.

Belmaker, M. (2006). Large mammalian community structure through time: Ubeidiya as a case study. Ph.D. dissertation. Hebrew University.

Binford, L. R. (2001). *Constructing frames of reference: an analytical method for archaeological theory building using hunter–gatherer and environmental data sets.* Berkeley: University of California Press.

Bousman, C. B. (1993). Hunter–gatherer adaptations, economic risk and tool design. *Lithic Technology, 18*, 59–86.

Brantingham, P. J., Olsen, J. W., Rech, J. A., & Krivoshapkin, I. A. (2000). Raw material quality and prepared core technologies in Northeast Asia. *Journal of Archaeological Science, 27*, 255–271.

Braun, D. R. (2006). Ecology of Oldowan Technology: Koobi Fora and Kanjera South. Ph.D. dissertation. Rutgers University.

Braun, D. R., & Harris, J. W. K. (2001). Oldowan technology: new insights from digital image analysis of Koobi Fora lithic assemblages. *Journal of Human Evolution, 40*, A4–A4.

Braun, D. R., & Harris, J. W. K. (2003). Technological developments in the Oldowan of Koobi Fora: innovative techniques of artifact analysis. In J. M. Moreno, R. M. Torcal, & I. T. Sainz (Eds.), *Oldowan: rather more than smashing stones* (pp. 117–144). Barcelona: University of Barcelona Press.

Braun, D. R., Plummer, T., Ditchfield, P., Ferraro, J. V., Maina, D., Bishop, L. C., Potts, R. (2008a). Oldowan behavior and raw material transport: perspectives from the Kanjera Formation. *Journal of Archaeological Science, 35*, 2329–2345.

Braun, D. R., Rogers, M. J., Harris, J. W. K., Walker, S. J. (2008b). Landscape scale variation in hominin tool use: evidence from the developed Oldowan. *Journal of Human Evolution, 55*, 1053–1063.

Carbonell, E., Mosquera, M., Rodriguez, X. P., Sala, R., & van der Made, J. (1999). Out of Africa: the dispersal of the earliest technical systems reconsidered. *Journal of Anthropological Archaeology, 18*, 119–136.

Chiu, C., Ku, Y., Chang, Y., & Chang, S. (1973). Newly discovered *Sinanthropus* remains and stone artifacts from Choukoutien. *Vertebrata PalAsiatica, 11*, 109–131.

Christenson, A. L. (1982). Maximizing clarity in economic terminology. *American Antiquity, 47*, 419–426.

Clark, D. J. (1998). The Early Paleolithic of the eastern region of the Old world in comparison to the West. In M. D. Petraglia & R. Korisettar (Eds.), *Early human behavior in global context. The rise and diversity of the Lower Paleolithic record* (pp. 437–450). London: Routledge.

Clark, D. J., & Schick, K. D. (1988). Context and content: impressions of Paleolithic sites and assemblages in the People's Republic of China. *Journal of Human Evolution, 17*, 439–448.

de Heinzelin, J., Clark, J. D., White, T. D., Hart, W. S., Renne, P. R., Woldegabriel, G., et al. (1999). Environment and behavior of 2.5-million-year-old Bouri hominids. *Science, 284*, 625–629.

de la Torre, I. (2004). Omo revisited – evaluating the technological skills of Pliocene hominids. *Current Anthropology, 45*, 439–465.

de la Torre, I., Mora, R., Dominguez-Rodrigo, M., de Luque, L., & Alcalá, L. (2003). The Oldowan industry of Peninj and its bearing on the reconstruction of the technological skills of Lower Pleistocene hominids. *Journal of Human Evolution, 44*, 203–224.

Delagnes, A., & Roche, H. (2005). Late Pliocene hominid knapping skills: the case of Lokalalei 2C, West Turkana, Kenya. *Journal of Human Evolution, 48*, 435–472.

Dennell, R. W. (1998). Grasslands, tool making and the hominid colonization of southern Asia: a reconsideration. In R. Korisettar & M. D. Petraglia (Eds.), *Early human behavior in global context: the rise and diversity of the Lower Paleolithic record* (pp. 280–303). London: Routledge.

Dennell, R. W. (2001). *From Sangiran to Olduvai, 1937–1960: the quest for 'centres' of hominid origins in Asia and Africa. Studying Human Origins: Disciplinary History and Epistemology* (pp. 45–66). Amersterdam: Amsterdam University Press.

Dennell, R. (2003). Dispersal and colonisation, long and short chronologies: how continuous is the Early Pleistocene record for hominids outside East Africa? *Journal of Human Evolution, 45*, 421–440.

Dominguez-Rodrigo, M., Pickering, T. R., Semaw, S., & Rogers, M. J. (2005). Cutmarked bones from Pliocene archaeological sites, at Gona, Afar, Ethiopia: implications for the function of the world's oldest stone tools. *Journal of Human Evolution, 48*, 109–121.

Dowsett, H., Thompson, J., Barron, J., Cronin, T., Fleming, F., Ishman, S., et al. (1994). Joint investigations of the Middle Pliocene climate I: PRISM paleoenvironmental reconstrcutions. *Global and Planetary Change, 9*, 169–195.

Foley, R., & Lahr, M. M. (2003). On Stony Ground: lithic technology, human evolution, and the emergence of culture. *Evolutionary Anthropology, 12*, 109–122.

Gao, X., & Norton, C. J. (2002). A critique of the Chinese 'Middle Palaeolithic'. *Antiquity, 76*, 397–412.

Gao, X., Wei, Q., & Keates, S. G. (2005). New light on the earliest hominid occupation in East Asia. *Current Anthropology, 46*, 115.

Goldman, T., Feibel, C. S., Hovers, E. (2006). *Lithic raw material selection at the late Pliocene site AL894 (Hadar, Ethiopia)*. 71st Annual meeting of the society for American archaeology. Puerto Rico: San Juan.

Goren-Inbar, N., Sharon, G., Melamed, Y., & Kislev, M. (2002). Nuts, nut cracking, and pitted stones at Gesher Benot Ya'aqov, Israel. *Proceedings of the National Academy of Sciences of the United States of America, 99*, 2455–2460.

Gould, R. A. (1980). *Living archaelology*. Cambridge: Cambridge University Press.

Harmand, S. (2006) Oldowan intersite diversity in the Nachukui Formation, West Turkana, Kenya. 71st Annual meeting of the society for American archaeology. Puerto Rico: San Juan.

Harris, J. W. K., Braun, D. R., McCoy, J. T., Pobiner, B. L., & Rogers, M. J. (2002). New directions and preliminary results from a landscape approach to the study of archaeological traces for the behavior of Plio-Pleistocene hominids at Koobi Fora. *Journal of Human Evolution, 42*, A15–A15.

Hayden, B. (1979). *Palaeolithic reflections: lithic technology and ethnographic excavations among Australian aborigines*. Canberra: Australian Institute of Aboriginal Studies.

Hayden, B. (1989). From chopper to celt: the evolution of resharpening techniques. In R. Torrence (Ed.), *Time, energy and stone tools* (pp. 30–42). Cambridge: Cambridge University Press.

Haynes, G. (1982). Prey bones and predators: potential ecologic information from analysis of bone sites. *OSSA: International Journal of Skeletal Research, 7*, 75–97.

Haynes, G. (1988). Longitudinal studies of African elephant death and bone deposit. *Journal of Archaeological Science, 15*, 131–157.

Holmes, K. M. (2007). Using Pliocene palaeoclimatic data to postulate dispersal pathways of early hominins. *Palaeogeography Palaeoclimatology Palaeoecology, 248*, 96–108.

Inizan, M.-L., Roche, H., & Tixier, J. (1992). *Technology of knapped stone: prehistoire de la Pierre Taillee*. Meudon: CREP.

Isaac, G. L. (1971). The diet of early man: aspects of archaeological evidence from Lower and Middle Plestocene sites in Africa. *World Archaeology, 2*, 278–299.

Isaac, G. L., & Harris, J. W. K. (1997). The stone artifact assemblages: a comparative study. In G. L. Isaac (Ed.), *Koobi Fora research project* (Plio-Pleistocene Archaeology, Vol. 5, pp. 262–362). Oxford: Clarendon.

Jeske, R. (1989). Economies in raw material use by prehistoric hunter–gatherers. In R. Torrence (Ed.), *Time, energy and stone tools* (pp. 34–45). Cambridge: Cambridge University Press.

Keates, S. G. (2000). *Early and Middle Pleistocene hominid behavior in Northern China*. Oxford: BAR Archaeopress.

Keates, S. G. (2004). Home range size in Middle Pleistocene China and human dispersal patterns in Eastern and Central Asia. *Asian Perspectives, 43*, 228.

Kibunjia, M. (1994). Pliocene archaeological occurrences in the Lake Turkana Basin. *Journal of Human Evolution, 27*, 159–171.

Kibunjia, M., Brown, F. H., Leakey, R. E., & Roche, H. (1992). Pliocene and Pleistocene archaeological sites west of Lake Turkana, Kenya. *Journal of Human Evolution, 23*, 431–438.

Kimura, Y. (2002). Examining time trends in the Oldowan technology at Beds I and II, Olduvai Gorge. *Journal of Human Evolution, 43*, 291–321.

Klein, R. G. (1986). Carnivore size and quaternary climatic change in southern Africa. *Quaternary Research, 26*, 153–170.

Leakey, M. D. (1971). *Olduvai Gorge*. Cambridge: Cambridge University Press.

Ludwig, B. V., & Harris, J. W. K. (1998). Towards a technological reassessment of East African Plio-Pleistocene lithic assemblages. In M. Petraglia & K. Paddaya (Eds.), *The rise and diversity of the Lower Paleolithic* (pp. 84–106). New York: Academic.

Lycett, S. J. (2007). Is the Soanian techno-complex a Mode 1 or Mode 3 phenomenon? A morphometric assessment. *Journal of Archaeological Science, 34*, 1434–1440.

Lycett, S. J., & Norton, C. J. (2010). A demographic model for Palaeolithic technological evolution: the case of East Asia and the Movius Line. *Quaternary International, 211*, 55–65.

McHenry, H. M., & Coffing, K. (2000). Australopithecus to Homo: transformations in body and mind. *Annual Review of Anthropology, 29*, 125–146.

Monahan, C. M. (1996). New zooarchaeological data from Bed II, Olduvai Gorge, Tanzania: implications for hominid behavior in the Early Pleistocene. *Journal of Human Evolution, 31*, 93–128.

Mora, R., & de la Torre, I. (2005). Percussion tools in Olduvai Beds I and II (Tanzania): implications for ealry human activities. *Journal of Anthropological Archaeology, 24*, 179–192.

Norton, C. J., & Gao, X. (2008). Hominin-carnivore interactions during the Chinese Early Paleolithic: Taphonomic perspectives from Xujiayao. *Journal of Human Evolution, 55*, 164–178.

Norton, C. J., Bae, K., Harris, J. W. K., & Lee, H. (2006). Middle Pleistocene handaxes from the Korean Peninsula. *Journal of Human Evolution, 51*, 527–536.

Peterson, C., Shen, C., Chen, C., Chen, W., & Tang, Y. (2003). Taphonomy of an Early Pleistocene Archaeofauna from Xiaochangliang, Nihewan Basin, North China. In C. Shen & S. G. Keates (Eds.), *Current Research in Chinese Pleistocene Archaeology*. Oxford: BAR Archaeopress.

Pope, G. G. (1988). Recent advances in Far Eastern paleoanthropology. *Annual Review of Anthropology, 17*, 43–77.

Pope, G. (1995). The influence of climate and geography on the biocultural evolution of the Far Eastern Hominids. In E. S. Vrba, G. H. Denton, T. C. Partridge, & L. H. Burckle (Eds.), *Paleoclimate and evolution, with emphasis on human origins* (pp. 493–506). New Haven: Yale University Press.

Pope, G., & Keates, S. G. (1994). The evolution of human cognition and cultural capacity: a view from the Far East. In R. S. Corruccini & R. L. Ciochon (Eds.), *Integrative paths to the past* (pp. 531–567). Englewood Cliffs: Prentice Hall.

Potts, R. (1991). Why the Oldowan?: Plio-Pleistocene toolmaking and the transport of resources. *Journal of Anthropological Research, 47*, 153–176.

Potts, R. (1994). Variables versus models of Early Pleistocene hominid land-use. *Journal of Human Evolution, 27*, 7–24.

Potts, R. (1996). Evolution and climate variability. *Science, 273*, 922–923.

Roche, H. (2000). Variability of Pliocene lithic productions in East Africa. *ACTA Anthropological Sinica, 19*, 98–103.

Rogers, M. J., Harris, J. W. K., & Feibel, C. S. (1994). Changing patterns of land-use by Plio-Pleistocene hominids in the Lake Turkana Basin. *Journal of Human Evolution, 27*, 139–158.

Rolland, N. (1998). The Lower Palaeolithic settlement of Eurasia, with special reference to Europe. In K. Paddaya & M. Petraglia (Eds.), *Early human behaviour in global context: the rise and diversity of the Lower Palaeolithic record* (pp. 187–220). London: Routledge.

Rosenzweig, M. L. (1968). Net primary production of terrestrial communities: preditction for climatological data. *American Naturalist, 102*, 67–74.

Schick, K. D., & Dong, Z. (1993). Early Paleolithic of China and Eastern Asia. *Evolutionary Anthropology, 2*, 22–35.

Schick, K. D., Qi, W., Clark, J. D., Etler, D., & Toth, N. (1991). Archaeological perspectives in the Nihewan Basin, China. *Journal of Human Evolution, 21*, 13–28.

Semaw, S. (2000). The world's oldest stone artefacts from Gona, Ethiopia: their implications for understanding stone technology and patterns of human evolution between 2.6–1.5 million years ago. *Journal of Archaeological Science, 27*, 1197–1214.

Semaw, S., Renne, P., Harris, J. W. K., Feibel, C. S., Bernor, R. L., Fesseha, N., et al. (1997). 2.5-million-year-old stone tools from Gona, Ethiopia. *Nature, 385*, 333–336.

Shen, C., & Chen, C. (2000). A use-wear study of the lithic artifacts from Xiaochangliang and hominid activities in the Nihewan Basin, northern China. *ACTA Anthropologica Sinica, 19*, 115–121.

Shen, C., & Chen, C. (2003). New evidence of hominid behavior from Xiaochangliang, Northern China: site formation and lithic technology. In C. Shen & S. G. Keates (Eds.), *Current Research in Chinese Pleistocene Archaeology*. Cambridge: BAR Archaeopress.

Shen, C., & Qi, W. (2004). Lithic technological variability of the Middle Pleistocene in the Eastern Nihewan Basin, Northern China. *Asian Perspectives, 43*, 282.

Shott, M. J., & Sillitoe, P. (2001). The mortality of things: correlates of use life in Wola material culture using age-at-census data. *Journal of Archaeological Method and Theory, 8*, 269–302.

Stout, D., & Chaminade, T. (2007). The evolutionary neuroscience of tool making. *Neuropsychologia, 45*, 1091–1100.

Stout, D., Quade, J., Semaw, S., Rogers, M. J., & Levin, N. E. (2005). Raw material selectivity of the earliest stone toolmakers at Gona, Afar, Ethiopia. *Jounral of Human Evolution, 48*, 365–380.

Teague, R., & Potts, R. (2007). Faunal and behavioral context of the earliest Pleistocene Homo in east Asia. *American Journal Physical Anthropology Supplement, 44*, 230.

Torrence, R. (1989a). Retooling: towards a behavioral theory of stone tools. In R. Torrence (Ed.), *Time, energy, and stone tools* (pp. 57–66). Cambridge: Cambridge University Press.

Torrence, R. (1989b). Tools as optimal solutions. In R. Torrence (Ed.), *Time, energy and stone tools* (pp. 1–6). Cambridge: Cambridge University Press.

Torrence, R. (2001). Hunter–gatherer technology: macro- and microscale approaches. In C. Panter-Brick, R. H. Layton, & P. Rowley-Conwy (Eds.), *Hunter–gatherers: an interdisciplinary perspective* (pp. 73–98). Cambridge: Cambridge University Press.

Turner, A. (1992). Large carnivores and earliest European hominids: changing determinants of resource availability during the Lower and Middle Pleistocene. *Journal of Human Evolution, 22*, 109–126.

Williams, M. A. J., Williams, F. M., Gasse, F., Curtis, G. H., & Adamson, D. A. (1979). Plio-Pleistocene environments at Gadeb prehistoric site, Ethiopia. *Nature, 282*, 29–33.

Wood, B., & Collard, M. (1999). The changing face of genus Homo. *Evolutionary Anthropology, 8*, 195–207.

Wynn, J. G. (2004). Influence of Plio-Pleistocene aridification on human evolution: evidence from paleosols of the Turkana Basin, Kenya. *American Journal of Physical Anthropology, 123*, 106–118.

Zhang, S. S. (1985). The Early Palaeolithic of China. In R. K. Wu & J. W. Olsen (Eds.), *Palaeoanthropology and Palaeolithic Archaeology in the People's Republic of China* (pp. 147–186). New York: Academic.

Zhu, R. X., Potts, R., Xie, F., Hoffman, K. A., Deng, C. L., Shi, C. D., et al. (2004). New evidence on the earliest human presence at high northern latitudes in northeast Asia. *Nature, 431*, 559–562.

Chapter 5
Inter-continental Variation in Acheulean Bifaces

Ceri Shipton and Michael D. Petraglia

Abstract This chapter uses euclidean distance measurements of biface length, width and thickness to assess inter-continental variability in Acheulean biface morphology. Assemblages from eastern Asia and the Arabian peninsula are tested for their similarity to Acheulean assemblages in eastern Africa and the Indian sub-continent. Our expectation that the Arabian bifaces are part of the Acheulean tradition is confirmed. On the basis of metric ratios we conclude that the bifaces from Imjin and Hantan river basins in Korea and the Bose basin in China, are not part of the Acheulean tradition, suggesting independent development of bifacial forms. However the bifaces from the Luonan basin in China do resemble those of the Acheulean, indicating that there may have been intermittent dispersals of populations manufacturing Acheulean bifaces into eastern Asia.

Keywords Movius Line • Elongation • Refinement

Introduction

Acheulean bifaces such as handaxes and cleavers were the principal tools of hominins for much of the Pleistocene. In Africa, Acheulean bifaces were used for over a million years from 1.7 million years ago (Asfaw et al. 1992) to around 0.2 million years ago. During the Pleistocene bifaces are also found in many parts of Eurasia from Britain to south India.

Movius (1948) noted that Acheulean bifaces do not occur north of the Himalayas or east of the Brahmaputra, a dichotomy which became known as the Movius Line. More recently bifacial artifacts have been reported in East Asia (e.g. Yi and Clark 1983; Huang 1989; Schick 1994; Lin 1994; Hou et al. 2000; Wang 2005; Norton et al. 2006), although disagreement remains as to whether these constitute genuine Acheulean assemblages (Petraglia 1998, 2001; Wang 2001; Ranov 2001; Corvinus 2004; Lycett and Cramon-Taubadel 2007).

Here we use published data and our own measurements of biface length to breadth ratio (elongation) and biface thickness to breadth ratio (refinement), to assess inter-continental variability in biface morphology. These variables have previously been shown to be the principal sources of bifacial variation and therefore should be useful for characterising assemblages (Callow 1976; Wynn and Tierson 1990; White 1998; McPherron 2006). We have chosen sites and regions to represent the range of biface variation during the early and middle Pleistocene of Asia. These include the well established Acheulean region of the Indian sub-continent, the understudied Arabian peninsula, Eastern Asia where the presence of the Acheulean occupation remains controversial, and Eastern Africa where the oldest Acheulean sites are found (see Fig. 5.1). Our goal is to determine whether it is possible to discern geographical trends in biface variation.

Previous studies of biface elongation have found allometric patterns (Crompton and Gowlett 1993; McPherron 2000), longer bifaces being relatively narrower. This relationship has been attributed to ergonomic constraints, differential rejuvenation, and regional differentiation (Crompton and Gowlett 1993; McPherron 2000; Wynn and Tierson 1990). In this chapter we will assess the variability in biface elongation between assemblages to determine if there are regional differences.

Norton et al. (2006) found that they could distinguish Korean bifaces from Acheulean ones on the basis of thickness. Here we will conduct a similar comparison between Eastern Asian and Acheulean assemblages using biface refinement, as the refinement ratio is a more comprehensive measure of biface shape than thickness alone.

C. Shipton (✉)
Garreglefain, Porthyrhyd, Llanwrda, Sir Gar SA19 8PE, UK
e-mail: ceri.shipton@cantab.net

M.D. Petraglia
Research Laboratory for Archaeology and the History of Art, University of Oxford, Dyson Perrins BuildingSouth Parks Road, Oxford, OX1 3QY, UK
e-mail: michael.petraglia@rlaha.ox.ac.uk

C.J. Norton and D.R. Braun (eds.), *Asian Paleoanthropology: From Africa to China and Beyond*, Vertebrate Paleobiology and Paleoanthropology, DOI 10.1007/978-90-481-9094-2_5, © Springer Science+Business Media B.V. 2010

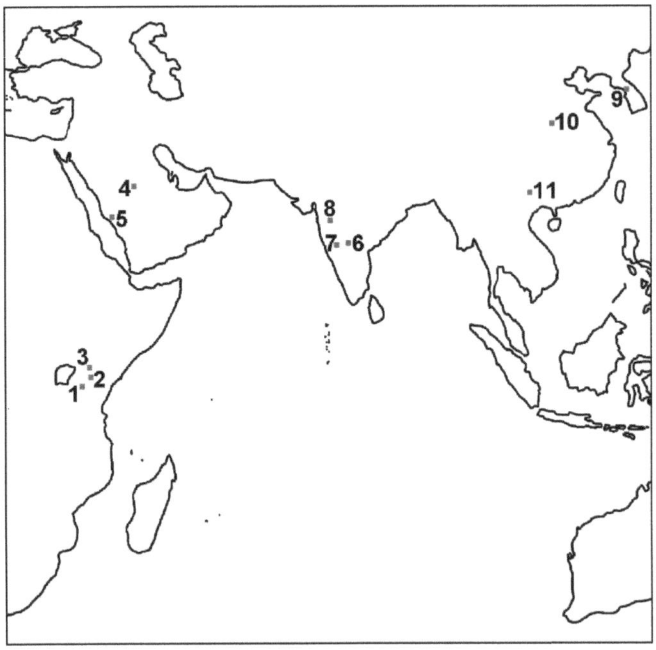

Fig. 5.1 Locations of the sites used in this study. 1 = Olduvai Gorge; 2 = Olorgesailie; 3 = Kariandusi; 4 = Dawadmi 206–76; 5 = Wadi Fatimah; 6 = Hunsgi-Baichbal; 7 = Anagwadi; 8 = Godavari; 9 = IHRB; 10 = Luonan; 11 = Bose

Table 5.1 Mean and median refinement for the sites in this study with the total scores for each region in bold

	N	Mean Refinement	Median Refinement
Olduvai Gorge Bed II	21	0.6839	0.6634
Kariandusi	58	0.4702	0.4596
DE89A Olorgesailie	63	0.4469	0.4314
H9AM Olorgesailie	13	0.3674	0.3717
I3 Olorgesailie	62	0.5551	0.5446
FB Olorgesailie	16	0.5843	0.5486
DE89C Olorgesailie	69	0.5545	0.506
Africa	**232**	**0.5083**	**0.4894**
Dawadmi 206–76	49	0.5457	0.5339
Wadi Fatimah	35	0.4692	0.4588
Arabia	**84**	**0.5138**	**0.5**
Hunsgi V	151	0.5352	0.5
Hunsgi II	34	0.5232	0.5
Gulbal II	17	0.4999	0.5
Mudnur VIII	9	0.5825	0.5455
Yediyapur I	21	0.4346	0.4143
Yediyapur IV	20	0.5412	0.5
Yediyapur VI	66	0.4924	0.5
Fatehpur V	31	0.4434	0.4365
Teggihalli II	31	0.4711	0.467
Anagwadi	25	0.5762	0.575
Godavari	10	0.5298	0.5278
India	**310**	**0.5139**	**0.5**
Acheulean Total	**531**	**0.5132**	**0.5**
Chongokni	25	0.6814	0.681
Chuwoli/Kawoli	19	0.6129	0.6
Kumpari	13	0.597	0.5854
IHRB	**57**	**0.6393**	**0.63**
Luonan Basin	89	0.524	No data
Bose/Baise	35	No data	No data

Geographic Regions

Sites from several regions of the Old World were selected in order to compare traditionally accepted Acheulean occupation areas to less well known or more controversial assemblages. Data was collected from published literature and the collections in the Cambridge Archaeology and Anthropology Museum, UK, the Department of Antiquities and Museums, Saudi Arabia and the Deccan College, India. A full list of the sites used and the number of artifacts from each locality is given in Tables 5.1–5.3.

Eastern Africa

The Acheulean originates in East Africa, this region having all the earliest examples of Acheulean bifaces. Three East African sites were examined for the present study: Olduvai Gorge Bed II, Kariandusi and Olorgesailie, all three of these sites are dated to the Early and Middle parts of the Acheulean period.

The Olduvai Gorge is cut through the south-east of the Serengeti plains in northern Tanzania. Bed I has yielded only Oldowan artifacts, while in the upper part of Bed II the earliest Acheulean bifaces at Olduvai Gorge are found, dated to c. 1.6 mya (Delson and Van Couvering 2000). The Bed II bifaces are typically made on large cobbles of quartz and basalt. A total of twenty Bed II Acheulean bifaces from the Cambridge Archaeology and Anthropology Museum, were sampled for this study; nine are from unlabelled localities, the remainder are from TK II (n = 1), FLK II (n = 3), DK II (n = 1), BWK II (n = 2), MLK II (n = 3), VEK II (n = 1), MK II (n = 1).

Kariandusi is in Kenya, in the Gregory Rift Valley. The faunal guild, together with the form of the bifaces, indicate that Kariandusi is broadly contemporaneous with Olduvai Gorge Bed IV (Cole 1954), dated to c. 0.8 mya (White 2000). K-Ar dates taken by Evernden and Curtis (1965) at Kariandusi pointed to just under 1 million years. Palaeomagnetism (Gowlett and Crompton 1994) and $^{40}Ar/^{39}Ar$ (Deino et al. 2004) have put upper and lower limits on this time frame, so that Kariandusi can be assigned an age bracket of 0.98–0.78 mya. Underlying the Kariandusi sediments is the Gilgil trachyte (McCall 1967), an outcrop of which just 80 m from the site may have provided the raw material on which some bifaces were manufactured. The 66 trachyte bifaces examined in this study derive from the Upper or Main Site at Kariandusi, and where it is possible to tell they were all made on flakes.

Table 5.2 Mean and median elongation for the sites in this study with the total scores for each region in bold

	N	Mean Elongation	Median Elongation
Olduvai Gorge Bed II	21	2.0087	1.968
Kariandusi	58	1.7693	1.7475
Olorgesailie DE89A	63	1.7521	1.7241
Olorgesailie H9AM	13	1.9376	2.0672
Olorgesailie I3	62	1.5425	1.5309
Olorgesailie FB	16	1.5964	1.6036
Olorgesailie DE89C	23	1.7625	1.7442
East Africa	**232**	**1.7247**	**1.7209**
Dawadmi 206–76	49	1.7711	1.7
Wadi Fatimah	35	1.4795	1.5238
Arabia	**84**	**1.6496**	**1.6111**
Hunsgi V	151	1.6324	1.625
Hunsgi II	34	1.6835	1.625
Gulbal II	17	1.5863	1.6458
Mudnur VIII	9	2.1567	2.2727
Yediyapur I	21	1.5462	1.4722
Yediyapur IV	20	1.6608	1.6667
Yediyapur VI	66	1.5369	1.5
Fatehpur V	31	1.4946	1.4773
Anagwadi	25	1.7008	1.6522
Godavari	10	1.3191	1.2933
Teggihalli II	31	1.5282	1.5233
India	**302**	**1.609**	**1.6**
Acheulean Total	**650**	**1.6618**	**1.6479**
Chongokni	25	1.5859	1.5603
Chuwoli/Kawoli	19	1.781	1.7194
Kumpari	13	1.569	1.6224
IHRB	**57**	**1.6471**	**1.6408**
Luonan Basin	89	1.474	No data
Bose/Baise	35	No data	No data

The Olorgesailie basin is located in the southern Kenya rift valley. The Acheulean localities at Olorgesailie occur in Members 1 to 8 which have been dated by $^{40}Ar/^{39}Ar$ to between 0.99–0.79, so that they are broadly contemporary with Kariandusi (Potts et al. 1999). The bifaces used in this study come from localities DE89A, DE89C, H9AM, I3 and FB and were measured by Noll (2000). Where possible to determine two thirds of the bifaces were made on flake blanks and one third were made on slabs and cobbles. Most of the bifaces were made on lavas and occasionally obsidian was used.

The Arabian Peninsula

Arabian assemblages have not been included in previous inter-continental comparisons of biface technology (e.g. Wynn and Tierson 1990). Given the position of Arabia at the juncture between Africa and Eurasia, it is important to situate the technology of Arabian hominins in a comparative context.

Table 5.3 A comparison of the mean length and mean weight of Acheulean and eastern Asian assemblages

	N	Mean length (mm)	Mean weight (g)
Olduvai Gorge Bed II	21	195.39	1349.28
Kariandusi	58	157.94	552.74
DE89A Olorgesailie	63	180.76	847.78
H9AM Olorgesailie	13	199.77	814
I3 Olorgesailie	62	97.95	253
FB Olorgesailie	16	98.81	212.13
DE89C Olorgesailie	69	158.7	778.26
Dawadmi 206–76	49	162.87	No data
Wadi Fatimah	35	141.86	No data
Hunsgi V	151	143.51	638.14
Hunsgi II	34	162.9	994.35
Gulbal II	17	147.14	868.21
Mudnur VIII	9	227.78	1302.22
Yediyapur I	21	123.13	424.69
Yediyapur IV	20	132.94	570
Yediyapur VI	66	127.89	555.39
Fatehpur V	31	126.82	469.09
Teggihalli II	31	121.54	350.77
Anagwadi	25	137.24	No data
Godavari	10	114	No data
Acheulean Total	**531**	**145.54**	**634.18**
Chongokni	25	146.42	No data
Chuwoli/Kawoli	19	169.47	No data
Kumpari	13	148.08	No data
IHRB	**57**	**154.48**	**No data**
Luonan Basin	89	149.08	851.75
Bose/Baise	35	180	1534

Bifaces from two sites in Arabia were measured for this analysis.

A sample of 49 bifaces from the locality of Dawadmi 206–76 were measured. Dawadmi 206–76 is a large site in central Arabia, measuring approximately 150 × 200 m (Whalen et al. 1984). A total of 3,256 artifacts were collected in a surface collection from a 30 × 30 m grid. Dawadmi 206–76 is part of a group of localities located along an andesite dyke and appears to have been a manufacturing area (Petraglia 2005). Most of the bifaces were produced on andesite from the dyke, although locally derived quartz, granite and rhyolite were occasionally used. The bifaces show large, deep, expanding flake scars as though struck with a hard hammer, sinuous edges, irregular cross-sections and cortical butts, leading the excavator to describe them as Middle Acheulean (Whalen et al. 1984).

In western Arabia between the Asir Mountains and the Red Sea flows the Wadi Fatimah. A total of 31 biface yielding localities and 2,227 artifacts have been discovered along the north side of this water course (Whalen et al. 1988). Similar to Dawadmi the majority of artifacts are made on andesite and infrequently quartzite was also used. A sample of 35 bifaces from several of the Wadi Fatimah localities were measured for the present study.

The Indian Sub-continent

The Indian Sub-continent is the eastern most extent of dense Acheulean occupation (Petraglia 2006). It is distinctive in the high proportion of cleavers among its assemblages which frequently constitute over half of all bifaces in an assemblage. Assemblages from three areas in southern India are used in this study: the Hunsgi-Baichbal valley, Anagwadi and Godavari.

The Hunsgi-Baichbal Valley in southern India contains over 200 Acheulean localities (Paddayya 2007). One of the most unusual features of the Hunsgi-Baichbal bifaces is the preference for limestone as the raw material, although several other materials including dolerite and quartzite were also used. Bifaces from eight localities were measured for the present study. Hunsgi II (Hunsgi valley) a surface locality of limestone flake bifaces; Hunsgi V (Hunsgi valley), a large excavated locality yielding around 150 bifaces, made on limestone slabs on flakes and quartzite cobbles (Paddayya 1977); Gulbal II (Hunsgi valley) a surface locality containing limestone bifaces; Mudnur VIII (Baichbal valley), a surface locality, possibly a cache, where several large limestone bifaces were found with no other cultural material in the vicinity; Yediyapur I (Baichbal valley) a surface locality where bifaces were made on several raw materials including limestone, dolerite and granite; Yediyapur IV (Baichbal valley) a surface locality containing two giant bifaces among other smaller specimens (Paddayya 1987); Yediyapur VI (Baichbal valley) an excavated locality where bifaces were produced on a variety of raw materials, including quartzite cobbles (Paddayya 1987); and Fatehpur V (Baichbal valley), a surface collection of eight sub-localities (Paddayya 1989).

Anagwadi is located in the Ghatprabha Basin, which similarly to the Hunsgi-Baichbal Valley empties into the River Krishna. A total of 252 bifaces were collected from Anagwadi both as surface finds and during excavation (Pappu 1974). The vast majority of bifaces are made on quartzite cobbles although chert was also used occasionally. Data on 25 Anagwadi bifaces taken from Pappu (1974) was added to the database for this study.

At Locality I in the upper Godavari Basin a handful bifaces were recovered prior to the construction of a dam (Sankalia 1952). The tools were found in a pebbly gravel matrix and are made on basalt. The measurements of 10 Godavari bifaces taken from Sankalia's 1952 publication were included in this analysis.

Eastern Asia

Several biface yielding localities have been discovered in the Bose or Baise basin in South China (Hou et al. 2000). Two thirds of the large cutting tools in the Bose assemblage are unifacial, however the present study only incorporated the bifacial implements. These bifaces are made on quartz, quartzite, sandstone and chert. Around a quarter of the bifaces were made on flakes with the remainder produced on thick cobbles and larger cores. The area where bifaces were recovered coincides with the location of the largest class in the basin, which are cobbles greater than 20 cm in length. The bifaces have clearly defined tip and butt ends, so their form has been argued to represent a target morphology similar to Acheulean bifaces, rather than an unintentional by-product of flaking. The association of some of the excavated bifaces with tektites from the Australasian asteroid impact strewnfield, has been used to date the Bose bifaces to 800 kya (Hou et al. 2000), making them by far the oldest in East Asia. However it has been contended that the tektites and the associated bifaces represent material accumulated on a palaeo-erosional surface and may not be contemporaneous (Langbroek 2004). Data on the Bose bifaces was taken from Hou et al. (2000) for the following analysis.

In the Luonan Basin near the city of Xi'an, China, a total of 89 bifaces have been recovered from 50 open air localities (Wang 2005). The bifaces are mostly made on quartzite, although greywacke and sandstone are also present. The majority of bifaces were made on flake blanks although pebbles and larger cores were used. The Luonan bifaces include cleavers which are a rare tool type in Eastern Asia. Based on the presence of handaxes and cleavers, the investigator has argued that the Luonan sites are part of the Acheulean tradition. A preliminary estimate of the date of occupation in the Basin based on faunal comparison, TL dates and Loess stratigraphy suggests between 500 and 250 kya (Wang 2005).

In the Imjin and Hantan river basins (IHRB), Korea, a number of biface-bearing localities have been discovered (Norton et al. 2006). Chongokni next to the Hantan River, a tributary of the Imjin, was the first of these sites to be discovered and contains low percentages of quartz and quartzite bifaces. At Chuwoli and Kawoli next to the Imjin River more bifaces were discovered as part of a larger lithic assemblage made on quartz and quartzite river cobbles. Further downstream is the site of Kumpari where it appears manufacturing of lithics was carried out including occasional bifaces (Bae 2002). Chongokni has been dated to 350,000–300,000 years old, and stratigraphic concordance puts the other sites in a similar age bracket (Danhara et al. 2002). Length, width and thickness measurements of the bifaces from these Korean localities were taken from Norton et al. (2006) and used in the following analysis.

Analyses

Different methodologies have found that elongation and refinement are the most important components of biface variability (Callow 1976; Wynn and Tierson 1990; White

Table 5.4 Kruskall-Wallis test comparing the refinement of Acheulean bifaces between regions

	N	Mean rank	Chi-square	Significance
Africa	232	299.95		
Arabia	84	314.46		
India	310	323.38	2.232	N.S.

Table 5.5 Mann-Whitney U test comparing the elongation of all Indian bifaces with all bifaces from Arabia

	N	Mean rank	Mann-Whitney U	Significance
India	310	196.33		
Arabia	84	201.83	12656.5	N.S.

Table 5.6 Mann-Whitney U test comparing the elongation of Indian bifaces those from East Africa

	N	Mean rank	Mann-Whitney U	Significance
India	310	240.99		
Africa	232	312.27	26501	P < 0.001

Table 5.7 Mann-Whitney U test comparing the elongation of all the Acheulean bifaces with those from the IHRB Basin in Korea

	N	Mean rank	Mann-Whitney U	Significance
Acheulean	650	354.57		
IHRB	57	347.51	18155	N.S.

1998). Elongation and refinement are thus useful variables for characterising and discriminating between biface assemblages.

Table 5.1 shows that the three Acheulean regions (Eastern Africa, the Indian Sub-Continent and the Arabian peninsula) have similar refinement scores. To test whether or not there is statistical homogeneity in biface refinement across these regions the non-parametric Kruskall-Wallis test was used, as the data was not normally distributed (see Table 5.4 for results). Table 5.2 shows that African and Arabian bifaces were more elongated than the Indian bifaces. This trend was tested statistically by comparing all the Indian bifaces to all the African bifaces and all the Indian bifaces to all the Arabian bifaces. Elongation data was not normally distributed so the non-parametric Mann-Whitney U scores were used in these tests (see Tables 5.5 and 5.6 for results).

The bifaces from the Korean IHRB sites were grouped together and compared with all of the bifaces from assemblages west of the Movius line, identified as Acheulean. Mann-Whitney U tests were used to compare the elongation and refinement of the Korean and Acheulean bifaces as the data were not normal (see Tables 5.7 and 5.8 for results).

For the Luonan bifaces only mean, rather than individual refinement scores were available, so the analysis was limited to a t-test. Mean refinement of the Luonan bifaces was compared with the total mean refinement of all the Acheulean

Table 5.8 Mann-Whitney U test comparing the refinement of all the Acheulean bifaces with those from the IHRB Basin in Korea

	N	Mean rank	Mann-Whitney U	Significance
Acheulean	650	339.07		
IHRB	57	524.2	8823.5	P<0.001

Table 5.9 T-test comparing the mean refinement of all the Acheulean bifaces with those from the Luonan Basin in China

	Mean refinement	Standard deviation	t	Significance
Acheulean	0.5132	0.1263		
Luonan	0.524	0.1351	0.7156	N.S.

Table 5.10 Unequal variances t-test comparing the mean weight of Acheulean bifaces from Olduvai Gorge Bed II, Mudnur VIII, Olorgesailie DE89A and Olorgesailie H9AM, with the mean weight of bifaces from the Bose Basin

	Mean weight (g)	Standard deviation	t	Significance
Acheulean	981.58	523.083		
Bose	1534	766	3.972	P<0.001

assemblages west of the Movius Line (see Table 5.9 for results). Elongation data was not available for the Luonan bifaces.

For the Bose bifaces neither elongation nor refinement data were available, although mean length and weight data were. Table 5.3 shows that the Bose bifaces are the heaviest in the sample though they are not the longest. The mean weight of the four Acheulean assemblages with longer bifaces than Bose (Olduvai Gorge Bed II, Mudnur VIII, Olorgesailie DE89A and Olorgesailie H9AM), was compared with the mean weight of the Bose bifaces using a t-test. A preliminary F test showed the two samples had unequal variances, therefore the unequal variance t-test was used (see Table 5.10 for the results).

Discussion

This study attempted to identify inter-regional differences within the Acheulean and examined whether Eastern Asian biface assemblages could be described as belonging to the Acheulean tradition. This study only used Euclidean distance measurements and weight, so its conclusions must be regarded as tentative, but we believe some informative aspects of shape variation have been captured. We look forward to more detailed technological studies employing three dimensional morphometrics and comparisons of reduction sequences in order to examine inter-regional variation in Early and Middle Pleistocene large cutting tools in greater detail.

With respect to the Acheulean assemblages, the results show that the African bifaces are significantly more elongated than the Indian ones at the $P < 0.001$ level, while there is no significant difference between the Arabian and Indian bifaces (see Tables 5.5 and 5.6). The age of the sites sampled to represent Africa may explain why the African bifaces are the most elongated. The three African sites are all older than 600,000 years and bifaces from before this time are usually larger than younger bifaces (McBrearty 2001). Since larger bifaces tend to be more elongated (Crompton and Gowlett 1993) this also accounts for the greater elongation of the African bifaces.

The Kruskall-Wallis test showed no significant difference between the refinement of African, Arabian and Indian bifaces, while the Arabian and Indian bifaces were also statistically similar in terms of elongation (see Tables 5.4 and 5.5). Tables 5.1, 5.2 and 5.3 show that the Arabian bifaces fall within the range of variation of African and Indian Acheulean assemblages, for refinement, elongation, length and weight. The Arabian assemblages can thus be ascribed to the Acheulean tradition, as would be expected from their geographical position and observed artifact types and technology.

For the Eastern Asian bifaces, Table 5.7 shows that there is no significant difference in elongation between the Korean bifaces and the Acheulean bifaces. However Table 5.8 shows that the relationship between width and thickness is significantly different between the Korean bifaces and the Acheulean bifaces, such that the Korean bifaces are thicker relative to constant width. All three IHRB assemblages have larger mean and median refinement scores than any Acheulean assemblage with the exception of Olduvai Gorge Bed II (see Table 5.2). Olduvai Gorge Bed II is exceptional because it is one of the earliest Acheulean sites and its bifaces cannot be regarded as typical of later Acheulean artifact assemblages. While they have comparable refinement ratios Tables 5.2 and 5.3 show that the Olduvai Gorge Bed II bifaces are far longer and more elongated than the IHRB bifaces. Though we do see some overlap in biface measures, we concur with Norton et al. (2006) who conclude that the IHRB bifaces do not belong within the Acheulean tradition. Indeed it is not yet established whether the IHRB bifaces are finished implements, or cores for the extraction of flakes.

An argument has been that the Bose bifaces are similar in many respects to Acheulean assemblages (Hou et al. 2000). However, the Bose bifaces are significantly heavier than Acheulean bifaces, despite the fact that the Acheulean bifaces tested are longer than those from Bose (see Table 5.10). This suggests that the Bose bifaces have a different shape to Acheulean ones and cannot be ascribed to the Acheulean industry. It is possible that the Bose bifaces are made from a particularly heavy stone. However this is unlikely to account for the shorter Bose bifaces being half as heavy again as the Acheulean bifaces (see Tables 5.3 and 5.10).

Biface assemblages from IHRB and Bose are metrically distinctive compared to Acheulean assemblages west of the Movius Line. We suggest that these East Asian biface assemblages are independent inventions of handaxe-like bifaces, rather than typical Acheulean industries. Handaxe-like forms appear many times in human history such as in the MTA of Middle Palaeolithic Europe, the Pacitan industry of southeast Asia and the British Neolithic, thus we suggest such forms can be arrived at without continuity of descent.

The Luonan bifaces were less refined than the mean for all the Acheulean bifaces, though this difference was not significant (see Tables 5.1 and 5.9). The Luonan assemblage falls comfortably within the range of Acheulean assemblages for mean refinement, length and weight (see Tables 5.1 and 5.3). This finding, coupled with the fact that the distinctive cleaver tool type is well represented among the Luonan bifaces (Wang 2005), has lead us to reaffirm the conclusion that the Luonan bifaces are part of the Acheulean tradition (Petraglia and Shipton 2009). Such a conclusion has major implications for hominin dispersal patterns, suggesting the possibility of a dispersal of Acheulean hominins into Eastern Asia. It is notable that Rightmire (2001) has proposed that the Chinese hominins from Dali and Jinniushan be attributed to *Homo heidelbergensis*, perhaps indicating a later Pleistocene expansion of this species into Eastern Asia. It is possible that the Luonan Acheulean assemblage represents the cultural signature of this expansion, though based on our study and the high frequency of Acheulean sites west of the Movius Line, we do not believe the Acheulean tradition was widespread or prolonged in Eastern Asia.

Acknowledgements The authors would like to thank the authorities of the Deccan College, the Cambridge Museum of Archaeology and Anthropology and the Department of Antiquities and Museums of Saudi Arabia for access to artifact collections. The Fulbright Senior Specialists Program Fellowship, the Allen, Meek and Read Scholarship and the Leakey Foundation are gratefully acknowledged for their financial support. Christopher Norton, David Braun and three anonymous reviewers are thanked for their comments on earlier draft of this chapter.

References

Asfaw, B., Beyene, Y., Suwa, G., Walter, R. C., White, T. D., WoldeGabriel, G., et al. (1992). The earliest Acheulean from Konso-Gardula. *Nature, 360*, 732–735.

Bae, K. D. (2002). Chongokni Paleolithic site: current understandings. In K. D. Bae & J. C. Lee (Eds.), *Paleolithic Archaeology in Northeast Asia* (pp. 55–75). Seoul: Yeoncheon County and The Institute of Cultural Properties.

Callow, P. (1976). The lower and middle Palaeolithic of Britain and adjacent areas of Europe. Unpublished Ph.D. dissertation, Cambridge University.

Cole, S. M. (1954). *The Prehistory of East Africa.* Harmondsworth: Pelican Books.

Corvinus, G. (2004). *Homo Erectus* in East and Southeast Asia, and the questions of the age of the species and its association with stone artifacts, with special attention to handaxe-like tools. *Quaternary International, 117*, 141–151.

Crompton, R. H., & Gowlett, J. A. J. (1993). Allometry and multidimensional form in Acheulean bifaces from Kilombe, Kenya. *Journal of Human Evolution, 25*, 175–199.

Danhara, T., Bae, K. D., Okada, T., Matsufuji, K., & Hwang, S. H. (2002). What is the real age of the Chongokni Paleolithic site? In K. D. Bae & J. C. Lee (Eds.), *Paleolithic Archaeology in Northeast Asia* (pp. 77–116). Seoul: Yeoncheon County and The Institute of Cultural Properties.

Deino, A., Trauth, M.H., Bergner, A. G., Potts, R. (2004). $^{40}Ar/^{39}Ar$ Age Calibration of the Lacustrine Sediments at Kariandusi, Central Kenya Rift. American Geophysical Union, Fall Meeting 2004.

Delson, E., & Van Couvering, J. A. (2000). Composite stratigraphic chart of Olduvai Gorge. In E. Delson, I. Tattersall, & J. A. Van Couvering (Eds.), *Encyclopedia of Human Evolution and Prehistory* (2nd ed., p. 488). New York: Garland.

Evernden, J. F., & Curtis, G. H. (1965). Potassium-argon dating of late Cenozoic rocks in East Africa and Italy. *Current Anthropology, 6*, 343–385.

Gowlett, J. A. J., & Crompton, R. H. (1994). Kariandusi: Acheulean morphology and the question of allometry. *African Archaeological Review, 12*, 1–40.

Hou, Y. M., Potts, R., Yuan, B., Guo, Y., Deino, Z. T., Wang, A., et al. (2000). Mid-Pleistocene Acheulean-like stone technology of the Bose Basin, South China. *Science, 287*, 1622–1626.

Huang, W. W. (1989). Bifaces in China. *Human Evolution, 4*, 87–92.

Langbroek, M. (2004). *'Out of Africa': an investigation into the earliest occupation of the Old World*. BAR International Series 1244, Oxford.

Lin, S. (1994). Restudy of nine hand-axe specimens and the applicability of Movius' theory. *Acta Anthropologica Sinica, 13*, 205–220.

Lycett, S. J., & Cramon-Taubadel, N. V. (2007). Acheulean variability and hominin dispersals: a model-bound approach. *Journal of Archaeological Science, 35*, 553–562.

McBrearty, S. (2001). The Middle Pleistocene of East Africa. In L. Barham & K. Robson-Brown (Eds.), *Human roots: Africa and Asia in the Middle Pleistocene* (pp. 81–98). Bristol: Westbury Press.

McCall, G. J. H. (1967). *Geology of the Nakuru-Thomson's Falls-Lake Hannington Area*. Nairobi: Geological Survey of Kenya.

McPherron, S. P. (2000). Handaxes as a measure of the mental capabilities of early hominids. *Journal of Archaeological Science, 27*, 655–663.

McPherron, S. P. (2006). What typology can tell us about Acheulian handaxe production. In N. Goren-Inbar & G. Sharon (Eds.), *Axe Age: Acheulian Toolmaking from Quarry to Discard* (pp. 267–286). London: Equinox.

Movius, H. L. (1948). The Lower Paleolithic cultures of southern and eastern Asia. *Transactions of the American Philosophical Society, 38*, 329–420.

Noll, M. P. (2000). Components of Acheulean lithic assemblage variability at Olorgesailie, Kenya. Unpublished Ph.D. dissertation, University of Illinois at Urbana-Champaign.

Norton, C. J., Bae, K., Harris, J. W. K., & Lee, H. (2006). Middle Pleistocene handaxes from the Korean Peninsula. *Journal of Human Evolution, 51*, 527–536.

Paddayya, K. (1977). An Acheulian occupation site at Hunsgi, peninsular India: a summary of the results of two seasons of excavation (1975–76). *World Archaeology, 8*, 344–355.

Paddayya, K. (1987). Excavation of a new Acheulean occupation site at Yediyapur, Peninsula India. *Anthropos, 82*, 610–614.

Paddayya, K. (1989). The Acheulian culture localities along the Fatehpur Nullah, Baichbal Valley, Karnataka (Peninsular India). In J. M. Kenoyer (Ed.), *Old Problems and New Perspectives in the Archaeology of South Asia, Madison* (pp. 21–28). Wisconsin: University of Wisconsin Press.

Paddayya, K., (2007). The Acheulean of Peninsular India with special reference to the Hunsgi and Baichbal Valleys of the Lower Deccan. In: Petraglia, M.D., Allchin, B. (Eds.) *The Evolution and History of Human Populations in South Asia*, 97–119.

Pappu, R. S. (1974). *Pleistocene Studies in the Upper Krishna Basin*. Pune: Deccan College.

Petraglia, M. D. (1998). The Lower Palaeolithic of India and its bearing on the Asian record. In M. D. Petraglia & R. Korisettar (Eds.), *Early human behaviour in global context: the Rise and Diversity of the Lower Palaeolithic Record* (pp. 343–390). London: Routledge.

Petraglia, M. D. (2001). The Lower Palaeolithic of India and its behavioural significance. In L. Barham & K. Robson-Brown (Eds.), *Human roots: Africa and Asia in the Middle Pleistocene* (pp. 217–233). Westbury Press: Bristol.

Petraglia, M. D. (2005). Hominin responses to Pleistocene environmental change in Arabia and South Asia. In M.J. Head, P.L. Gibbard (Eds.), *Early-Middle Pleistocene Transitions: The Land-Ocean Evidence. Geological Society, London, Special Publications, 247*, 305–319.

Petraglia, M. D. (2006). The Indian Acheulian in global perspective. In N. Goren-Inbar & G. Sharon (Eds.), *Axe Age: Acheulian Toolmaking from Quarry to Discard* (pp. 389–414). London: Equinox.

Petraglia, M. D., & Shipton, C. (2009). Erratum to "Large cutting tool variation west and east of the Movius Line" J. H. Evol. 55 (2008) 962–966. *Journal of Human Evolution, 57*, 326–330.

Potts, R., Behrensmeyer, A. K., & Ditchfield, P. (1999). Palaeolandscape variation and Early Pleistocene hominid activities: members 1 and 7, olorgesailie formation, Kenya. *Journal of Human Evolution, 37*, 747–788.

Ranov, V. A. (2001). Cleavers: their distribution, chronology and typology. In S. Milliken & J. Cook (Eds.), *A very Remote Period Indeed: Papers on the Palaeolithic Presented to Derek Roe* (pp. 105–113). Oxford: Oxbow Books.

Rightmire, G. P. (2001). Comparison of Middle Pleistocene hominids from Africa and Asia. In L. Barham & K. Robson-Brown (Eds.), *Human roots: Africa and Asia in the Middle Pleistocene* (pp. 123–133). Bristol: Westbury Press.

Sankalia, H. D. (1952). *The Godavari Palaeolithic Industry*. Pune: Deccan College.

Schick, K. D. (1994). The Movius Line reconsidered: perspectives on the earlier Paleolithic of eastern Asia. In R. S. Corruccini & R. L. Ciochon (Eds.), *Integrative Paths to the Past* (pp. 569–596). New Jersey: Prentice Hall.

Wang, Y. P. (2001). The Middle Pleistocene archaeology of the valley of the Yangtze River. In L. Barham & K. Robson-Brown (Eds.), *Human Roots: Africa and Asia in the Middle Pleistocene* (pp. 149–158). Bristol: Westbury Press.

Wang, S. (2005). *Perspectives on Hominid Behaviour and Settlement Patterns: a Study of the Lower Palaeolithic Sites in the Luonan Basin, China*. Oxford: Archaeopress.

Whalen, N. M., Siraj-Ali, J. S., & Davis, W. (1984). Excavation of Acheulean sites near Saffaqah, Saudi Arabia, 1403 AH 1983. *Atlatl, 8*, 9–24.

Whalen, N. M., Siraj-Ali, J. S., Sindi, H. O., Pease, D. W., & Badein, M. A. (1988). A complex of sites in the Jeddah – Wadi Fatimah area. *Atlatl, 11*, 77–85.

White, M. J. (1998). On the significance of Acheulean biface variability in southern Britain. *Proceedings of the Prehistoric Society, 64*, 15–44.

White, T. D. (2000). Olduvai Gorge. In E. Delson, I. Tattersall, & J. A. Van Couvering (Eds.), *Encyclopedia of Human Evolution and Prehistory* (2nd ed., pp. 486–489). New York: Garland.

Wynn, T., & Tierson, F. (1990). Regional comparison of the shapes of Later Acheulean handaxes. *American Anthropologist, 92*, 73–84.

Yi, S. B., & Clark, G. A. (1983). Observations on the Lower Palaeolithic of Northeast Asia. *Current Anthropology, 24*, 181–202.

Chapter 6
Cranial Shape in Asian *Homo erectus*: Geographic, Anagenetic, and Size-Related Variation

Karen L. Baab

Abstract A strong focus on the morphological differences between African and Asian *H. erectus* has generally overshadowed variation among populations of Asian *H. erectus*. This study explored variation in Asian *H. erectus* using 3D geometric morphometric methods, examining the shape of the neurocranium, frontal bone, occipital bone and temporal base. Analyses focused on the elucidation of geographic, temporal and size-related patterns of cranial shape variation in a representative sample from Zhoukoudian, Sangiran, Ngandong, Sambungmacan and Ngawi. In regards to the neurocranium, geographic differences explained the greatest proportion of variation, followed by geochronological age (these two factors explained similar shape differences within the neurocranium) and size. The temporal base, frontal and occipital bones were strongly influenced by geography and size. Although the later Javanese and Zhoukoudian specimens were generally distinct, there was some overlap between Sangiran 2 and the northern Chinese specimens. This may suggest that isolation between the two regions did not occur until the Middle or Late Pleistocene or that the Sangiran hominins are morphologically close to the common ancestor of the Zhoukoudian and later Javanese *H. erectus*. The late Indonesian fossils did group together in the principal components analyses. Yet there were subtle shape differences between the Sambungmacan/Ngawi fossils and those from Ngandong, indicating the presence of two similar but slightly different potentially contemporaneous morphs. The Sambungmacan specimens do not appear to be morphological intermediaries between the Sangiran and Ngandong fossils in regards to overall cranial shape.

Keywords Shape variation • 3D geometric morphometrics • China • Indonesia • Middle Pleistocene

K.L. Baab (✉)
Department of Anatomical Sciences, Stony Brook University Medical Center, Stony Brook, NY 11794, USA
e-mail: KarenBaab@stonybrook.edu

Introduction

Despite the implicit assumption of broad homogeneity within Asian *H. erectus*, there is variation in the cranial morphology of the Javanese and Chinese *H. erectus* samples. This diversity has implications for the interpretation of the taxonomy and evolutionary history of *H. erectus*. Although *H. erectus* s.l. is now known from sites across the Old World, the initial *H. erectus* discoveries were from sites on the Indonesian island of Java (Trinil; Dubois 1894), and from mainland China (Zhoukoudian; Black 1929). A preoccupation with the distinctions between African and Asian *H. erectus* has generally overshadowed potential variation within Asian *H. erectus* (e.g., Wood 1984, 1994; Andrews 1984; Stringer 1984; Villmoare 2005). Yet, those studies that have explored differences within and among Asian *H. erectus* fossil assemblages have uncovered complex and sometimes conflicting patterns of geographic, temporal, and size-related cranial variation (von Koenigswald and Weidenreich 1939; Weidenreich 1943, 1951; Jacob 1975, 1978; Sartono 1975; Rightmire 1990; Kidder and Durband 2000, 2004; Antón 2002; Durband et al. 2005; Kaifu et al. 2005, 2008; Liu et al. 2005; Baab 2007).

Some researchers have reported a robust relationship between morphology and geography within Asian *H. erectus*. These studies show that the Javanese and northern Chinese samples are each relatively homogenous but clearly distinct from one another in regards to their cranial shape (Weidenreich 1951; Picq 1990; Kidder and Durband 2000, 2004; Antón 2002; Durband et al. 2005), despite the greater time depth of the Indonesian fossil sample. Differences have been documented in the overall shape of the cranial vault, including the frontal bone, occipital bone, and the temporomandibular joint (TMJ). In contrast, other studies have documented greater overlap between the Indonesian and Chinese morphologies (Rightmire 1990; Bilsborough 2000; Liu et al. 2005).

Cranial shape may have changed over time both within Asian *H. erectus* as a whole as well as within regional samples (e.g., Antón 2002). It has been well-established that the average brain size increased over time in *H. erectus* s.l.

C.J. Norton and D.R. Braun (eds.), *Asian Paleoanthropology: From Africa to China and Beyond*, Vertebrate Paleobiology and Paleoanthropology, DOI 10.1007/978-90-481-9094-2_6, © Springer Science+Business Media B.V. 2010

(Wolpoff 1984; Leigh 1992; Rightmire 2004) and within Asian *H. erectus* s.s. in particular (Sartono 1975; Antón and Swisher 2001). However, it is unclear exactly how this relates to changes in cranial shape within this taxon, although Antón (2002) has speculated that an increase in cranial height and minimum frontal breadth were the result of increasing endocranial volume in the Javanese *H. erectus*. Santa Luca (1980) identified several evolutionary trends in the shape of the cranium within Asian *H. erectus* which he also attributed to an increase in brain size. These changes include an increase in the breadth across the temporal squamae, a decrease in the size of the sagittal keel, a reduction of the supraorbital sulcus and a more vertically oriented occipital plane. More recently, Kaifu et al. (2008) suggested that increases in external cranial dimensions in the more recent Indonesian fossils correspond to increases in endocranial capacity.

Differences within the sites of Sangiran, Ngandong, and Zhoukoudian have also been attributed to sexual variation (e.g., Weidenreich 1943; Santa Luca 1980; Rightmire 1990; Wolpoff 1999). Interpretations of sexual variation must be viewed cautiously, however, due to well-documented difficulties in assigning sex to individual fossils (Armelagos and Van Gerven 1980). To avoid the risks associated with assigning sex to the fossils, this study focused on size-related variation; within Asian *H. erectus*, and within limited temporogeographic samples (such as Ngandong), this probably corresponds to sexual variation. However, size variation outside of these circumscribed samples must be viewed as a combination of sexual and allometric variation. Recent work has highlighted the effect of size on certain aspects of cranial shape often used to define *H. erectus*, such as occipital angulation and degree of postorbital constriction (Spoor et al. 2005).

In addition to questions which apply to the Asian *H. erectus* sample as a whole, there are also areas of ambiguity related specifically to the Javanese *H. erectus* sample. For example, it is unclear whether the Sambungmacan fossils form a single, homogenous sample and, if so, whether they are a good morphological intermediary between the Sangiran and Ngandong samples (Frayer et al. 1993; Wolpoff 1999; Baba et al. 2003; Kaifu et al. 2008), or whether they are more closely related to the Ngandong specimens (Sartono 1975; Jacob 1975, 1978; Delson et al. 2001; Antón 2002). The less widely-known Ngawi specimen has also been grouped with the Ngandong and Sambungmacan fossils, although only limited morphometric comparisons have been made between Ngawi and other Indonesian *H. erectus* (Sartono 1990; Widianto and Zeitoun 2003).

Jacob (1975, 1978) and Sartono (1975) have further linked the Ngandong/Sm 1 group to S 17 which is considerably older (1.3 Ma: Larick et al. 2001) on the basis of overall size and details of cranial form such as the large supraorbital tori. By excluding other Sangiran hominins from this group, these workers were implicitly supporting the presence of two or more separate lineages within Javanese *H. erectus* for over 1 Myr. Jacob (1978) extended this lineage back in time even further to >1.5 Ma by arguing for an ancestor–descendent relationship between "*Pithecanthropus modjokertensis*" (which includes S 4) and "*P. soloensis*" (Ngandong, Sm 1, and S 17).

This study complements previous efforts to better characterize and understand the degree and pattern of variation within Asian *H. erectus* based on linear measurements and discrete characters by focusing on an alternate source of information: three dimensional (3D) cranial shape. Three dimensional geometric morphometrics has advantages over standard linear measurements in that it retains the original geometry of the specimens and can be used to visualize shape differences related to external factors such as geography or size. Although discrete traits have proven central to the debate about the distinctiveness of Asian vs. African *H. erectus*, they have also proven quite controversial. For instance, disagreements concerning individual character definitions, independence of characters, and taxonomic distribution of characters remain unsettled (e.g., Stringer 1984; Andrews 1984; Wood 1984, 1991, 1994; Hublin 1986; Bräuer 1990; Kennedy 1991; Bräuer and Mbua 1992; Villmoare 2005). Cranial shape, as quantified by geometric morphometric techniques, provides a less subjective source of information about cranial morphology and allows for the study of the shape of biological structures, rather than just a limited set of linear measurements. In this study the relationship between cranial shape variation and several external variables, specifically geography, geological age and cranial size, was explored using principal components and multiple regression analyses. In addition, the affinities of individual specimens within the Javanese sample and possible evolutionary relationships were also assessed using this dataset.

Materials

In this study 3D landmark data were acquired from Indonesian and Chinese *H. erectus* cranial fossils. Original specimens were used when available, but casts were substituted for the Zhoukoudian, S 12, S 17 and Ngawi fossils (Table 6.1). Although data was collected from the well-preserved Perning fossil calvaria, it was not included in any analysis as a result of its very young age. The age of Zkd 3 is unclear. It may be a young subadult (8–9 years of age) (Weidenreich 1943), an older subadult (Black 1929, 1931; Antón 2001), or an adult (Mann 1971). It was therefore included in analyses when possible, but its possible subadult status must be kept in mind when discussing this specimen.

Table 6.1 List of fossil specimens used in this study and their respective cranial capacities, as well as centroid sizes from two of the neurocranial analyses

Region	Specimen*	Endocranial capacity (cm³)	Centroid size/log centroid size[g]	
			Max. Indonesian	Max. Chinese
China	Zkd 3	915[a]	N/A	425.57/6.05
	Zkd 5	1140[a]	N/A	459.14/6.13
	Zkd 11	1015[a]	405.37/6.00	430.97/6.07
	Zkd 12	1030[a]	411.62/6.02	437.32/6.08
Early Indonesia	S 2	813[b]	379.82 / 5.94	N/A
	S 4	908[b]	N/A	N/A
	S 12	1059[b]	N/A	N/A
	S 17	1004[b]	416.11/6.03	448.43/6.11
Late Indonesia	Sm 1	1100[c]	421.31/6.04	N/A
	Sm 3	917[d]	394.82/5.98	422.73/6.05
	Ng 1	1172[e]	N/A	N/A
	Ng 6	1251[e]	434.28/6.07	468.49/6.15
	Ng 7	1013[e]	N/A	N/A
	Ng 10	1135[e]	N/A	N/A
	Ng 11	1231[e]	427.20/6.06[h]	458.74/6.13
	Ng 12	1090[e]	418.27/6.04	446.30/6.10
	Ngawi	1000[f]	391.36/5.97	418.62/6.04

[a]Weidenreich (1943)
[b]Holloway (1981)
[c]Jacob (1975)
[d]Marquez et al. (2001)
[e]Holloway (1980)
[f]Sartono (1990)
[g]Centroid size is calculated as the square root of the sum of squared distances from the centroid to each landmark and is therefore dependent upon the number and type of landmark used. Hence, centroid sizes in different analyses are not directly comparable
[h]Ng 11 had all necessary landmarks so centroid size could be computed, but it was not included in the analysis to maintain more equivalent temporogeographic sample sizes
*Zkd (Zhoukoudian), S (Sangiran), Sm (Sambungmacan), Ng (Ngandong)

Although clearly relevant to the broader questions pertaining to *H. erectus*, non-Asian *H. erectus* specimens, including those from Africa and Dmanisi, were excluded from this study in order to focus on variation within a more limited geographic region; details regarding cranial shape variation in *H. erectus* as a whole can be found in Baab (2007, 2008). Although the transition among groups of Pleistocene *Homo* in China is not well understood, later specimens from sites such as Dali, Maba and Jinniushan were also not analyzed because they are most commonly assigned to "archaic" *H. sapiens* based on their large endocranial volume and cranial morphology (e.g., Rightmire 2004; Baab 2007). In contrast, numerous studies have confirmed the attribution of the younger Ngandong fossils to *H. erectus* (Santa Luca 1980; Rightmire 1990; Antón 2002; Baab 2007). Unfortunately, most of the oldest fossils in both Indonesia and China are quite distorted due to diagenetic forces (e.g., S 27 and Gongwangling), and several of the Chinese *H. erectus* fossils are not widely accessible to researchers. Therefore, information about the morphology in the earliest Asian *H. erectus* crania is limited to that of S 4, and it was not possible to evaluate claims for intra-Chinese variation (Dong 1989; Li and Etler 1992; Antón 2002; Kidder and Durband 2004; Durband et al. 2005).

Methods

Data Acquisition and Processing

A total of 131 midline and bilateral 3D landmarks (such as standard osteometric points like bregma) were recorded from each cranial specimen with a Microscribe 3D/3D-X digitizer. In addition, three midline and nine bilateral curves were recorded from each specimen when possible. Each curve is a series of landmarks (termed semilandmarks) taken along a cranial contour like the midsagittal plane or along the temporal squama. For all specimens, each curve was then resampled to the same number of semilandmarks using weighted linear interpolation. The linear interpolation worked as follows: (1) the total length of the curve was calculated as the sum of the distances between the original neighboring semilandmarks, (2) the total distance was divided by the number of desired resampled semilandmarks, (3) the resampled semilandmarks were evenly spaced along this curve, and their coordinates were calculated as averages of the two closest original semilandmarks (weighted by proximity to the resampled landmark position). The endpoints of the curves remained unchanged in all cases. The landmark protocol included landmarks and curves from all regions of the vault, temporal base, and occipital (see Table 6.2) and was similar to landmark sets employed by Frost et al. (2003), Harvati et al. (2004) and McNulty et al. (2006). Not all landmarks collected

Table 6.2 Landmarks and curves: definitions and abbreviations

Landmark/curve	Abbr.	Definition	Ref.
Dorsal landmarks			
Inion	IN	Point at which the superior nuchal lines merge in the external occipital protuberance (or linear tubercle)	b
Lambda	LA	The apex of the occipital bone at its junction with the parietals, in the midline	a
Bregma	BR	Posterior border of the frontal bone in the midline	a
Midline post-toral sulcus	PTS	Minima of concavity on midline post-toral frontal squama	f
Glabella	GL	Most anterior midline point on the frontal bone, usually above the frontonasal suture	b
Nasion	NA	The intersection of the fronto-nasal suture and the median plane	a
Alveolare	AL	Midline point at the inferior tip of the bony septum between the upper central incisors	b
Alare	ALR	The most lateral point on the margin of the nasal aperture	b
Zygoorbitale	ZO	Point where the orbital rim intersects the zygomaticomaxillary suture	b
Dacryon	DA	Point where the lacrimomaxillary suture meets the frontal bone	b
Supraorbital notch	SON	Point of greatest projection of notch into the orbital space, taken on the medial side of the notch	f
Frontomalare- temporale	FMT	Point where the frontozygomatic suture crosses the temporal line (or outer orbital rim)	b
Frontomalare-orbitale	FMO	Point where the frontozygomatic suture crosses the inner orbital rim	b
Mid-torus inferior	MTI	Point on inferior margin of supraorbital torus roughly at the middle of the orbit (on superior margin of orbit)	c
Mid-torus superior	MTS	Point on superior aspect of supraorbital torus, directly above mid-torus inferior on anterior aspect of torus	c
Anterior pterion	AP	Where coronal suture intersects sphenofrontal or sphenoparietal suture	
Jugale	JG	The point in the depth of the notch between the temporal and frontal processes of the zygomatic	b
Porion	PO	Uppermost point on the margin of the external auditory meatus	b
Auriculare	AU	A point vertically above the center of the external auditory meatus at the root of the zygomatic process	b
Malar root origin	MR	Point where zygomatic root arises from the maxilla (often a point of convexity between molar juga and zygomatic root)	f
Frontotemporale	FT	The point where the temporal line reaches its most anteromedial position on the frontal	b
Parietal notch	PN	On postero-superior border of the temporal where the squamosal and parietomastoid sutures meet	b
Asterion	AS	The common meeting point of the temporal, parietal, and occipital bones, on either side	a
Ventral Landmarks			
Opisthion	OP	Midline point at the posterior margin of the foramen magnum	b
Tympanomastoid junction	TM	Point where tympanic tube and mastoid fissure meet laterally	
Medial petro-tympanic crest	MPT	Most medial point of petrotympanic crest at level of carotid canal	e
Lateral petro-tympanic crest	LPT	Lateral origin of petrotympanic crest	e
Stylomastoid foramen	SF	Posterior border of stylomastoid foramen	d
Postglenoid	PG	Infero-lateral most point posterior to glenoid fossa and anterior to ectotympanic tube (corresponds to postglenoid process)	c
Entoglenoid	EG	Most inferior point on the entoglenoid pyramid	d
Lateral articular fossa	LAF	Deepest point on the lateral margin of the articular eminence	d

(continued)

Table 6.2 (continued)

Landmark/curve	Abbr.	Definition	Ref.
Center of mandibular fossa	CMF	The intersection of the perpendicular axes though the longest and widest parts of the mandibular fossa	
Medial squamo-tympanic fissure	MST	Intersection of squamotympanic fissure with medial edge of mandibular fossa	
Lateral squamo-tympanic fissure	LST	Intersection of squamotympanic fissure with lateral edge of mandibular fossa	
Foramen spinosum	FS	Point on posterior margin of foramen spinosum	
Temporosphenoid suture	TS	Point where temporosphenoid suture passes from squama to cranial base (often on infratemporal crest)	
Lingual canine margin	LC	Most lingual aspect of canine alveolar process	f
M1–M2 contact	M1–2	Point of contact projected onto buccal alveolar margin	c
M2–M3 contact	M2–3	Point of contact projected onto buccal alveolar surface	f
Metopion	ME	Point midway between glabella and bregma in midline, calculated *a postieri*	b
Mid-parietal	MP	Point midway between bregma and lambda in midline, calculated *a postieri*	
Mid-temporal squama	MTS	Point midway along temporal squama between TS and PN, calculated *a postieri*	
Curves			
Midsagittal Frontal Profile	MFP	From bregma to glabella along midline	
Midsagittal Occipital Profile	MOP	From lambda to opisthion along midline	
Superior Frontal Torus	SFT	From frontomalaretemporale to midline, along the supero-anterior edge of the frontal torus	

a Howells (1973)
b White and Folkens (2000)
c Frost (2001)
d Harvati (2001)
e Harvati (2003)
f McNulty (2003)

were analyzed. Four landmarks (metopion, mid-parietal, and mid-temporal squama left and right) were calculated as the midpoint of a single curve (see Table 6.2 for details).

While most Asian *H. erectus* calvaria and crania preserve the cranial vault and the lateral basicranium (particularly the mastoid and TMJ regions), the exact structures preserved varied among individuals. Bilateral landmarks missing unilaterally were reconstructed by reflection of their antimeres across the midsagittal plane (e.g., McNulty et al. 2006; Gunz and Harvati 2007). Landmark configurations for all individuals were then superimposed using generalized Procrustes analysis (GPA) in order to remove the effects of location, orientation and scale (although shape variation related to size is not removed) (Rohlf 1990; Rohlf and Marcus 1993; O'Higgins 2000; Adams et al. 2004). GPA works by first superimposing specimens' centroids (geometric center) at the origin, scaling configurations to a unit centroid size (centroid size is the square root of the sum of the squared distances between each landmark and the centroid) and then rotating them until the residual sum-of-squares across all landmarks and specimens falls below a set tolerance level (Gower 1975; Rohlf and Slice 1990). In all cases the landmarks were averaged with their reflected equivalents in order to minimize the effects of bilateral asymmetry, particularly essential for fossil specimens (Bookstein 1996) and the superimposed landmarks were then treated as shape variables in standard multivariate statistical analyses (Kent 1994). Morpheus et al. (Slice 1998) was used for generalized Procrustes analysis as well as for visualization purposes, and statistical analysis was performed in SAS 8 (SAS Institute 1999–2001).

In the case of Zkd 5, portions of both temporal bones were recovered in the mid-1930s and described by Weidenreich (1935, 1936–1937, 1943). In 1966, members of the Institute of Vertebrate Paleontology and Paleoanthropology recovered both a frontal and an occipital bone that also belong to Zkd 5 (but see Schwartz and Tattersall 2003). Together these form a nearly complete cranium lacking its superior aspect. While most of the frontal squama is present, the posterior border is not complete. This specimen has been reconstructed in the area of bregma but no sutures are preserved. The midsagittal curve was recorded from glabella to the general region of bregma on Zkd 5, and Zkd 2, 3, 10, 11 and 12 were used to estimate the position of bregma along this original curve.

Study Design

Two geographic groups were recognized in this analysis: Indonesia and China. In order to statistically investigate the effects of time on the fossil sample, specimens were assigned to one of four time bands of 400 kyr each from 1.6 Ma to the present. While narrower time bands (e.g., 200 Ka; Wood 1993) would have provided more fine-grained temporal resolution, the 400 kyr time bands still permitted a reasonable degree of resolution while also allowing for wider confidence limits around the age of each specimen (necessary due to well-documented uncertainties surrounding dating of Asian, particularly Indonesian, *H. erectus* sites). Each specimen was assigned to a single time band based on evaluation of the available data (see Baab 2007 for discussion) as summarized in Table 6.3. In each analysis the centroid size was calculated for all individuals (see Table 6.1) and acted a multivariate proxy for size (Bookstein 1996).

Several sets of landmarks were designed to maximize inclusion of incomplete fossil specimens. The three neurocranial landmark sets captured the overall shape of the vault including the supraorbital torus. There were also three regional analyses which focused on the frontal bone, temporal base, and occipital bone. The fossil samples for each landmark set differed based on individual fossil preservation. Table 6.4 provides details about the composition of samples and landmarks in each analysis. Results were reported for each landmark set separately, but interpretations of geographic, anagenetic and size-related variation were based on the combined results of these analyses. One of the neurocranial analyses was performed on only the Indonesian fossil hominins ("Indonesians Only"), in order to focus on questions specific to this more temporally variable sample.

Both the frontal and occipital bone analyses included landmarks and semilandmarks (see Table 6.4). For the frontal bone analysis, the semilandmarks were resampled from three curves: the midsagittal frontal profile (MFP) and the superior frontal (supraorbital) torus (SFT left and right). MFP was resampled to eight semilandmarks, including glabella and bregma. SFT right and left were resampled to five semilandmarks each, which included frontomalaretemporale as the first landmark. Because the supraorbital curves ended at the midline which was also captured by the midsagittal line, the last of these five semilandmarks was dropped from each of the supraorbital curves. Combined, a total of 25 landmarks and semilandmarks were analyzed in the frontal bone analysis. The occipital bone analysis was focused exclusively on the curvature of the bone in the midline and the position of asterion. The midline occipital profile (MOP) curve was taken from lambda to opisthion in the mid-sagittal plane. In addition to asterion, lambda, and opisthion, eight semilandmarks were analyzed that were evenly spaced

Table 6.3 Geochronological time bands for Asian Homo erectus sites/specimens (preferred ages in Ma outlined)

Site/specimen	Time band 1: ≤1.6–1.2	Time band 2: <1.2–0.8	Time band 3: <0.8–0.4	Time band 4: <0.4
S 4	>1.5[a]	1.07–0.99[b, c]		
S 17	1.3–1.24[d]	0.99–0.78[e, c]		
S 12		1.2–0.98[d]	~0.78[b]	
S 2		1.2–0.98[d]	~0.78[b]	
Zhoukoudian (all but 5)*			0.6–0.4[f] or >0.5 (0.8–0.6)[g, o]	
Zkd 5			0.4–0.5[g]	~0.35[f], 0.29[h]
Sambungmacan		~1.0 (Kabuh or Jaramillo)[e, i]	≤0.78 (Brunhes)[e, i–j]	0.05–0.032[k]
Ngawi		Kabuh[l]	Kabuh[l]	High Terraces (same age as Ngandong)[l]
Ngandong			Middle Pleistocene[m]	0.05–0.032[k] or 0.1[n]

[a]Antón (2003)
[b]Itihara et al. (1994)
[c]Hyodo et al. (1993, 2002)
[d]Larick et al. (2001)
[e]Jacob et al. (1978)
[f]Grün et al. (1997)
[g]Shen et al. (2001)
[h]Yuan et al. (1991)
[i]Matsu'ura et al. (2000) for Sm 2
[j]Baba et al. (2003) for Sm 4
[k]Swisher et al. (1996)
[l]Sartono (1990)
[m]Santa Luca (1980)
[n]Bartstra et al. (1988)
[o]Shen et al. (2009).
*Zkd 3 is the oldest of the Zhoukoudian fossils and may be older than 0.8 Ma, but this date is based on extrapolations from approximating sedimentation rates (Shen et al. 2001)

Table 6.4 Neurocranial and regional analyses: composition of samples and landmark sets

	Fossils		
Analysis	China	Indonesia	Landmarks[a]
Maximum Landmarks	Zkd 11 & 12	S 17, Sm 3, Ng 6, 11 &12, Ngawi	OP, IN, LA, MP, BR, MET, PTS, GL, SON, FMT, FMO, AP, MTS, PO, AU, FT, PN, AS, TM, EG, TS[b]
Maximum Indonesian	Zkd 11 & 12	S 2, 17, Ngawi, Ng 6, 11, 12,Sm 1 & 3	OP, IN, LA, MP, BR, FMO, FMT, FT, AP, MTS, PO, AUR, TM, EG, PN, AS, TS
Maximum Chinese	Zkd 3, 5, 11 & 12	S 17, Ng 6, 11 & 12, Sm 3, Ngawi	IN, LA, MP, BR, ME, PTS, GL, SON, FMO, FMT, FT, AP, MTS, PO, AUR, TM, EG, PN, AS
Indonesians Only	None	S 2, 17, Ngawi, Ng 6, 11, 12, Sm 1 & 3	OP, IN, LA, MP, BR, FMO, FMT, FT, AP, MTS, PO, AUR, TM, EG, PN, AS, TS
Frontal Bone	Zkd 3, 5, 11 & 12	S 17, Sm 3[c], Ng 6 & 12, Ngawi	BR, GL, NA, FMT, FMO, MTI, AP, FT, twelve semilandmarks[d]
Temporal Base	Zkd 3 & 5	S 2, 4 & 17, Sm 1, Ng 6 & 12	TM, MPT, LPT, SF, PG, EG, LAF, CMF, MST, LST, TS[e]
Occipital Bone	Zkd 11 & 12	S 2, 4, 12 & 17, Sm 1 & 3, Ngawi, Ng 1, 6, 7, 11 & 12	LA, OP, eight semilandmarks[f]

[a] Abbreviations refer to Table 6.2
[b] See Fig. 6.1 for an illustration of all neurocranial landmarks used set
[c] Nasion was taken at the broken inferior edge of the frontal bone but may be slightly superior to the actual frontonasal suture
[d] Six evenly spaced semilandmarks were resampled from the MFP curve and three each were resampled from the left and right SFT curves (see Fig. 6.7)
[e] See Fig. 6.9
[f] Eight semilandmarks were resampled from the MOP curve between lambda and opisthion (see Fig. 6.12)

between these lambda and opisthion, for a total of 12 semilandmarks and landmarks.

Statistical Analysis

TpsSmall (Rohlf 2003) was used to test the correspondence between the datasets in Kendall's shape space and the Euclidean space which is tangent to Kendall's shape space at the mean configuration (Marcus et al. 2000). The very tight correlations (R^2 = 0.99999 – 1.0) between the pairwise Procrustes distances (calculated in Kendall's shape space) and the pairwise Euclidean distances for all landmark sets described in Table 6.4 permitted statistical analyses to be carried out in GPA space (sensu Slice 2001, 2005).

It was possible to address whether temporal / geographic variation or size variation had a greater influence on Asian *H. erectus* cranial shape. If anagenetic change, which is attributable to either gene flow or convergent selection pressures in different regions, exerted the strongest influence on cranial morphology, then temporal variation will be most tightly correlated with cranial shape variation. If, however, local evolutionary forces such as genetic drift / isolation or adaptation to the local environment had a more significant impact on cranial morphology, then geography would explain more of the shape variation in Asian *H. erectus*. Finally, if some combination of sexual selection and species-specific allometric constraints are the primary determiners of cranial morphology, then cranial shape will be closely linked to size variation.

Regressions of the coordinates of the superimposed landmarks on the two geographic groups (coded as dummy variables: Indonesia = 1 and China = 0), geological age measured as millions of years before present (ln-transformed, based on the midpoint of each band) and centroid size (ln-transformed) were performed. The total variance remaining in the sample after the shape variables were adjusted for each independent variable was calculated and presented as a proportion of the total variance of the sample. Regressions were performed for all landmark sets. The angles (and cosines of the angles) between the vectors of regression coefficients for each pair of the three independent variables [geography, log geological age, and log centroid size] were reported in order to quantify the magnitude of correlation between each set of factors. The coordinates of the superimposed landmarks were regressed on the three independent factors separately (in three multiple regression analyses), which resulted in three vectors of regression (beta) coefficients, one for each factor. The larger the cosine value, the stronger the correlation between the two factors. If landmarks were removed in order to accommodate individual specimens, new regressions were performed and all results were reported.

Principal components analysis (PCA) was performed on each data set. PCA reduces the dimensionality of the data by rotating the GPA-aligned data (shape coordinates) so that the new axes (principal components) are orthogonal to one another and are ordered by the amount of variance explained.

PCA ordinations highlight patterns of shape covariation in the data. Only plots of the first two components were presented because higher order components were not statistically significantly correlated with geography, age, or size of the fossils.

Differences between geographic groups were tested over a larger number of PCs (accounting for at least 97% of the total variance) using multivariate analysis of variance (MANOVA). Principal component (PC) 1 and 2 scores were also regressed on geography, log geological age, and log centroid size. In order to also examine the interaction among the independent variables, the general linear model was used. The analysis results in both a probability associated with the F value for the overall model, as well as t-tests for the independent variables. The probabilities associated with statistically significant interactions were also presented.

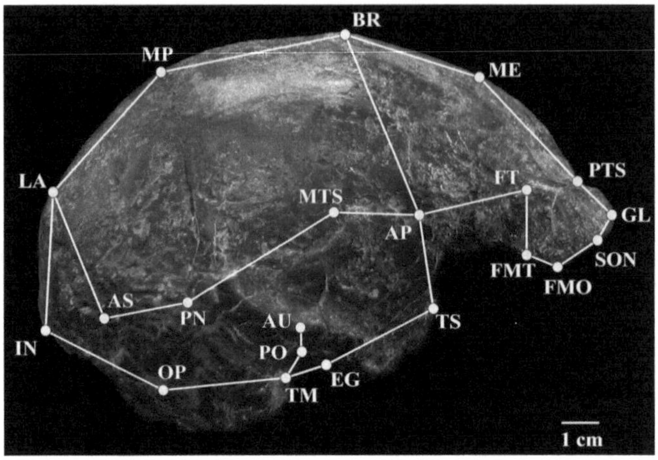

Fig. 6.1 Illustration of all landmarks used in three neurocranial analyses. This particular set of landmarks corresponds to those used in the maximum landmarks analysis whereas both the maximum Indonesian and maximum Chinese analyses used slightly different subsets of these landmarks. The *lines* connecting the landmarks are for visualization purposes only and do not represent actual data. Again, the *wireframes* used in the maximum Indonesian and maximum Chinese analyses differ slightly from the one shown here for the maximum landmark set. Landmarks OP and EG cannot be seen in this view but their position is estimated. Landmark definitions and abbreviations can be found in Table 6.2

Visualization

Shape variation along each of the first two principal components was illustrated at the positive and negative ends of the axes for each analysis (by adding/subtracting the eigenvectors of each component to/from the consensus configuration) scaled by the maximum and minimum values observed in the data set. In some cases comparisons were also made between the average shapes from each of three restricted temporogeographic groups: Zhoukoudian, early Indonesian (Sangiran), and late Indonesian (Ngandong, Sambungmacan, and Ngawi). This approach was deemed preferable to a comparison of the mean from each group to the grand mean for all Asian *H. erectus* fossils because the former method emphasized the small scale variations among temporogeographic groups.

Results

Neurocranium

Maximum Landmarks Analysis

The first neurocranial landmark set includes the greatest density of landmarks, but the fewest specimens (see Table 6.4), and is illustrated in Fig. 6.1. This landmark set captures the shape of the entire neurocranium. MANOVA on the scores from the first six PCs, which together account for 98% of the total shape variance in the sample, indicates significant group differences between the Chinese and Indonesian fossils (Table 6.5). The first component in the PCA separates the Indonesian fossils (which score low) from the two Zhoukoudian specimens (which score high) as seen in Fig. 6.2. This corresponds well with the observation that geography accounted for the highest proportion of the total variance in cranial shape (Table 6.6). The first component is clearly related to differences between the Chinese and Indonesian fossils, while 75% of the variance in PC 2 scores was accounted for by differences in the size of the individual fossils (Table 6.7) with the larger specimens scoring higher than smaller ones. In all of the neurocranial analyses, the shape differences accounted for by geography and time overlap to a great extent as indicated by the high cosine values in Table 6.6.

As seen in Fig. 6.2, individuals with higher scores along PC 1 (i.e., Zhoukoudian) have more postero-superiorly positioned inion, more inferiorly positioned opisthion, longer but less outwardly expanded midline frontal squamae, narrower occipital and frontal bones, greater width across the basal temporal bone, and the lateral ends of the supraorbital torus are less down-turned. Also illustrated in Fig. 6.2 are the shape differences associated with an increase in size toward the positive end of PC 2. High scoring individuals have longer neurocrania due to greater expansion of the anterior frontal bone, a lower temporal squama, and a more posteriorly located inion and lambda. The basal temporal structures are more medially positioned but auriculare is more laterally projecting relative to porion, and the sphenosquamosal suture is more anteriorly located. The frontal bone also differed in being slightly narrower with a more superiorly positioned frontotemporale.

The differences between the Chinese and Javanese averages correspond closely to those described for PC 1 above

Table 6.5 MANOVA test of group differences between Zhoukoudian and Indonesian samples on the first n principal components that explain ≥97% of the total variance in the neurocranial shape analyses

Analysis	PCs scores used for MANOVA	% of total variance explained by PCs	Wilks' Lambda F value	Probability associated with F value	Significant difference in group means on PCs
Maximum Landmarks	1–6	97.7%	1399.56	0.0205*	PC 1
Maximum Indonesian	1–8	98.8%	73.20	0.0902	-
Maximum Zhoukoudian	1–7	97.6%	39.06	0.0252*	PC 1

*Indicates significant result at $p < 0.05$

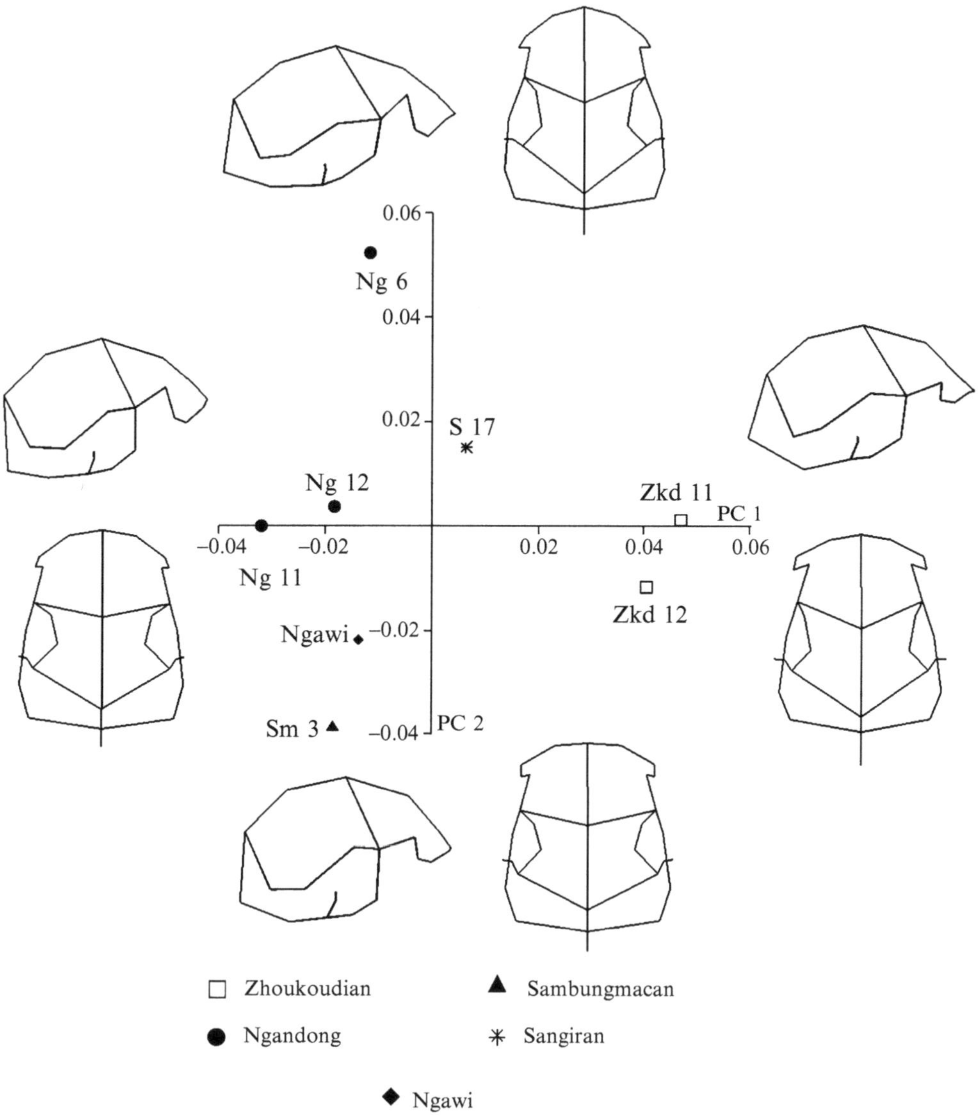

Fig. 6.2 PCs 1 and 2 of the maximum landmark data set. The shape changes seen from the negative to the positive ends of the first two components illustrated below in *right lateral* and *superior* views are also described in the text. The wireframe corresponds with that shown in Fig. 6.1. For this and subsequent PCA graphs, the proportion of variance captured by each component is listed in Tables 6.7 and 6.10

(Fig. 6.3). Relative to the younger Indonesian fossils, the Zhoukoudian specimens are wider across the basal temporal bone structures, but narrower at all points along the vault (FT, AP, MTS, PN, and AS) (Fig. 6.3b, c). This pattern corresponds to the narrow frontal and occipital dimensions relative to the midvault in Zhoukoudian as noted by previous researchers (Antón 2002; Kidder and Durband 2004; Durband 2006). However, the greater width across auriculare is not indicative of greater width across the mid-vault in general; rather, the suprameatal crest appears to be more laterally projecting in the Chinese specimens (Fig. 6.3a, b). Similarly, despite the greater constriction at frontotemporale, the Chinese fossils are actually wider across anterior pterion compared to S 17. S 17 does not differ very much from the

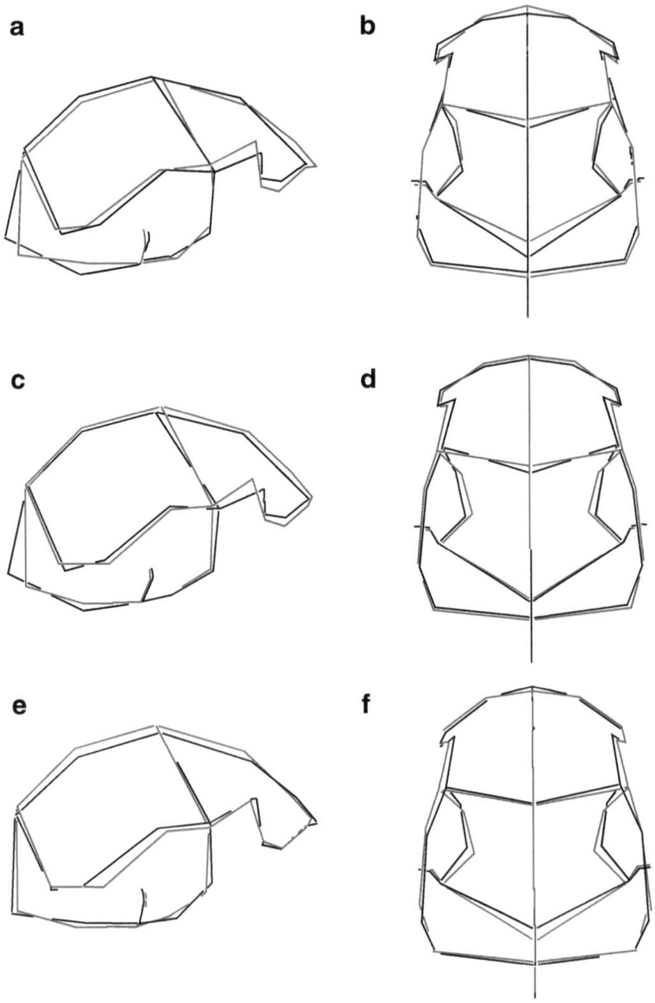

Fig. 6.3 Comparison of the average shapes of the Zhoukoudian, early Indonesian and late Indonesian groups. The shape differences are illustrated in *left lateral* and *superior* views between (**a**, **b**) Zhoukoudian (*black*) and S 17 (*gray*), (**c**, **d**) Zhoukoudian (*black*) and the younger Indonesian fossils (*gray*), and (**e**, **f**) S 17 (*black*) and the younger Indonesian fossils (*gray*)

younger Indonesian fossils, but does have a lower cranial profile in the midline, a taller mid-temporal squama, and a narrower frontal squama (see Fig. 6.3e, f).

Maximum Indonesians Analysis

By excluding several of the midline frontal bone landmarks and SON from the analysis, it was possible to include additional specimens from Java, specifically Sm 1 and S 2 (see Table 6.4), thus increasing the representation of older and smaller individuals in the analysis.

A MANOVA test for differences between these two regional groups for scores on the first eight PCs together was not significant (see Table 6.5), and geographic origin of fossils accounts for less of the total shape variance than in the previous maximum landmarks analysis (see Table 6.6). There is no clear Chinese–Indonesian division in regards to cranial shape on the first four PCs. On PCs 1, 2, and 4, both of the Zhoukoudian fossils scored higher than the majority of the Javanese specimens, but the two geographic groups overlapped one another on all PCs. S 2 in particular does not group with its closest geographic neighbors on PC 1 (see Fig. 6.4). There is a complex relationship between age, size, geography, and PC 1 scores, including statistically significant interaction between the geochronological age and size of each fossil (Table 6.7). Among the older fossils from Sangiran and Zhoukoudian, the larger specimens have lower scores, whereas the smaller specimens of the younger group (Ngandong, Sambungmacan, and Ngawi) score even lower. On PC 1, S 2 and the Zhoukoudian fossils differ from the other, lower scoring, Indonesian fossils, in having a narrower frontal bone with greater constriction across frontotemporale, a lower cranial vault with a more posteriorly projecting inion, and a shorter parietomastoid suture (see Fig. 6.4).

The absence of landmarks from the central portion of the frontal squama may have contributed to the lack of separation between Indonesian and Zhoukoudian samples in this analysis as this region did vary consistently between the two geographic groups. However, S 2 also differed from S 17 and the other fossils from Java in several other features of neurocrania shape (as described above), and it is certainly possible that it exhibited further variations in frontal bone morphology.

Maximum Zhoukoudian Analysis

When OP and TS are excluded from the maximum landmark set, two additional Zhoukoudian fossils, Zkd 3 and 5, can be examined. This introduces some additional time depth to the sample as Zkd 3 and 5 represent the oldest and youngest specimens, respectively (see Table 6.3). These are also the largest (Zkd 5) and smallest (Zkd 3) of the Zhoukoudian specimens (see Table 6.1).

The inclusion of these additional fossils results in a strong Zhoukoudian–Indonesian dichotomy along the first component (Fig. 6.5). Geographic differences account for 40% of the total variance in neurocranial shape (Table 6.6), with most of these differences concentrated on the first component (Table 6.5). There is a relationship between size and PC 2 scores, which is modified by geography. Specifically, larger Indonesian fossils score lower on PC 2 while the opposite pattern holds true within the Chinese sample (i.e., smaller fossils score lower on PC 2). This indicates a different relationship between size and shape within the two geographic regions.

Most of the shape differences seen along PC 1, which separate the two geographic groups, are similar to those described in the maximum landmarks analysis along the first component. Fossils with higher scores along PC 2

Table 6.6 Neurocranial analyses: proportion of total variance explained by geography, log geological age, and log centroid size

	Maximum Landmarks		Maximum Indonesian		Maximum Zhoukoudian		Indonesians Only	
Factor	Variance	% of Total	Variance	% of Total	Variance	% of Total	Variance	% of Total
Total	0.00307	100.0	0.00347	100.0	0.00324	100.0	0.00338	100
Geography	0.00227	25.9	0.00287	17.1	0.00131	40.4	N/A	N/A
Log geological age	0.00237	22.6	0.00265	23.6	0.00086	26.5	0.00251	25.6
Log centroid size	0.00242	21.1	0.00296	14.6	0.00032	9.9	0.00275	18.6
Geog. & ln(geol. age)	0.00171	44.4	0.00220	36.6	0.00167	51.5	N/A	N/A
Geog. & ln(cent. size)	0.00165	46.0	0.00238	31.2	0.00162	49.8	N/A	N/A
ln(geol. age) & ln(cent. size)	0.00173	43.6	0.00223	35.7	0.00118	36.4	0.00194	42.5
Geog., ln(cent. size) & ln(geol. age)	0.00110	64.2	0.00176	49.4	0.00197	60.7	N/A	N/A
Angle/cosine of angle between Geography & ln(geol. age)	52.04°/0.62		52.21°/0.61		30.65°/0.86		N/A[a]	
Angle/cosine of angle between Geography & ln(cent. size)	72.17°/0.31		63.47°/0.45		74.34°/0.27		N/A[a]	
Angle/cosine of angle between ln(geol. age) & ln(cent. size)	86.72°/0.06		58.42°/0.52		82.49°/0.13		63.08°/0.45	

Larger cosine values indicate stronger correlations

[a]There were no geographic differences in the Indonesians only analysis, so this was not considered as a separate variable in this analysis

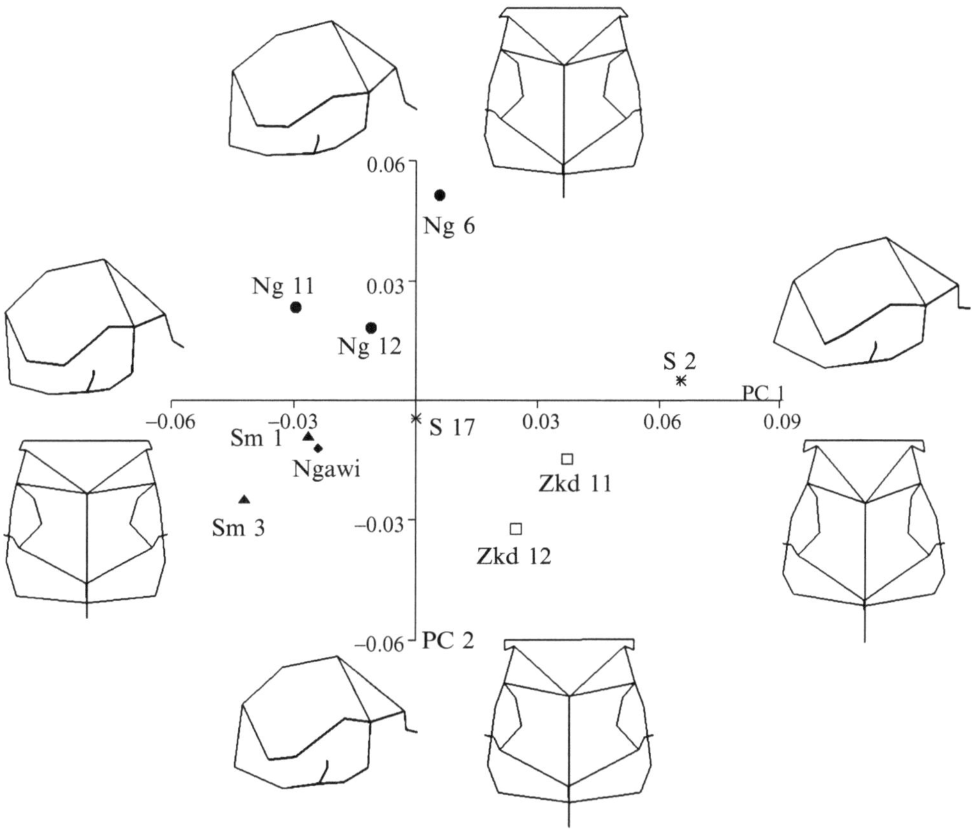

Fig. 6.4 PCs 1 and 2 of the maximum Indonesian data set. The shape changes along the first two components are illustrated below in *right lateral* and *superior* views. The exclusion of numerous landmarks from the anterior frontal bone resulted in its somewhat truncated appearance in the wireframe illustrations. See Fig. 6.2 for symbol legend

Table 6.7 Neurocranial analyses: results of PCA and multiple regression of PC scores on geography, log geological age, and log centroid size

			Overall Model ANOVA		Parameter estimates (Pr > |t|)			
Analysis	PC	% of Variance	R^2	Pr (R^2) > F	Geography	ln (geol. age)	ln(centsize)	Interaction
Maximum Landmarks	1	27.7%	0.84	0.0013	0.0013	N/S	N/S	N/S
	2	23.9%	0.75	0.0055	N/S	N/S	0.0055	N/S
Maximum Indonesian	1	33.2%	0.92	0.0065	0.0398	0.0071	0.0093	(size & age) 0.0073
	2	18.3%	-	N/S	N/S	N/S	N/S	N/S
Maximum Zhoukoudian	1	45.3%	0.86	0.0001	0.0001	N/S	N/S	N/S
	2	15.3%	0.71	0.0461	0.0323	N/S	0.0129	(geog. & size) 0.0324
Indonesians Only	1	37.6%	0.87	0.0307	N/A[a]	0.0214	0.0403	(size & age) 0.0218
	2	19.6%	0.62	0.0196	N/A[a]	N/S	0.0196	N/S

[a]There was only one geographic group in the Indonesians only analysis, therefore geography was not included as a separate variable in the analysis

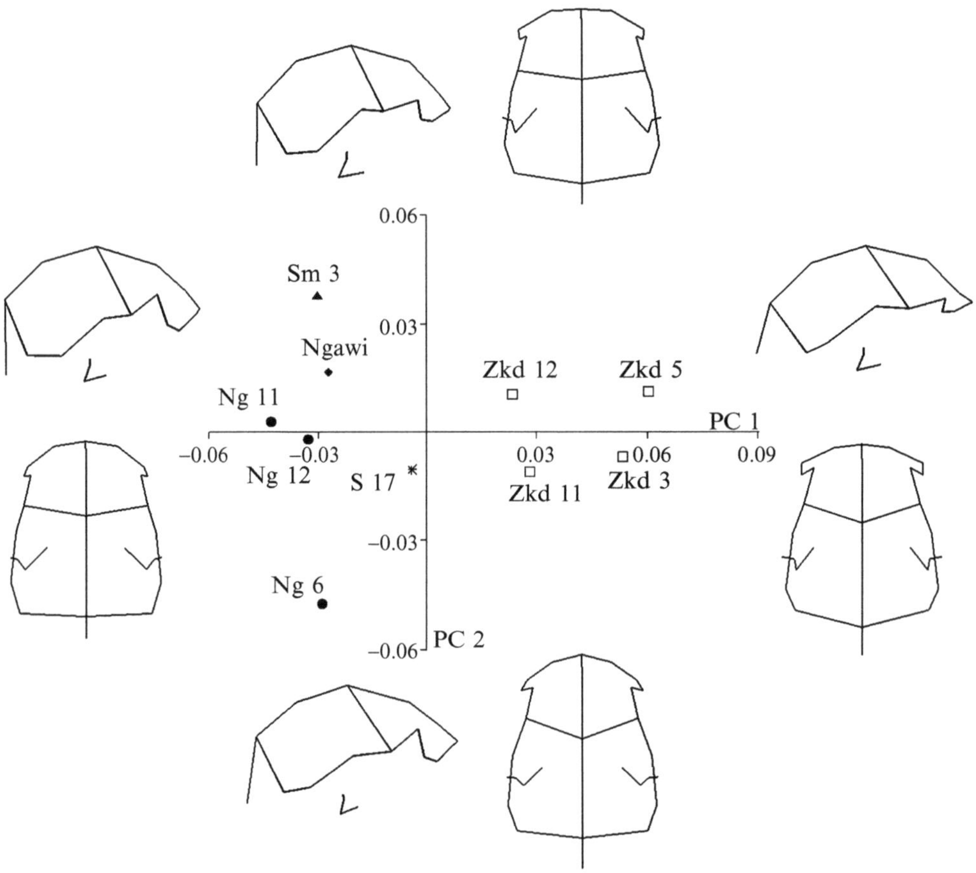

Fig. 6.5 PCs 1 and 2 of the maximum Chinese data set. The shape changes along the first two components are illustrated below in *right lateral* and *superior* views. The absence of opisthion and the temporosphenoid suture landmarks from this analysis result in the "floating" landmarks on the temporal base (EG, TM, PO, and AU). See Fig. 6.2 for symbol legend

have relatively shorter and (slightly) taller cranial vaults (due to less posterior projection at inion and less expansion at glabella) and wider frontal squamae with more inferiorly positioned frontotemporale. Also, there is greater antero-posterior distance between the entoglenoid process and the tympanomastoid junction, both of which are more laterally positioned.

Indonesians Only Analysis

This analysis utilized the same landmark set as the maximum Indonesian data set described above (see Table 6.4), but included only the Indonesian specimens in order to address questions specific to the southeast Asian *H. erectus* fossil record. Geochronological age accounted for the greatest proportion of the total shape variance in this sample, explaining about a quarter of shape differences (Table 6.6). Size

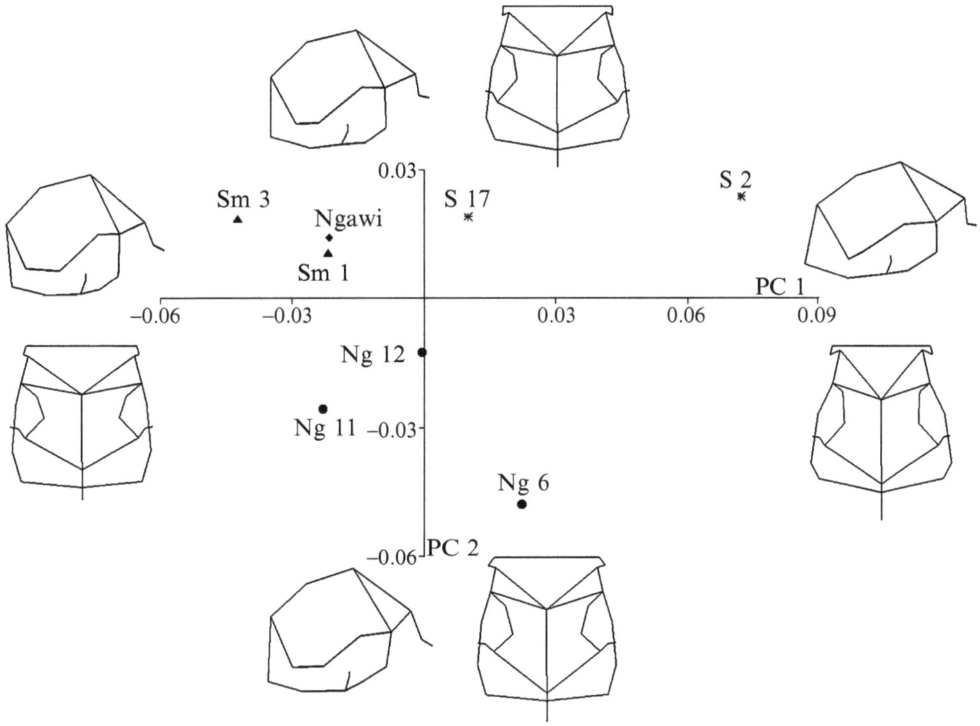

Fig. 6.6 PCs 1 and 2 of the Indonesians only data set. The shape changes along the first two components are illustrated below in *right lateral* and *superior* views. See Fig. 6.2 for symbol legend

explained slightly less of the total variance (19%), and there was some overlap in the shape variance explained by these two factors (cosine = 0.45).

There was a complex relationship between age and size of the fossil on PC 1 (Table 6.7), where, in the older age group (i.e., Sangiran), the larger S 17 specimen scored lower than S 2, but in the younger age group the pattern was the opposite (the larger specimens scored higher). Overall, the most variable aspects of neurocranial shape within this sample of Indonesian *H. erectus* separates S 2 from the other specimens, as seen in Fig. 6.6. Those specimens which scored lower on PC 1 (Sm 1, 3, and Ngawi) had much wider frontal bones, less projecting occipital bones, longer parietomastoid sutures, and more posteriorly positioned frontotemporale.

On the second component, which is influenced by overall size of the fossils (Table 6.7), the Ngandong specimens scored lower than any of the other Indonesian fossils. The negative end of PC 2 is associated with more medially located temporal base structures and more height at bregma. The temporosphenoid suture is more superiorly positioned while frontotemporale is more supero-posteriorly positioned, and anterior pterion is located more anteriorly.

Frontal Bone Analysis

The frontal bone landmarks and semilandmarks from the MFP and SFT curves are illustrated in Fig. 6.7. There is no overlap between the two geographic groups on the first component. Although there is not a significant group (Java vs. Zhoukoudian) difference on PCs 1–7 (accounting for 98% of total variance) (Table 6.8), geography accounts for the greatest proportion of shape variance in this sample (see Table 6.9), and PC 1 is strongly influenced by the geographic origin of the fossils (see Table 6.10). The high cosine of the angle between the vectors of correlation coefficients for both geography and log centroid size of 0.85 indicates that these factors

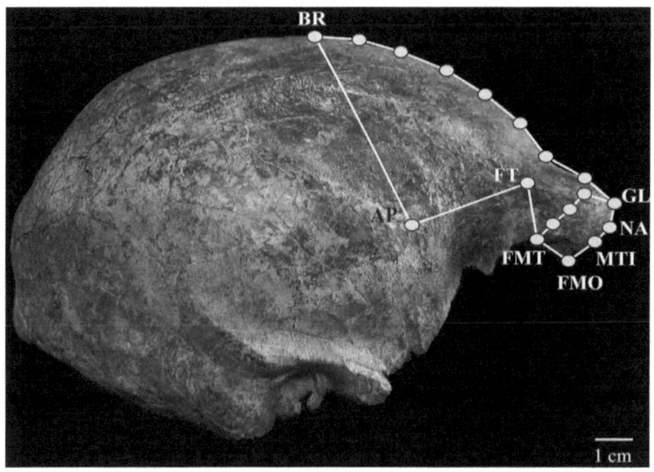

Fig. 6.7 Landmarks (labeled) and semilandmarks used in the frontal bone analysis, illustrated on Sm 1. The semilandmarks are from the MFP and SFT curves. See Table 6.2 for landmark and curve abbreviations

Table 6.8 MANOVA test of group differences between Zhoukoudian and Indonesian samples on the first n principal components that explain ≥97% of the total variance in the regional analyses

Analysis	PC scores used for MANOVA	% of total variance explained by PCs	Wilks' Lambda F value	Probability associated with F value	Significant difference in group means on PCs
Frontal Bone	1–7	98.5%	39.34	0.1222	–
Temporal Base	1–7	98.1%	4.72	0.3407	–
Occipital Bone	1–6	97.0%	1.95	0.1873	–

Table 6.9 Regional analyses: proportion of total variance explained by geography, log geological age, and log centroid size

Factor	Frontal Bone		Temporal Base		Occipital Bone	
	Variance	% of Total	Variance	% of Total	Variance	% of Total
Total	0.00544	100.0	0.03204	100.0	0.00565	100.0
Geography	0.00355	34.8	0.02447	23.6	0.00493	12.8
Log geological age	0.00433	20.4	0.02692	16.0	0.00513	9.2
Log centroid size	0.00428	21.4	0.02823	11.9	0.00442	21.8
Geog. & ln(geol. age)	0.00253	53.5	0.01965	38.7	0.00446	21.1
Geog. & ln(cent. size)	0.00304	44.2	0.02183	31.8	0.00383	32.3
ln(geol. age) & ln(cent. size)	0.00321	41.0	0.02394	25.3	0.00393	30.6
Geog., ln(cent. size) & ln(geol. age)	0.00198	63.6	0.01755	45.2	0.00345	38.9
Angle / cosine of angle between Geography & ln(geol. age)	54.19°/0.59		72.39°/0.30		65.82°/0.41	
Angle / cosine of angle between Geography & ln(cent. size)	32.29°/0.85		130.27°/–0.65[a]		121.24°/–0.52[a]	
Angle / cosine of angle between ln(geol. age) & ln(cent. size)	68.17°/0.37		126.61°/–0.60		74.81°/0.26	

[a]To obtain the vector of regression coefficients, geography was coded as a dummy variable with the Indonesian specimens coded as "1" and the Chinese specimens coded as "0". Therefore an angle greater than 90° (or a negative cosine) implies that changes in shape from Chinese to Indonesian follow the opposite pattern to those observed for increasing size (or decreasing age) to the extent that they account for similar patterns of shape variation

Table 6.10 Regional analyses: results of PCA and multiple regression of PC 1 and PC 2 scores on geography, log geological age, and log centroid size

Analysis	PC	% of Variance	Overall Model ANOVA		Parameter estimates (Pr > |t|)			
			R^2	Pr (R^2) > F	Geography	ln (geol. age)	ln(centsize)	Interaction
Frontal Bone	1	42.4%	0.75	0.0027	0.0027	N/S	N/S	N/S
	2	20.7%	0.77	0.0019	N/S	0.0019	N/S	N/S
Temporal Base	1	30.0%	0.55	0.0231	0.0231	N/S	N/S	N/S
	2	25.5%	-	N/S	N/S	N/S	N/S	N/S
Occipital Bone	1	48.4%	0.42	0.0090	N/S	N/S	0.0090	N/S
	2	17.2%	0.42	0.0093	0.0093	N/S	N/S	N/S

accounted for much of the same shape change (see Table 6.9). Most of the variation in PC 2 scores is explained by the geochronological age of the specimens (see Table 6.10), but S 17 is the only specimen older than 1 Ma in the analysis. It is not clear whether this specimen is broadly representative of older Asian *H. erectus*.

There is a fairly tight grouping of the younger fossils from Indonesia in the upper right quadrant of the PC plot in Fig. 6.8. The Zhoukoudian fossils also form a fairly cohesive group on PCs 1 and 2, particularly Zkd 3, 5, and 11. The general temporal trend seen on PC 2 is not followed within the Zhoukoudian sample.

Specimens at the negative end of PC 1 (Zhoukoudian) have a stronger frontal tuberosity, a marked and anteriorly positioned post-toral sulcus on the midline, more antero-inferiorly positioned frontotemporale, and the supraorbital torus is straighter in anterior view (less arched). Several of these traits were described previously by Weidenreich (1943, 1951), Santa Luca (1980) and Antón (2002). These fossils also did not have strong development of the lateral "corners" of the

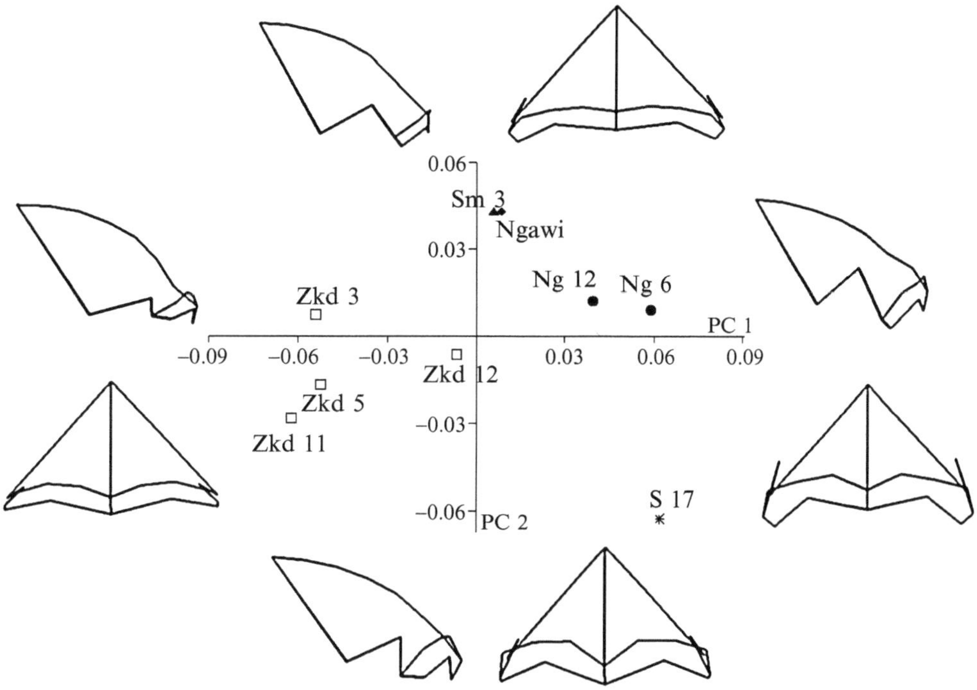

Fig. 6.8 PCs 1 and 2 of the frontal bone data set. The shape changes along the first two components are illustrated below in right lateral and anterior views. See Fig. 6.2 for symbol legend

supraorbital torus, which has been described in the Ngandong (frontal trigone: Weidenreich 1951) and Sambungmacan (angular trigone: Wolpoff 1999) specimens. Lower scoring (thus older) fossils on PC 2 have taller and more arched supraorbital tori that are more anteriorly projecting at glabella and frontotemporale is located more anteriorly.

The differences among the mean temporogeographic samples are illustrated in Fig. 6.9. Both the older and younger Indonesian fossils share certain features not found in the Chinese fossils, which generally correspond to the differences described above (e.g., the absence of a midline post-toral sulcus). However, S 17 is also different from both the Zhoukoudian and recent Indonesian averages in having a much taller and more anteriorly projecting supraorbital torus with a greatly expanded glabellar region.

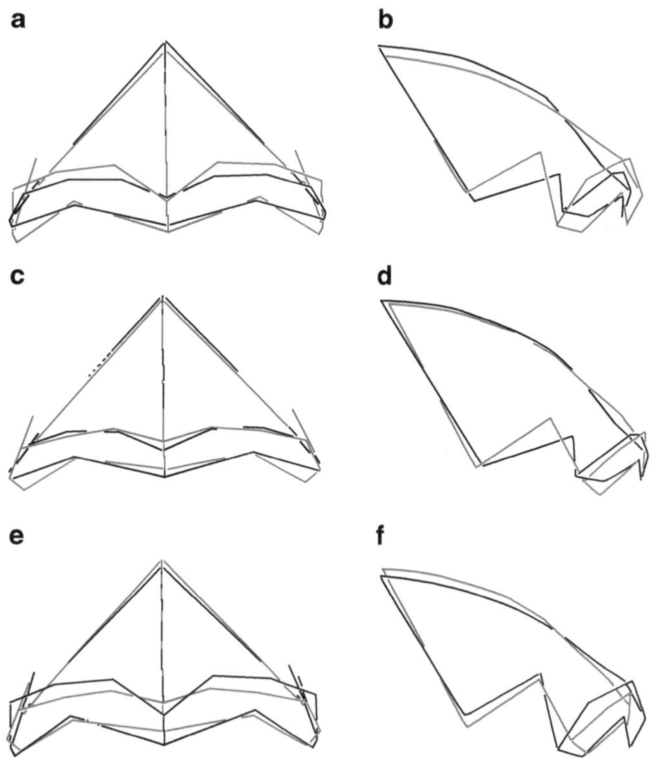

Fig. 6.9 Comparison of the average frontal bone shape in the Zhoukoudian, older Indonesian (i.e., Sangiran), and younger Indonesian (i.e., Ngandong, Sambungmacan, and Ngawi) samples. The shapes are illustrated in right lateral and anterior views between (**a**, **b**) Zhoukoudian (*black*) and older Indonesian (*gray*), (**c**, **d**) Zhoukoudian (*black*) and younger Indonesian (*gray*), and (**e**, **f**) older Indonesian (*black*) and younger Indonesian (*gray*)

Temporal Base Analysis

A more detailed analysis of the temporal base morphology was also undertaken (Table 6.4 and Fig. 6.10). Geographic differences explain the greatest proportion of variation in the shape of the temporal base (24%), followed by differences in age (16%) (Table 6.9). Although there is not a significant difference between the Indonesian and Zhoukoudian groups on the first seven components, which together accounted for 98% of the total shape variance (Table 6.8), the Zhoukoudian fossils do have significantly higher scores on PC 1 (Table 6.10).

The Zhoukoudian specimens have mandibular fossae that are both narrower (medio-laterally) and longer (antero-posteriorly) and the lateral aspect of the fossa is more superiorly located (Fig. 6.11). The antero-posteriorly long mandibular fossa was also noted by Picq (1990). In addition, the stylomastoid foramen is more medially positioned, the anterior face of the tympanic tube is longer, and the petrotympanic crest is less inferiorly projecting. Higher scoring individuals on PC 2 have a less postero-laterally angled petrotympanic crest, the medial end of the squamotympanic fissure is more anteriorly located, and the mandibular fossa is slightly wider. Also, both the lateral end of the squamotympanic fissure and the postglenoid process are more inferiorly positioned, and the petrotympanic crest is more supero-laterally oriented.

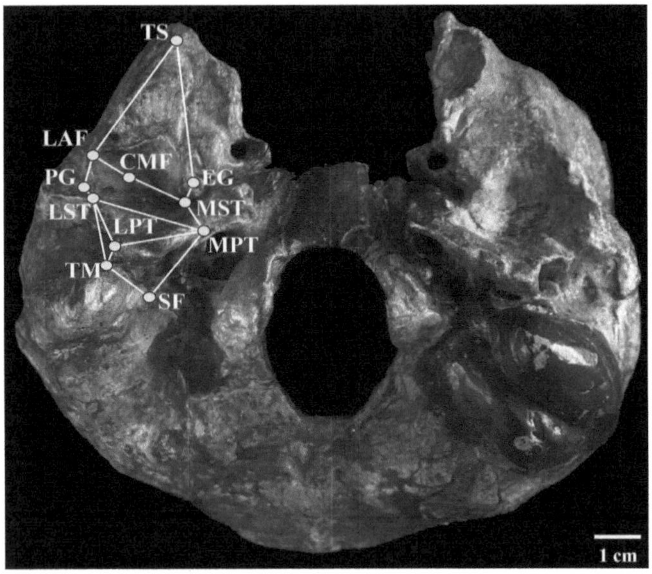

Fig. 6.10 The temporal base landmarks illustrated on the base of S 4. See Table 6.4 for details about the temporal base analysis and Table 6.2 for landmark abbreviations

This analysis is not directly comparable to the analysis of 3D temporal bone shape in *H. erectus* conducted by Terhune et al. (2007) due to more focused landmark sampling here (vs. the entire temporal bone) and differences in sample composition. That said, Terhune et al. also found a certain degree of separation between the Zhoukoudian and recent Javanese *H. erectus*, but with greater overlap between the two groups. Several observations regarding morphological differences between the two groups also appear to correspond between the two studies, such as the more laterally positioned postglenoid process, the more anteriorly angled tympanic process (in lateral view), and the greater distance between the center of the mandibular

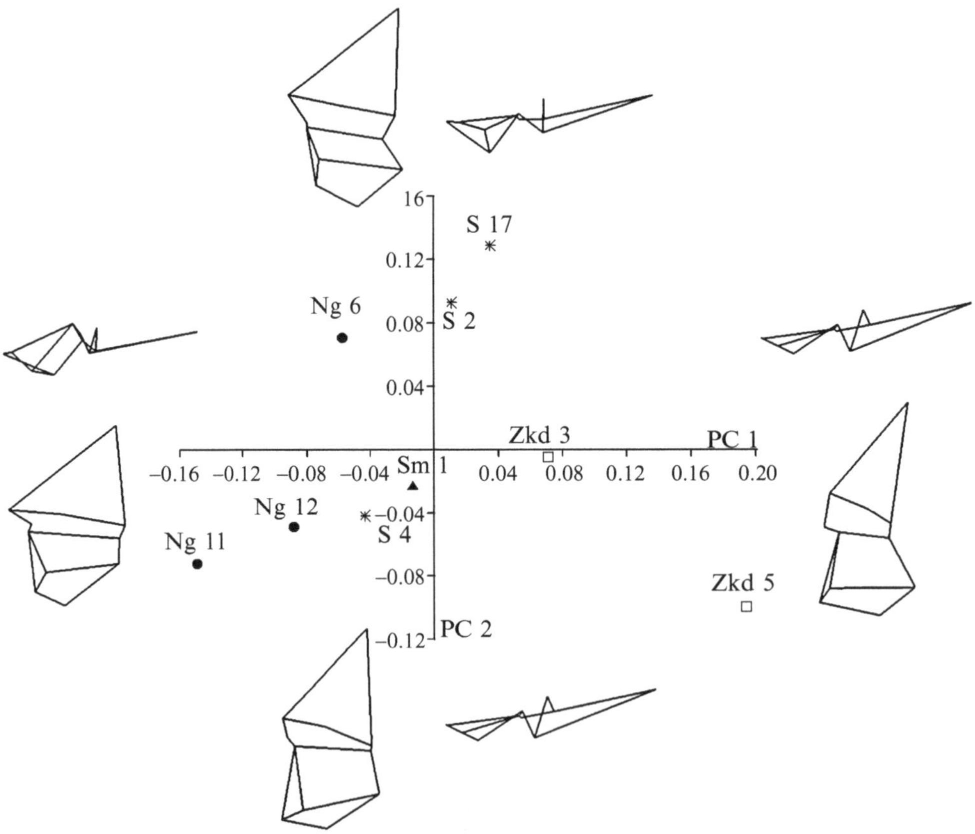

Fig. 6.11 PCs 1 and 2 of the temporal base data set. The shape changes along the first two components are illustrated below in *inferior* and *right lateral* views on the right temporal bone. See Fig. 6.2 for symbol legend

fossa and the postglenoid process in the majority of Indonesian *H. erectus*. The greater breadth of the TMJ noted here for the Javanese specimens was not apparent in their analysis, which may be attributable to different landmark choice as they did not record the lateral extent of the mandibular fossa.

Occipital Bone Analysis

The occipital bone landmark/semilandmark set is described in Table 6.4 and illustrated in Fig. 6.12. Despite a statistically significant relationship between PC 2 scores and geography, the geographic signal in the occipital bone shape is weak overall (Tables 6.8 and 6.9), and there is overlap between the specimens from Zhoukoudian and Java on all principal components. Size of the occipital bone explains the greatest proportion variance in shape (Table 6.9) and strongly influenced the position of specimens along PC 1 (Fig. 6.13 and Table 6.10). The occipital torus is slightly more superiorly positioned in higher scoring (i.e., larger) specimens, so that the nuchal plane is longer than the occipital plane. The occipital angle is also more open in these fossils and asterion is more postero-superiorly located.

Those specimens at the positive end of PC 2 (the majority of the Indonesian fossils) are relatively wider across asterion and have a more inferiorly positioned opisthocranion. In addition, the occipital plane is longer and less rounded than that seen in the lower scoring fossils.

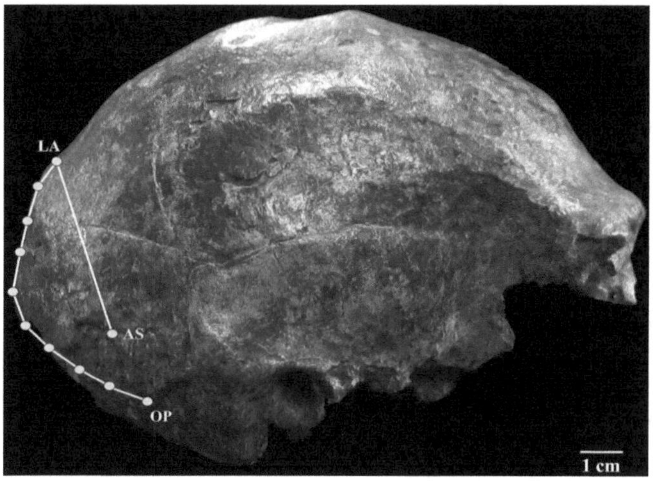

Fig. 6.12 Occipital landmarks (*labeled*) and semilandmarks. The four landmarks (LA, OP, and AS, *left* and *right*) as well as the eight evenly spaced (resampled) semilandmarks from the MOP curve are shown on Ng 7. The last few semilandmarks and OP are not visible in lateral view but are estimated here. The *lines* between the *points* are for visualization purposes only. See Table 6.2 for landmark and curve abbreviations

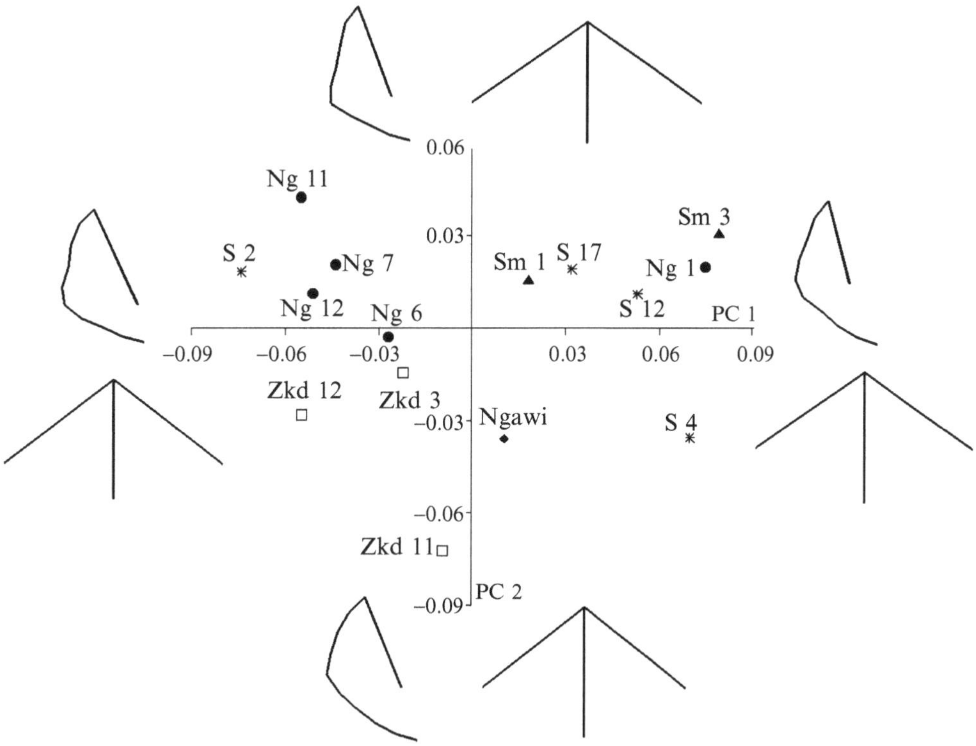

Fig. 6.13 PCs 1 and 2 of the occipital bone data set. The shape changes along the first and second components are illustrated below in *right lateral* and *posterior* views. See Fig. 6.2 for symbol legend

Discussion:

This study explored temporal, geographic, and size-related patterns of intraspecific variation within Asian *H. erectus* using the shape of both the overall neurocranium as well as several more restricted cranial regions (i.e., the frontal, temporal base, and occipital). Overall, geography accounted for the greatest proportion of variation in cranial shape, but geography, age, and size each explained a similar proportion of the total variance (between 21% and 26%). Although Rightmire (1990) suggested that there was not a consistent geographic pattern of cranial morphology within *H. erectus*, the findings of this study supported regional differentiation as an important aspect of morphological variability in Asian *H. erectus* (c.f. Weidenreich 1951; Picq 1990; Wolpoff 1999; Antón 2002), albeit with some overlap between early Indonesian (specifically S 2) and Zhoukoudian individuals (Liu et al. 2005; but see Antón 2002, 2007).

Generally, the differences in cranial shape between Indonesian and Zhoukoudian fossils are also consistent with previous descriptions (particularly Weidenreich 1943, 1951; Rightmire 1990; Picq 1990; Frayer et al. 1993; Wolpoff 1999; Antón 2002, 2003; Kidder and Durband 2004; Durband 2006). Specifically, the relatively narrower frontal and occipital bones relative to biauriculare breadth of the Zhoukoudian specimens corresponds to previous descriptions by Weidenreich (1943), Antón (2002) and Kidder and Durband (2004). Likewise, the more pronounced frontal tuberosity of the Zhoukoudian fossils, the lack of a midline post-toral sulcus and more posteriorly positioned frontotemporale in Indonesian *H. erectus* are well-documented elsewhere (Weidenreich 1943, 1951; Wolpoff 1999; Antón 2002). Both Weidenreich (1951) and Picq (1990) noted the antero-posteriorly wide mandibular fossa of the Zhoukoudian fossils, while the more vertical occipital squama in the Indonesian fossils was described by Santa Luca (1980). More subtle differences reported within the recent Indonesian sample are not as well documented and will be discussed in more detail below.

Geographic Patterns of Variation

In agreement with previous studies, the Zhoukoudian specimens present a fairly homogenous cranial shape, grouping together in all analyses. The results obtained here, however, suggest that there was not a single Indonesian cranial *Bauplan*, instead demonstrating that S 2 is quite distinct from other Indonesian hominins in terms of relative frontal bone breadth at anterior pterion, constriction across frontotemporale, and the projection of the occipital bone at inion. The imperfect distinction between the Zhoukoudian and Sangiran fossils suggest that there was less genetic isolation in the Early Pleistocene or, that the Sangiran fossils resemble the common ancestor of both the Zhoukoudian and later Indonesian populations of *H. erectus*. On the other hand, the clear separation between the Zhoukoudian and *later* Indonesian fossils could reflect genetic isolation of northern China and island Southeast Asia in the Middle or Late Pleistocene (Antón 2002, 2007). Any interpretation must be viewed with caution, however, due to the relatively small sample sizes, and the possibility that other fossils from China may not fit the Zhoukoudian pattern (see below).

The finding that there was not a consistent cranial shape in Java across time contradicted some earlier studies, particularly that of Antón (2002) and Antón et al. (2002). However, Liu et al. (2005) reported a similar result, finding that S 2, Sangiran IX (Tjg-1993.05; Arif 2005) and Trinil overlapped the Zhoukoudian / Nanjing *H. erectus* range of variation, although they did not focus on this finding in their report. While metric approaches have generally emphasized how distinct the Zhoukoudian (and Nanjing) *H. erectus* were from Indonesian *H. erectus*, these same studies have also found that an expanded Chinese sample is more heterogeneous (Kidder and Durband 2000, 2004; Antón 2002; Antón et al. 2002; Durband et al. 2005; Liu et al. 2005). In particular, the eastern Chinese specimen from Hexian shares closer metric affinities with Indonesian fossils than with Zhoukoudian and Nanjing (Dong 1989 cited in Brown 2001). Although specimens from the two geographic groups were not always distinct in their cranial shape, it is still possible that non-metric features, such as the presence of a supramastoid sulcus and the position of the squamotympanic fissure, may distinguish between Javanese and northern Chinese *H. erectus* (Antón 2002). These observations of heterogeneity in both the Chinese and Indonesian samples suggest that Zhoukoudian may not be typical of all Chinese *H. erectus* (Antón 2002), nor is any single specimen representative of all Indonesian *H. erectus*.

Size-Related and Temporal Variation

It is difficult to comment on the presence of a clear temporal pattern within the Asian *H. erectus* sample due to issues of sampling and dating of individual specimens. This study did not uncover a pattern of overarching anagenetic change in the Asian *H. erectus* neurocranium. While this result may have been affected by decisions regarding which time band to assign particular fossils, there is general agreement that the Zhoukoudian fossils are intermediate in age between Sangiran and Ngandong/Sambungmacan/Ngawi. However, the Zhoukoudian fossils were not morphologically intermediate between these two sets of Indonesian fossils. Even if the Sambungmacan specimens were older than the Zhoukoudian sample, there would not be a clear temporal pattern. The only statistically significant correlation was between age and frontal bone shape (PC 2 in Fig. 6.8), but this correlation was strongly influenced by the position of S 17, the only fossil older than 1 Ma. Although Santa Luca (1980) suggested that both the minimum frontal breadth and the depth of the supraorbital sulcus became reduced over time in Asian *H. erectus*, these traits appeared to vary more according to geography than time in this study. However, neither S 2 nor Trinil, which were both included in the Santa Luca (1980) study, were complete enough to analyze here. Until additional, older, specimens can be included in this analysis, the evidence for anagenetic change in the frontal bone remains ambiguous. The sequence of evolutionary changes in the occipital bone outlined by Santa Luca (1980), which included a lengthening and verticalization of the upper scale and a shortening of the lower scale, was also not clear in this analysis. However, this study did confirm that S 4 and 12 have longer nuchal planes compared to most Ngandong specimens, although this pattern was neither restricted to the Sangiran fossils (a similar pattern was observed in Ng 1 and Sm 3), nor was it ubiquitous in the early Javanese fossils as S 2 was very distinct in its shape from the other Sangiran specimens.

Although size influenced the shape of the neurocranium in general and the occipital bone in particular, there was generally not a strongly allometric pattern present within Asian *H. erectus* because the effect of size on neurocranial shape was always mediated by either the geochronological age or the geographic origin of the fossils. The shape of the occipital bone was related to overall size of this bone, but if this pattern is indicative of a general allometric trend within Asian *H. erectus*, it did not emerge until after 1 Ma. Although the size of the occipital bone did account for the greatest proportion of variation in occipital bone shape, several fossils did not fit this pattern. In particular, S 2 scored much lower than predicted for its small size.

The relationship between size and shape in the occipital bone may also have been related to sexual variation in the Zhoukoudian and later Indonesian populations, but not in the Sangiran sample. Presumed female specimens (Ng 1, Sm 1, Zkd 3 and 11; Weidenreich 1943, 1951; Santa Luca 1980; Wolpoff 1999; Delson et al. 2001) generally scored higher on PC 1 than male specimens within their respective groups. The most obvious exception to this trend was Ng 7, another supposed female, who scored among the other Ngandong fossils (which are either male or of unknown sex). However, within the Sangiran sample, the only widely accepted female (S 2) scored much lower than the considerably more robust S 4 and S 17. If this pattern truly captures male–female differences in occipital bone shape, then patterns of sexual variation varied over time.

Variation Within Javanese Homo erectus

Despite smaller sample sizes, the older Javanese fossils from the Sangiran dome were quite variable, a finding consistent with previous studies (Antón 2002, 2003; Kaifu et al. 2008). The younger Indonesian fossils from Ngandong, Sambungmacan, and Ngawi consistently plotted together at one end of the first component (c.f. Fig. 6.2), indicating that they all shared certain similarities in cranial shape. This pattern is mostly likely related to either genetic drift in the later populations or sampling from more restricted time intervals (and therefore more limited gene pools) in the both the northern Chinese and later Indonesian samples (Antón 2002). Genetic isolation has also been implicated in the evolution of *H. floresiensis* from *H. erectus* on the more easterly located island of Flores (Brown et al. 2004; Morwood et al. 2004, 2005). While the Flores cranium does generally resemble *H. erectus*, it is not particularly similar to the later Indonesian populations (Baab et al. 2007; Baab and McNulty 2009). However, the greater variability observed in early Javanese specimens leaves open the possibility of a Javanese ancestry for this small-bodied and small-brained taxon.

Kaifu et al. (2005) attributed mandibulo-dental variation within the early Indonesian sample to multiple migrations to the region or to marked anagenetic changes within the Indonesian *H. erectus* lineage. The focus of the Kaifu et al. (2005) study was on differences between the older Sangiran fossils that were recovered from or below the Grenzbank zone (the lowest layer of the Bapang Formation) and those fossils that derived from the Bapang Formation above the Grenzbank layer. According to their stratigraphic categorization of individual Sangiran dome fossils, the only specimen in this study that is definitely older is S 4. Although most workers consider S 2 to derive from Bapang Formation sediments above the Grenzbank zone (e.g., Itihara et al. 1985, 1994; Larick et al. 2001), the fluorine content of S 2 is more consistent with a Grenzbank derivation (Matsu'ura 1982). Regardless of whether S 2 is older or younger than S 17 (which is in turn older than S 12), there is not a clear temporal pattern within the Sangiran sample in any of the analyses, and S 4 (the only specimen > 1.5 Ma) does not appear to be an outlier within this sample. However, older Indonesian specimens such as S 27 and 31 were not included in this analysis due to issues of preservation, and previous descriptions suggest that they, along with S 4 and several mandibular fragments (e.g., S 5, 6a, 8, 22 and *Meganthropus* D), present a more robust morphology than later Sangiran fossils (Tyler 1991, 2001; Tyler et al. 1995; Antón 2003). Therefore, the lack of temporal change documented in this study should be viewed with caution until a greater sampling of crania from the earlier time periods is possible.

Relationships Among Javanese Homo erectus

In contrast to some previous analyses (e.g., Frayer et al. 1993; Wolpoff 1999; Baba et al. 2003; Kaifu et al. 2008), the results of this study do not provide strong support for a linear progression in neurocranial shape from Sangiran to Ngandong via Sambungmacan/Ngawi. Rather, this study found greater support for a Ngandong/Sambungmacan/Ngawi group (Jacob 1975, 1978, 1981; Sartono 1990; Delson et al. 2001; Widianto and Zeitoun 2003). The Sambungmacan/Ngawi group is often at least as distinct from the Sangiran fossils as are the Ngandong specimens in their neurocranial shape. Moreover, the Sambungmacan/Ngawi specimens are never intermediate between Sangiran and Ngandong along PC 1, which captures the most variable aspects of cranial shape. However, some aspects of the Sambungmacan/Ngawi neurocranial shape are intermediate between S 17 and Ngandong along PC 2 in the Indonesian only analysis (e.g., cranial height, cranial base breadth, and the length of the parietomastoid suture; see Fig. 6.6), but this may be related to the size of these specimens (see Table 6.7).

Although Sm 1 and Sm 3 were not found at the same locality (Delson et al. 2001), and their genetic relatedness has been questioned (Schwartz and Tattersall 2003), the overall similarity of their neurocranial shapes may support a close populational relationship. Moreover, Ngawi shared greater affinities with these specimens than with the Ngandong fossils, possibly highlighting two slightly different morphs on Java from approximately the same time period (Swisher et al. 1996). The Sambungmacan/Ngawi specimens are distinct from both S 17 and Ngandong in regards to their vault length, breadth of the frontal squama, and projection of glabella (see Figs. 6.4–6.6), although this may be influenced by the smaller size of the Sambungmacan/Ngawi specimens.

Evidence for Separate Lineages in Java > 1 Ma

There was no clear evidence for multiple long-lived lineages within Java. The greater resemblance of S 17 (compared to S 2) to the Ngandong specimens in its overall neurocranial shape (see Fig. 6.4), may have been influenced by the much larger size of S 17 as it did not particularly resemble Ngandong in the more restricted regions of the neurocranium. This pattern highlights the influence of both geological age and size on cranial form in the Indonesian fossils. Kaifu et al. (2008, this volume) also emphasized certain similarities between S 17 and the Ngandong fossils, but did not attribute these resemblances to allometric trends within this taxon.

The evidence is more ambiguous regarding a hypothesized relationship between S 4 on one hand and later Indonesian *H. erectus* on the other. Jacob (1975, 1978) hypothesized a separate S 4 ("*Pithecanthropus modjokertensis*") – Ngandong ("*P. soloensis*") lineage due to similarities in the base and occipital bone. Although S 4 does plot near Ng 1 in the occipital bone analysis, it does not group with the majority of Ngandong fossils (see Fig. 6.13). In the temporal bone analysis, S 4 plots slightly closer to Ngandong and Sambungmacan specimens than to S 2 and S 17 on the first two components (see Fig. 6.11). Neither of these results is particularly convincing evidence for a separate lineage encompassing S 4 and Ngandong/Sm 1.

Conclusions

Quantification of intraspecific variation has bearing not only on taxonomy but on past events in the evolutionary history of *H. erectus*. The results of the analyses reported here confirmed that Asian *H. erectus* exhibited variation in terms of overall neurocranial shape as well as frontal, temporal base, and occipital shape. Much of this variation was correlated with geographic, temporal, and size-related differences among members of *H. erectus*, as observed by many previous workers (e.g., Weidenreich 1943, 1951; Le Gros Clark 1964; Jacob 1975; Santa Luca 1980; Rightmire 1990; Wolpoff 1999; Antón 2002, 2003; Antón et al. 2002; Kidder and Durband 2004; Liu et al. 2005), while inter-individual differences also accounted for some of the observed shape variation. The use of geometric morphometrics was an improvement over past studies, because 3D landmark data preserve the original geometric relationships among the landmarks and shape differences could be clearly visualized.

The most morphologically distinct sets of specimens are those from Zhoukoudian, Ngandong, and Sambungmacan/Ngawi, groups which likely correspond to both geographically and temporally circumscribed populations. The greater separation between Zhoukoudian and the younger Javanese fossils further confirms the influence of microevolutionary forces (genetic drift, local adaptation) in shaping cranial morphology within Asian *H. erectus*. Although the later samples of Indonesian fossils from Ngandong, Sambungmacan, and Ngawi were fairly coherent with respect to cranial shape, this analysis did uncover subtle differences between Ngandong and Sambungmacan/Ngawi. There was no clear temporal pattern within more morphologically variable Sangiran fossils, leaving open the possibility that the greater variation was the result of multiple migrations to Java in the terminal Pliocene or early Pleistocene (Kaifu et al. 2005). This supports the idea that the Sangiran fossils should not be treated as a single population. Some of the earlier Indonesian fossils from the Sangiran dome resembled the later hominins from Java, in agreement with previous comparative analyses (Jacob 1975, 1978; Sartono 1975). There was not, however, strong support for separate, long-lived lineages within Javanese *H. erectus*. It seems more likely that certain combinations of traits seen in the variable Sangiran group were concentrated in later populations of *H. erectus* (Antón 2002).

The Zhoukoudian fossils formed a very homogenous group in terms of their cranial shape. Yet, the Indonesian fossils did not group together as consistently and sometimes overlapped the Zhoukoudian range. Although not addressed here, previous analyses have also highlighted greater variation in a more broadly sampled northern Chinese sample (Antón 2002; Kidder and Durband 2004; Durband et al. 2005). A complex pattern of migration, gene flow, and isolation likely characterized the evolutionary history of *H. erectus* within Asia.

Acknowledgements I am grateful to Dr. Christopher Norton, Dr. David Braun and Dr. Jack Harris for inviting me to participate in this volume and the original symposium from which this volume originated. I acknowledge Dr. Eric Delson and Dr. Ryan Raaum for their help in the preparation of this manuscript, as well as the comments of the editors, Dr. Yousuke Kaifu and two anonymous reviewers. My gratitude is also extended to the late Dr. Teuku Jacob and Dr. Etty Indriati at Gadjah Mada University; Dr. Friedemann Schrenk at the Senckenberg Research Institute and Natural History Museum; and Dr. Ian Tattersall, Dr. Ken Mowbray and Gisselle Garcia at the American Museum of Natural History for allowing me to study casts and original fossils under their care.

References

Adams, D. C., Rohlf, F. J., & Slice, D. E. (2004). Geometric morphometrics: Ten years of progress following the 'revolution'. *Italian Journal of Zoology, 71*, 5–16.

Andrews, P. (1984). An alternative interpretation of the characters used to define *Homo erectus*. *Courier Forschungs-Institut Senckenberg, 69*, 167–175.

Antón, S. C. (2001). Cranial growth in *Homo erectus*. In N. Minugh-Purvis & K. J. McNamara (Eds.), *Human evolution through developmental change* (pp. 349–380). Baltimore: Johns Hopkins University Press.

Antón, S. C. (2002). Evolutionary significance of cranial variation in Asian *Homo erectus*. *American Journal of Physical Anthropology, 118*, 801–828.

Antón, S. C. (2003). Natural history of *Homo erectus*. *Yearbook of Physical Anthropology, 46*, 126–170.

Antón, S. C. (2007). Climatic influences on the evolution of early *Homo*? *Folia Primatologica, 78*, 365–388.

Antón, S.C., & Swisher III, C.C. (2001). *Evolution and variation of cranial capacity in Asian Homo erectus. A scientific life: Papers in honor of Professor Dr. Teuku Jacob* (pp. 25–39). Yogyakarta, Indonesia: Bigraf Publishing.

Antón, S. C., Marquez, S., & Mowbray, K. (2002). Sambungmacan 3 and cranial variation in Asian *Homo erectus*. *Journal of Human Evolution, 43*, 555–562.

Arif, J. (2005). The skull IX (Tjf-1993.05) as one of the variant examples of Asian *Homo erectus* from Indonesia. *PaleoAnthropology, A49*(Abst.).

Armelagos, G. J., & Van Gerven, D. P. (1980). Sexual dimorphism and human evolution: An overview. *Journal of Human Evolution, 9*, 437–446.

Baab, K.L. (2007). Cranial shape variation in *Homo erectus*. Ph.D. dissertation, City University of New York.

Baab, K. L. (2008). The taxonomic implications of cranial shape variation in *Homo erectus*. *Journal of Human Evolution, 54*, 827–847.

Baab, K. L., & McNulty, K. P. (2009). Size, shape, and asymmetry in fossil hominins: The status of the LB1 cranium based on 3D morphometric analyses. *Journal of Human Evolution, 57*, 608–622.

Baab, K.L., McNulty, K.P., Brown, P. (2007). Allometric scaling of craniofacial shape: Implications for the Liang Bua hominins. *PaleoAnthropology, A2*(Abst.).

Baba, H., Aziz, F., Kaifu, Y., Suwa, G., Kono, R. T., & Jacob, T. (2003). *Homo erectus* calvarium from the Pleistocene of Java. *Science, 299*, 1384–1388.

Bartstra, G. J., Soegondho, S., & van der Wijk, A. (1988). Ngandong man: Age and artifacts. *Journal of Human Evolution, 17*, 325–337.

Bilsborough, A. (2000). Chronology, variability and evolution in *Homo erectus*. *Variability and Evolution, 8*, 5–30.

Black, D. (1929). Preliminary notice of the discovery of an adult *Sinanthropus* skull at Chou Kou Tien. *Bulletin of the Geological Society of China, 8*, 15–32.

Black, D. (1931). On an adolescent skull of *Sinanthropus pekinensis* in comparison with an adult skull of the same species and with other hominid skulls, recent and fossil. *Palaeontologica Sinica, Ser. D7*, 1–144.

Bookstein, F. L. (1996). Combining the tools of geometric morphometrics. In L. F. Marcus, M. Corti, A. Loy, G. J. P. Naylor, & D. E. Slice (Eds.), *Advances in morphometrics* (pp. 131–151). The Netherlands: Kluwer.

Bräuer, G. (1990). The occurrence of some controversial *Homo erectus* cranial features in the Zhoukoudian and East African hominids. *Acta Anthropologica Sinica, 9*, 352–358.

Bräuer, G., & Mbua, E. (1992). Homo erectus features used in cladistics and their variability in Asian and African hominids. *Journal of Human Evolution, 22*, 79–108.

Brown, P. (2001). Chinese Middle Pleistocene hominids and modern human origins. In L. Barham & K. R. Brown (Eds.), *Human roots – Africa and Asia in the Middle Pleistocene* (pp. 135–145). Bristol: Western Academic and Specialist Press.

Brown, P., Sutikna, T., Morwood, M. J., Soejono, R. P., Jatmiko, Saptomo, E. W, & Due Rokus Awe (2004). A new small-bodied hominin from the Late Pleistocene of Flores, Indonesia. *Nature, 431*, 1055–1061.

Delson, E., Harvati, K., Reddy, D., Marcus, L. F., Mowbray, K., Sawyer, G. J., et al. (2001). The Sambungmacan 3 *Homo erectus* calvaria: A comparative morphometric and morphological analysis. *The Anatomical Record, 262*, 380–397.

Dong, X. (1989). *Homo erectus* in China. In R. Wu, X. Wu, & S. Zhang (Eds.), *Early humankind in China* (pp. 9–23). Beijing: Science Press.

Dubois, E. (1894). *Pithecanthropus erectus, eine menschenaehnliche Übergangsform aus Java*. Landsdrukkerij: Batavia.

Durband, A. C. (2006). Craniometric variation within the Pleistocene of Java: The Ngawi 1 cranium. *Human Evolution, 21*, 193–201.

Durband, A. C., Kidder, J. H., & Jantz, R. L. (2005). A multivariate examination of the Hexian calvaria. *Anthropological Science, 113*, 147–154.

Frayer, D. W., Wolpoff, M. H., Thorne, A. G., Smith, F. H., & Pope, G. G. (1993). Theories of modern human origins: The paleontological test. *American Anthropologist, 95*, 14–50.

Frost, S.R. (2001). Fossil Cercopithecidae of the Afar depression, Ethiopia: Species systematics and comparison to the Turkana Basin. Ph.D. dissertation, City University of New York.

Frost, S. R., Marcus, L. F., Bookstein, F. L., Reddy, D. P., & Delson, E. (2003). Cranial allometry, phylogeography, and systematics of large-bodied Papionins (Primates: Cercopithecinae) inferred from geometric morphometric analysis of landmark data. *The Anatomical Record Part A, 275A*, 1048–1072.

Gower, J. C. (1975). Generalized Procrustes analysis. *Psychometrika, 40*, 33–51.

Gunz, P., & Harvati, K. (2007). The Neanderthal "chignon": Variation, integration, and homology. *Journal of Human Evolution, 52*, 262–274.

Harvati, K. (2001). The Neanderthal problem: 3-D geometric morphometric models of cranial shape variation within and among species. Ph.D. dissertation, City University of New York.

Harvati, K. (2003). The Neanderthal taxonomic position: Models of intra-and inter-specific craniofacial variation. *Journal of Human Evolution, 44*, 107–132.

Harvati, K., Frost, S. R., & McNulty, K. P. (2004). Neanderthal taxonomy reconsidered: Implications of 3D primate models of intra-and interspecific differences. *Proceedings of the National Academy of Sciences of the United States of America, 101*, 1147–1152.

Holloway, R. L. (1980). Indonesian "Solo" (Ngandong) endocranial reconstructions: Some preliminary observations and comparisons with Neandertal and *Homo erectus* groups. *American Journal of Physical Anthropology, 53*, 285–295.

Holloway, R. L. (1981). The Indonesian *Homo erectus* brain endocasts revisited. *American Journal of Physical Anthropology, 55*, 503–521.

Howells, W. W. (1973). Cranial variation in man. *Papers of the Peabody Museum of Archaeology and Ethnology, 67*, 1–259.

Hublin, J.-J. (1986). Some comments on the diagnostic features of *Homo erectus*. *Anthropos (Brno), 23*, 175–187.

Hyodo, M., Watanabe, N., Sunata, W., & Susanto, E. E. (1993). Magnetostratigraphy of hominid fossil bearing formations in Sangiran and Mojokerto, Java. *Anthropological Science, 101*, 157–186.

Hyodo, M., Nakaya, H., Urabe, A., Saegusa, H., Shunrong, X., Yiyun, Y., et al. (2002). Paleomagnetic dates of hominid remains from Yuanmou, China and other Asian sites. *Journal of Human Evolution, 43*, 27–41.

Itihara, M., Wikarno, T., & Kagemori, Y. (1985). Tektites from the Sangiran Area. In N. Watanabe & D. Kadar (Eds.), *Quaternary geology of the hominid fossil bearing formation in Java* (pp. 125–134). Bandung: Geological research and development centre.

Itihara, M., Watanabe, N., Kadar, D., & Kumai, H. (1994). Quanternary stratigraphy of the hominid fossil bearing formations in the Sangiran Area, Central Java. *Courier Forschungs-Institut Senckenberg, 171*, 123–128.

Jacob, T. (1975). Morphology and paleoecology of early man in Java. In R. H. Tuttle (Ed.), *Paleoanthropology morphology and paleoecology* (pp. 311–325). The Hague: Mouton.

Jacob, T. (1978). The puzzle of Solo Man. *Modern Quaternary Reserach in Southeast Asia, 4*, 31–40.

Jacob, T. (1981). Solo man and Peking man. In B. A. Sigmon & J. S. Cybulski (Eds.), *Homo erectus: Papers in honor of Davidson Black* (pp. 87–104). Toronto: University of Toronto Press.

Jacob, T., Soejono, R. P., Freeman, L. G., & Brown, F. H. (1978). Stone tools from mid-Pleistocene sediments in Java. *Science, 202*, 885–887.

Kaifu, Y., Baba, H., Aziz, F., Indriati, E., Schrenk, F., & Jacob, T. (2005). Taxonomic affinities and evolutionary history of the early Pleistocene hominids of Java: Dentognathic evidence. *American Journal of Physical Anthropology, 128*, 709–726.

Kaifu, Y., Indriati, E., Jacob, T., Kurniawan, I., & Baba, H. (2008). Cranial morphology of Javanese *Homo erectus*: New evidence for continuous evolution, specialization, and terminal extinction. *Journal of Human Evolution, 55*, 551–580.

Kennedy, G. (1991). On the autapomorphic traits of *Homo erectus*. *Journal of Human Evolution, 20*, 375–412.

Kent, J. T. (1994). The complex Bingham distribution and shape analysis. *Journal of the Royal Statistical Society, Series B 56*, 285–299.

Kidder, J. H., & Durband, A. C. (2000). The question of speciation in *Homo erectus* revisited I: The metric evidence. *American Journal of Physical Anthropology, 30*, 194–195.

Kidder, J. H., & Durband, A. C. (2004). A re-evaluation of the metric diversity within *Homo erectus*. *Journal of Human Evolution, 46*, 299–315.

Larick, R., Ciochon, R. L., Zaim, Y., Sudijono, S., Rizal, Y., Aziz, F., et al. (2001). Early Pleistocene 40Ar/39Ar ages for Bapang Formation hominins, Central Jawa, Indonesia. *Proceedings of the National Academy of Sciences of the United States of America, 98*, 4866–4871.

Le Gros Clark, W. E. (1964). *The fossil evidence for human evolution*. Chicago: University of Chicago Press.

Leigh, S. R. (1992). Brain ontogeny and life history in *Homo erectus*. *Journal of Human Evolution, 50*, 104–108.

Li, T., & Etler, D. A. (1992). New Middle Pleistocene hominid crania from Yunxian in China. *Nature, 357*, 404–407.

Liu, W., Zhang, Y., & Wu, Z. (2005). Middle Pleistocene human cranium from (Nanjing), Southeast China: A new reconstruction and comparison of *Homo erectus* from Eurasia and Africa. *American Journal of Physical Anthropology, 127*, 253–262.

Mann, A. (1971). *Homo erectus*. In P. Dolhinow & V. Sarich (Eds.), *Background for man*. Boston: Little, Brown.

Marcus, L. F., Hingst-Zaher, E., & Zaher, H. (2000). Application of landmark morphometrics to skulls representing the orders of living mammals. *Hystrix, 11*, 27–47.

Marquez, S., Mowbray, K., Sawyer, G. J., Jacob, T., & Silvers, A. (2001). New fossil hominid calvaria from Indonesia – Sambungmacan 3. *The Anatomical Record, 262*, 344–368.

Matsu'ura, S. (1982). A chronological framing for the Sangiran hominids: Fundamental study by the fluorine dating method. *Bulletin of the National Science Museum of Tokyo, Ser D 8*, 1–53.

Matsu'ura, S., Kondo, M., Aziz, F., Sudijono, Narasaki, S., Watanabe, N. (2000). First known tibia of an early Javanese hominid. *Current Anthropology, 41*, 297–300.

McNulty, K.P. (2003). Geometric morphometric analyses of extant and fossil hominoid craniofacial morphology. Ph.D. dissertation, City University of New York.

McNulty, K. P., Frost, S. R., & Strait, D. S. (2006). Examining affinities of the Taung child by developmental simulation. *Journal of Human Evolution, 51*, 274–296.

Morwood, M. J., Soejono, R. P., Roberts, R. G., Sutikna, T., Turney, C. S. M., Westaway, K. E., et al. (2004). Archaeology and age of a new hominin from Flores in eastern Indonesia. *Nature, 431*, 1087–1091.

Morwood, M. J., Brown, P., Jatmiko, Sutikna, T., Wahyu Saptomo, E., Westaway, K. E., Due, R. A., Roberts, R. G., Maeda, T., Wasisto, S., Djubiantono, T. (2005). Further evidence for small-bodied hominins from the Late Pleistocene of Flores, Indonesia. *Nature, 437*, 1012–1017.

O'Higgins, P. (2000). The study of morphological variation in the hominid fossil record: Biology, landmarks and geometry. *Journal of Anatomy, 197*, 103–120.

Picq, P. G. (1990). *L'articulation temporo-mandibulaire des Hominides*. Paris: Le Centre national de la recherche scientifique.

Pope, G. G. (1988). Recent advances in Far Eastern paleoanthropology. *Annual Review of Anthropology, 17*, 43–77.

Rightmire, G. P. (1990). *The evolution of Homo erectus: Comparative anatomical studies of an extinct human species*. New York: Cambridge University Press.

Rightmire, G. P. (2004). Brain size and encephalization in early to Mid-Pleistocene *Homo*. *American Journal of Physical Anthropology, 124*, 109–123.

Rohlf, F. J. (1990). Morphometrics. *Annual Review of Ecology and Systematics, 21*, 299–316.

Rohlf, F. J. (2003). tpsSmall, v. 1.20. Ecology and Evolution. State University of New York, Stony Brook.

Rohlf, F. J., & Slice, D. (1990). Extensions of the Procrustes method for the optimal superimposition of landmarks. *Systematic Zoology, 39*, 40–59.

Rohlf, F. J., & Marcus, L. F. (1993). A revolution in morphometrics. *Trends in Ecology and Evolution, 8*, 129–132.

Santa Luca, A. P. (1980). The Ngandong fossil hominids: A comparative study of a far eastern *Homo erectus* group. Yale University Publications in Anthropology.

Sartono, S. (1975). Implications arising from Pithecanthropus VIII. In R. H. Tuttle (Ed.), *Paleoanthropology, morphology and paleoecology* (pp. 327–360). The Hague: Mouton.

Sartono, S. (1990). A new *Homo erectus* skull from Ngawi, east Java. *Indo-Pacific prehistory association bulletin, 11*, 14–22.

Schwartz, J. H., & Tattersall, I. (2003). *The human fossil record: Volume two: Craniodental morphology of the genus Homo (Africa and Asia)*. New York: Wiley.

Shen, G., Ku, T., Cheng, H., Edwards, R., Yuan, Z., & Wang, Q. (2001). High-precision U-series dating of Locality 1 at Zhoukoudian, China. *Journal of Human Evolution, 41*, 679–688.

Shen, G., Gao, X., Gao, B. Granger, D. E. (2009). Age of Zhoukoudian *Homo erectus* determined with $^{26}Al/^{10}Be$ burial dating. *Nature, 458*, 198–200.

Slice, D. E. (1998). Morpheus et al.: Software for morphometric research. Stony Brook, NY: State University of New York, Department of Ecology and Evolution.

Slice, D. E. (2001). Landmark coordinates aligned by Procrustes analysis do not lie in Kendall's shape space. *Systematic Zoology, 50*, 141–149.

Slice, D. E. (2005). Modern morphometrics. In D. E. Slice (Ed.), *Modern morphometrics in physical anthropology* (pp. 1–46). New York: Kluwer.

Spoor, F., Leakey, M. G., Leakey, L. N. (2005). A comparative analysis of the KNM-ER 42700 hominin calvaria from Ileret (Kenya). *American Journal of Physical Anthropology, 40*, 200–201(Abst.).

Stringer, C. B. (1984). The definition of *Homo erectus* and the existence of the species in Africa and Europe. *Courier Forschungsinstitut Senckenberg, 69*, 131–143.

Swisher III, C. C., Rink, W. J., Antón, S. C., Schwarcz, H. P., Curtis, G. H., Suprijo, A., Widiasmoro (1996). Latest *Homo erectus* of Java: Potential contemporaneity with *Homo sapiens*. *Science, 274*, 1870–1874.

Terhune, C. E., Kimbel, W. H., & Lockwood, C. A. (2007). Variation and diversity in *Homo erectus*: A 3D geometric morphometric analysis of the temporal bone. *Journal of Human Evolution, 53*, 41–60.

Tyler, D. E. (1991). A taxonomy of Javan hominid mandibles. *Human evolution, 6*, 401–420.

Tyler, D. E. (2001). Two new "Meganthropus" mandibles from Java. *Human Evolution, 16*, 151–158.

Tyler, D. E., Sartono, S., Krantz, G. S. (1995). *Homo erectus* mandible F (Sangiran 22) from Java: An announcement. *Palaeo-anthropology: Evolution and ecology of Homo erectus, Pithecanthropus centennial foundation* (pp. 203–206). Leiden, NL: Leiden University.

Villmoare, B. (2005). Metric and non-metric randomization methods, geographic variation, and the single-species hypothesis for Asian and African *Homo erectus*. *Journal of Human Evolution, 49*, 680–701.

von Koenigswald, G. H. R., & Weidenreich, F. (1939). The relationship between *Pithecanthropus* and *Sinanthropus*. *Nature, 144*, 926–929.

Weidenreich, F. (1935). The *Sinanthropus* population of Choukoutien (Locality 1) with a preliminary report on new discoveries. *Bulletin of the Geological Society China, 14*, 427–461.

Weidenreich, F. (1936–1937). The new discoveries of *Sinanthropus pekinensis* and their bearing on the *Sinanthropus* and *Pithecanthropus* problems. *Bulletin of the Geological Society of China, 16*, 439–471.

Weidenreich, F. (1943). The Skull of *Sinanthropus pekinensis:* A Comparative Study on a Primitive Hominid Skull. *Palaeontologia Sinica,* New Series D, *10*, 1–484.

Weidenreich, F. (1951). Morphology of solo man. *Anthropological Papers of the American Museum of Natural History, 43*, 222–288.

White, T. D., & Folkens, P. A. (2000). *Human osteology* (2nd ed.). New York: Academic.

Widianto, H., & Zeitoun, V. (2003). Morphological description, biometry and phylogenetic position of the skull of Ngawi 1 (East Java, Indonesia). *International Journal of Osteoarchaeolology, 13*, 339–351.

Wolpoff, M. H. (1984). Evolution in *Homo erectus*: The question of stasis. *Paleobiology, 10*, 389–406.

Wolpoff, M. H. (1999). *Paleoanthropology*. Boston: McGraw-Hill.

Wood, B. A. (1984). The origin of *Homo erectus*. *Courier Forschungs-Institute Senckenberg, 69*, 99–111.

Wood, B. A. (1991). *Koobi Fora research project: Hominid cranial remains*. New York: Oxford University Press.

Wood, B. A. (1993). Early *Homo*, how many species? In W. H. Kimbel & L. B. Martin (Eds.), *Species, species concepts, and primate evolution* (pp. 485–522). New York: Plenum.

Wood, B. A. (1994). Taxonomy and evolutionary relationships of *Homo erectus*. *Courier Forschungs-Institute Senckenberg, 171*, 159–165.

Yuan, S. X., Chen, T. M., Gao, S. J., & Hu, Y. Q. (1991). Study on uranium series dating of fossil bones from Zhoukoudian sites. *Acta Anthropologica Sinica, 10*, 189–193.

Chapter 7
Rethinking the Palearctic-Oriental Biogeographic Boundary in Quaternary China

Christopher J. Norton, Changzhu Jin, Yuan Wang, and Yingqi Zhang

Abstract This paper reviews the Chinese Quaternary biogeographic record, which has traditionally been divided into the Palearctic (North China) and Oriental (South China) biozones. Notable findings are: (1) Open-steppe taxa (e.g., *Equus sanmeniensis*, *E. yunnanensis*, *E. hemonius*, *Mammuthus*) are found in the Oriental region during the Early and Late Pleistocene; (2) *Beremendia* is found in large numbers at the Early Pleistocene Renzidong site located in the Oriental region, though it had previously been known only from the Palearctic biozone; (3) Oriental taxa are found in Middle Pleistocene deposits in northern China, though an increase in the winter monsoon intensity after 520 ka probably forced many of the warm climate taxa back southwards; (4) A higher number of primate taxa is present in the Oriental region, increasing in species diversity from the Early Pleistocene to the Holocene; and (5) Most of central-east China is less than 1,000 m above sea level and served as a continuous migration corridor between the Palearctic and Oriental biogeographic zones during the Quaternary. The general conclusion that we draw from this brief review is that the utility of a strict division between the Palearctic and Oriental biogeographic zones is not valid. In light of fluctuating paleoenvironmental pressures, Palearctic faunas often migrated southward during stadials and Oriental taxa were able to expand northward during interstadials. More detailed reconstructions of the eastern Asian biogeographic record, along with linking the data to the loess-soil, deep sea oxygen isotope, paleobathymetric, and pollen records, will facilitate a deeper understanding of how the paleoenvironment influenced hominin dispersals and evolution during the Quaternary.

Keywords China • Palearctic • Oriental • Quaternary • Biogeography • Migration corridor

Introduction

The initial hominin dispersals out of Africa and into Asia ca. 2 Ma triggered one of the most significant behavioral and ecological changes in human prehistory. The nature of early hominin dispersals and behavioral patterning is only beginning to be understood (Potts 1998; Antón and Swisher 2004; Dennell and Roebroeks 2005; Anton 2007). The paleoenvironmental setting, from which early *Homo erectus* sensu lato dispersed into and through China is critical to understand the dynamics involved in human evolution during the Asian Quaternary. Appreciating how the paleoenvironment influenced early hominin evolution is important, particularly because it is not until relatively late in our prehistory that humans were able to overcome major environmental stresses and occupy regions of the world that were previously beyond the range of hominin dispersals.

Currently, the number of studies that link Asian paleoenvironmental variability with the human evolutionary record is growing (e.g., Jablonski 1997; Wang et al. 1997; Zhu et al. 2003; Louys et al. 2007; Belmaker 2010; Dennell 2010). These studies have analyzed a variety of different proxies (e.g., isotopes, paleomagnetics, vertebrate paleontology) in order to develop an understanding of how changes in the paleoenvironment influenced hominin variability. In order to fully reconstruct the Chinese paleoenvironment, analysis of the deep sea oxygen isotope record, bathymetric fluctuation, pollen analysis, Red Clay and loess-soil deposition, and biogeography studies need to be fully synthesized. The purpose of this paper is to discuss one of these areas, biogeography, particularly in light of the apparently fluid boundary that separates the Palearctic and Oriental biogeographic zones, and how the boundary can be used to examine hominin dispersal patterning in Quaternary China.

Chinese Quaternary Environment

The present day East Asian environment can be broadly divided into northern and southern regions separated by the Himalayan Mountain Range in the west and the Qinling

C.J. Norton (✉)
Department of Anthropology, University of Hawaii, 2424 Maile Way, Saunders Hall 346, Honolulu, HI, 96822, USA
e-mail: cjbae@hawaii.edu

C.J. Norton and D.R. Braun (eds.), *Asian Paleoanthropology: From Africa to China and Beyond*, Vertebrate Paleobiology and Paleoanthropology, DOI 10.1007/978-90-481-9094-2_7, © Springer Science+Business Media B.V. 2010

Mountain Range, which runs through the central part of modern day China, in the east. In general, the modern day Northeast (NE) Asian region is characterized as being open grassland, with extensive Plio-Pleistocene loess-soil deposits indicative of a dynamic environment. The present day Southeast (SE) Asian region is known more for subtropical broadleaf deciduous forests, with large tracts of bamboo deposits and a more stable climate (Liu and Ding 1984; Sun and Wu 1988; Wang et al. 1997; Jin et al. 1999, 2000; Xie et al. 2004; Tougard and Montuire 2006; Louys et al. 2007).

Controlled by diverse climatic conditions, different areas are inhabited by distinct faunal elements, thus forming the present day biogeographic regions found in China. Affected by the summer monsoon, the climate in the southeast and southwest regions of China is warm and humid and inhabited by animals suited for that environment. The extensive Meng-Xin Plateau area (including Inner Mongolia and Uygur Autonomous Region), which is less affected by the southeast and southwest monsoon, is dominated by arid and/or semi-arid adapted taxa. The Qinghai-Tibetan Plateau, which has an average altitude ~4,500 m above sea level ("MASL"), has less circulating air and lower average temperature. The Qinghai-Tibetan Plateau is not suitable for many taxa, thus only hardy animals (e.g., goat, sheep) survive in this region.

The Qinling Mountain Range is currently 3,000–4,000 MASL, reaching its elevation probably sometime during the Middle to Late Miocene. Indirect support for this can be found in studies of the timing and nature of the uplift of the adjacent Himalayan Mountain Range and Tibetan Plateau, the latter geologic formation possibly having a more recent uplift (Fort 1996; Zheng et al. 2000; An et al. 2001; Spicer et al. 2003; but see Ji 1980; Wang and Deng 2005). Because most of the region east of the Qinling Mountain Range (central-east China) has always been less than 1,000 MASL, with no major geological barriers, it is feasible that the region served as a continuous faunal migration corridor (Jin et al. 1999, 2000). In particular, during stadials cold adapted faunas would have migrated southward, and during interstadials warm adapted taxa would have expanded northward.

Plio-Pleistocene Division

There has been a recent development in Quaternary research to move the Plio-Pleistocene boundary back to 2.6 Ma from its current position of 1.8 Ma (top of the Olduvai subchron). Even though it has not yet been universally accepted, it is recognized by many Quaternary scientists (e.g., see www.inqua.tcd.ie). In China, the 2.6 Ma Gauss-Matuyama magnetostratigraphic boundary is commonly considered the division between the Pliocene and Early Pleistocene (e.g., Ding et al. 1997; Jablonski et al. 2000). The primary events in East Asia that occurred at this boundary are: (1) the initial strengthening of the neotectonic movement and further uplift of the Tibetan Plateau; (2) an increase in the Northern Hemisphere glaciations; and (3) loess-paleosol accumulations in China (Heller and Liu 1982; Liu 1985; Ding et al. 1997) that are closely related to the increase in seasonal monsoon intensity (Derbyshire et al. 1995; Liu and Ding 1998; An 2000; An et al. 2001; Guo et al. 2002; Zhang et al. 2007). An event coinciding with this is the appearance of *Equus* in the Palearctic biogeographic region. *Equus* originated in North America and migrated into Eurasia across the Bering land bridge (Deng and Xue 1999).

Quaternary Biogeography

Quaternary East Asia is typically divided into two major faunal zones: Palearctic (NE Asia) and Oriental (South China and mainland SE Asia) (Fig. 7.1; Xue and Zhang 1991; Xie et al. 2004; Rink et al. 2008). The Palearctic fauna is represented by open-steppe species, while the Oriental fauna is represented by forest-dwelling, warm and humid adapted taxa (Tables 7.1 and 7.2). Although the climate is different in the two biogeographic zones, from the beginning of the Quaternary to the present day there is a general diminution of overall animal size in both, and many closely related species that were larger disappear over the course of the Pleistocene allowing the smaller species to survive and flourish (see for alternative views: Teilhard de Chardin 1941; Hooijer 1951; Pei 1957; Luchterhand 1978, 1984; Aigner 1981; Han and Xu 1985; Qi 1989, 1990; Xue and Zhang 1991; Dong 1997; Tougard and Montuire 2006; Louys et al. 2007; Rink et al. 2008).

A number of taxa are considered representative of each faunal zone and of each geologic period. Most of these representative taxa appear early in the Plio-Pleistocene paleontological record and flourish throughout most of the time period. For instance, *Hipparion* appears for the first time in NE Asia during the Pliocene. However, *Equus* enters NE Asia during the Early Pleistocene, eventually forcing the extinction of *Hipparion* by the end of the epoch. Evidence of overlap of *Hipparion* and *Equus* exists in Early Pleistocene deposits in the Nihewan Basin in northern China (Deng and Xue 1999). The extinction of *Hipparion* in NE Asia was not a sudden event, but occurred over the course of a million years. Other important open-steppe taxa, such as *Camelus* (*Paracamelus*), *Palaeoloxodon*, *Gazella*, *Rhinoceros*, and *Mammuthus* also appear in NE Asia during the Early Pleistocene and continue to thrive throughout the epoch. The most representative Early Pleistocene NE Asian faunal assemblages are from the Nihewan Basin, Xihoudu, Zhoukoudian Locality 18, and Gongwangling Lantian.

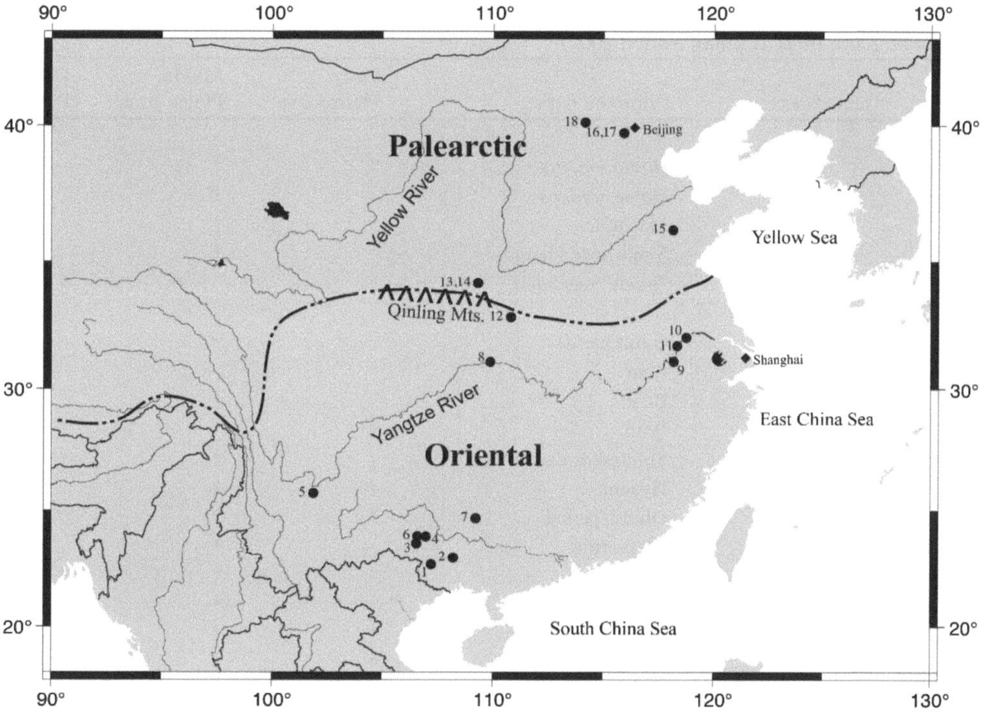

Fig. 7.1 Map of China divided into biogeographic zones along with primary Early and Middle Pleistocene sites yielding vertebrate paleontological materials in China. The boundary between the Palearctic and Oriental biogeographic zones lies between the Yangtze River and the Qinling Mountain Range. Sites: 1. Daxin; 2. Wuming; 3. Wuyun; 4. Upper Pubu; 5. Yuanmou; 6. Mohui; 7. Liucheng; 8. Longgupo; 9. Renzidong; 10. Tangshang; 11. Hexian; 12. Xuetangliangzi; 13. Gongwangling; 14. Chenjiawo; 15. Yiyuan; 16,17. ZKD; 18. Nihewan (see Table 7.3 for faunal distribution)

Overall, during the Early Pleistocene in NE Asia, there is a dearth of warm, humid, dense forest adapted taxa (e.g., *Ailuropoda*, *Macaca*, *Bubalus*) (Table 7.1; Teilhard de Chardin 1941; Wei 1978, 1988; Aigner 1981; Han and Xu 1985; Qi 1989, 1990; Pope and Keates 1994; Deng and Xue 1999; Jin et al. 1999, 2000; Wei et al. 2006).

Throughout the Pleistocene the SE Asian faunal group is referred to as the *Ailuropoda-Stegodon* faunal complex (Table 7.2; Pei 1957; Han and Xu 1985; Tougard 2001; Tougard and Montuire 2006; Louys et al. 2007; Wang et al. 2007; Rink et al. 2008). *Ailuropoda*, *Stegodon*, *Macaca*, and *Rhizomys* all appear early during the Pleistocene and flourish in the warm and humid climate of SE Asia. Important early *Ailuropoda-Stegodon* faunal accumulations include the *Gigantopithecus*-bearing assemblages of Liucheng and Jianshi (Han and Xu 1985). The Liucheng cave site is particularly important because excavations have produced three *Gigantopithecus* mandibles and over 1,000 associated teeth. One noticeable difference between the Early Pleistocene faunal composition of the Palearctic and Oriental regions is the presence of a greater diversity of primate genera in SE Asia (Tables 7.1 and 7.2). This distinction becomes even more prevalent through the Middle and Late Pleistocene and into the present day Holocene epoch.

The Middle Pleistocene of East Asia is characterized by a climate that became much warmer and more humid. Pronounced climatic change is further evident as many different warm and humid adapted species that originally appeared in the Oriental region make their appearance as far north as Zhoukoudian, Jinniushan, and the Korean Peninsula (Table 7.1). For instance, *Bubalus*, *Macaca*, *Rhinopithecus*, and *Ailuropoda* appear in Middle Pleistocene NE Asian deposits. The most representative faunal localities in NE Asia during the Middle Pleistocene are Zhoukoudian Locality 1, Chenjiawo, Kommunmoru, and Jinniushan (Aigner 1981; Liu and Ding 1984; Han and Xu 1985; Qi 1989, 1990; Xu 1997; Dong et al. 2000; Norton 2000).

The warmer Middle Pleistocene climate supported an expansion of primate groups in the Oriental region (Jablonski et al. 2000). Not only do *Homo erectus*, *Gigantopithecus blacki*, *Macaca*, and *Pongo* continue to be present, but *Hylobates*, *Rhinopithecus*, *Semnopithecus*, and *Trachypithecus* all appear for the first time in the paleontological record in this region (Table 7.2). Representative type sites for the SE Asian Middle Pleistocene are Daxin, Wuming, and Bama (Han and Xu 1985; Rink et al. 2008).

By the beginning of the Late Pleistocene the climate became cooler and the warm adapted species that appeared

Table 7.1 Macromammal diversity in Quaternary Northeast Asia (Pei 1957; Aigner 1981; Han and Xu 1985; Qi 1989, 1990; Xu et al. 1993; Wu and Poirier 1995; Dong et al. 2000; Jin et al. 2000; Norton 2000).[a]

Order	Genus/species	Common name	Early Pleistocene	Middle Pleistocene	Late Pleistocene	Holocene
PRIMATES						
	Homo erectus	*Homo erectus*	*	*		
	Homo sapiens	*Home sapiens*		?[b]	*	*
	Macaca robusta	Macaque		*		
	Rhinopithecus spp.	Snub-nosed langur		*		
	Procynocephalus cf. *wimani*	"giant" macaque	*			
CARNIVORA						
	Ailuropoda melanoleuca	Giant panda		?[c]		
	Ursus cf. *U. angustidens*	Bear	*			
	Ursus spp.	Bear	*	*	*	*
	Canis spp.	Wolf	**	*	*	*
	Crocuta crocuta ultima	Spotted or cave hyaena		*	*	
	Hyaena spp.	Hyaena	*	*	*	*
	Cuon alpinus	Dhole, jackal	*	*	*	*
	Gulo spp.	Wolverine		*		*
	Vulpes spp.	Fox	*	*	*	*
	Nyctereutes spp.	Raccoon dog	*	*	*	*
	Felis (*Panthera*) spp.	Tiger / leopard	*	*	*	*
	Megantereon (*Machairodus*) spp.	Sabretooth cat	*	*		
	Cynailurus spp.	Cheetah	*	*	*	
	Acinonyx (*Sivapanthera*) *pleistocaenicus*	Cheetah		*	*	
	Lynx spp.	Lynx	*		*	
	Mustela spp.	Polecat, ferret	*	*	*	*
	Lutra spp.	Otter	*	*		*
	Martes spp.	Marten, sable	*	*		
	Meles spp.	Old World badger	*	*	*	*
	Arctonyx spp.	Hog badger		*		
	Paguma spp.	Masked palm civet			*	*
	Viverricula spp.	Lesser Oriental civit	*	*		
PROBOSCIDEA						
	Archidiskodon planifrons	Yellow River elephant	*			
	Palaeoloxodon spp.	Elephant	*	*	*	
	Elephas spp.	Indian elephant	*	*	*	
	Stegodon spp.	Stegodont	*	*		
PERISSODACTYLA						
	Coelodonta antiquitatis	Woolly rhinoceros	*	*	*	
	Dicerorhinus spp.	Two horned rhinoceros		*	*	
	Elasmotherium spp.	Giant rhinoceros	*			
	Nestoritherium sinense	Rhinoceros	*	*		
	Rhinoceros spp.	Rhinoceros	**	*	*	*
	Equus spp.	Horse	*	*	*	*
	Hipparion sinense	Three toed horse	*			
	Megatapirus augustus	Giant tapir		*		
	Tapirus sinensis	Tapir		*		
ARTIODACTYLA						
	Camelus (*Paracamelus*) spp.	Camel	*	*	*	*
	Euctenoceros boulei	Deer	*			
	Cervulus sinensis	Chinese muntjak	*			
	Axis spp.	Spotted deer	*	*		
	Megaloceros (*Sinomegaceros*) spp.	Giant antlered deer	*	*	*	
	Cervus spp.	Deer	*	*	*	*
	Moschus spp.	Musk deer	*	*	*	
	Pseudaxis grayi	Deer		*	*	
	Elaphodus cephalophus	Tufted deer		*	*	
	Capreolus spp.	Roe deer		*	*	*

(continued)

Table 7.1 (continued)

Order	Genus/species	Common name	Early Pleistocene	Middle Pleistocene	Late Pleistocene	Holocene
	Hydropotes spp.	River deer		*	*	*
	Muntiacus spp.	Barking deer	*	*		
	Rusa spp.	Rusa deer	*	*		
	Gazella spp.	Gazelle	*	*	*	*
	Procapreolus spp.	Gazelle	*	*	*	
	Spiroceros spp.	Spiral horned antelope	*	*	*	
	Capricornis spp.	Serow		*	*	
	Ovis spp.	Sheep	*	*	*	*
	Pseudovis cf. *P. nayaur*	Blue sheep			*	
	Sus spp.	Wild boar/Pig	**	*	*	*
	Bison spp.	Bison	*	*		
	Leptobos spp.	Bison	*	*		
	Bos spp.	Aurochs	*	*	*	
	Bubalus sp.	Water buffalo		*		
	Bibos spp.	Wild ox	*	*		

[a]Data were gathered from different published sources. In order to avoid slight discrepancies between species identifications, when two or more members of the same genera appeared they were collapsed into one category and given a "spp." designation

[b]Early *Homo sapiens* is present in sites that may date to the Middle Pleistocene (e.g., Xujiayao)

[c]It is not clear whether *Ailuropoda* is present in Northeast Asia during the Middle Pleistocene. It has been suggested it may be present at Zhoukoudian Locality 1

north of the Qinling Mountain Range during the Middle Pleistocene retreated southward (Jablonski 1997; Jablonski et al. 2000). For instance, *Bubalus*, *Ailuropoda*, *Macaca*, and *Rhinopithecus* disappear from the NE Asian Late Pleistocene paleontological record (Aigner 1981; Han and Xu 1985). The most significant change in SE Asia during the Late Pleistocene is the absence of *Gigantopithecus blacki* from the paleontological record (Ciochon et al. 1990). The giant ape apparently went extinct during the later Middle Pleistocene (Rink et al. 2008), which may be related to the change in climate, but is also probably due to the increased competition over similar resources (e.g., food, sleeping areas) from the different competing primate taxa.

The more stable Holocene brought about the present day faunal diversity. This includes the majority of East Asian primate diversity located south of the Qinling Mountain Range, except for the Japanese macaques which were restricted to the southern region of the Japanese Archipelago when ocean bathymetry rose to its current level. The index genera of the Oriental, *Bubalus* and *Ailuropoda*, are currently limited to that particular region. Open-dwelling species, including *Equus* and *Gazella* are now restricted to the Palearctic (Tables 7.1 and 7.2). Due to the warmer and more humid climate in SE Asia, there continues to be a more diversified faunal complex vis-à-vis NE Asia. In addition, more taxa appear to have survived the transition from the Late Pleistocene to the Holocene in SE Asia. This is at least partially related to the more extreme climatic conditions in the north as evidenced from the loess-soil deposition in that region (Liu and Ding 1998; An 2000).

Central-East China: A Migration Corridor?

Much of the region just east of the Qinling Mountain Range is lower than 1,000 MASL and served as a bi-directional migration corridor during the Quaternary (Jin et al. 1999, 2000). A number of important Early and Middle Pleistocene sites exist in this region (e.g., Renzidong, Huludong [Tangshan], Longtandong [Hexian], Yiyuan [Table 7.3; Fig. 7.1; Wu and Poirier 1995; Jin et al. 2000; Wu and Li 2001; Zheng et al. 2001]). *Homo erectus* or hominin trace fossils (manuports, lithics, hominin-modified bone) have been found at these localities, in addition to a diversity of faunal remains. In support of the central-east China migration corridor argument, the taxonomic composition of the Early Pleistocene Renzidong site indicates that between 2.4 and 2.0 Ma, Palearctic (37%) and Oriental (34%) taxa are almost equally represented (29% of the faunal diversity is represented by taxa that crosscut both biogeographic zones) (Jin et al. 2000). In addition, *Beremendia*, a Palearctic-restricted shrew, has been identified in the Renzidong faunal assemblage (Jin et al. 2009), as were three typical Palearctic forms of arvicolids, *Villanyia fanchangensis, Mimomys* cf. *M. peii*, and *Cromeromys* cf. *C. gansunicus* (Zhang et al. 2008).

Further supporting data can be found in the Early Pleistocene Qipanshan Yinan cave, Shandong Province, where Oriental elements (e.g., *Hipposideros*, *Hylopetes*, and *Brachyrhizomys*) were found in the same context with Palearctic taxa, including *Kowalskia*, *Nannocricetus*, and *Ursus* (Jin et al. 1999). *Mammuthus* has also been identified

Table 7.2 Macromammal diversity in Quaternary mainland Southeast Asia (Data from Pei 1957; Han and Xu 1985; Wu and Poirier 1995; Dong et al. 2000; Jin et al. 2000; Tougard 2001; Louys et al. 2007; Rink et al. 2008)[a]

Order	Genus/species	Common name	Early Pleistocene	Middle Pleistocene	Late Pleistocene	Holocene
PRIMATES						
Homo erectus		*Homo erectus*	*	*	?[b]	
Homo sapiens		*Homo sapiens*		*[c]	*	*
Gigantopithecus blacki		Giant ape	*	**		
Macaca spp.		Macaque	*	**	*	*
Procynocephalus spp.		"giant" macaque	*			
Pongo spp.		Orangutan	*	*	*	*
Tarsius spp.		Tarsier				*
Hylobates spp.		Gibbon		*	*	*
Nasalis spp.		Proboscis monkey				*
Nycticebus spp.		Slow loris			*	*
Presbytis spp.		Leaf monkey			*	*
Pygathrix spp.		Douc langur				*
Rhinopithecus spp.		Snub nosed langur		*	*	*
Semnopithecus spp.		Hanuman langur		*		
Trachypithecus spp.		Silver leaf monkey		*	*	*
CARNIVORA						
Ailuropoda spp.		Giant panda	*	*	*	*
Ailurus spp.		Red panda		*	*	*
Ailuropoda microta		panda	*			
Ailuropoda melanoleuca fovealis		panda		**		
Ursus spp.		Bear	*	*	*	*
Ursus thibetanus Cuvier		Asiatic black bear		*		
Ursus cf. thibetanus Cuvier		Asiatic black bear	*			
Helarctos spp.		Sun bear			*	*
Sclenarctos spp.		Black bear			*	
Canis spp.		Wolf		*	*	*
Crocuta crocuta		Spotted or cave hyaena		*	*	
Hyaena brevirostris licenti		Short-faced hyena	*			
Hyaena spp.		Hyaena	*	*	*	
Cuon javanicus		Dhole		*		
Cuon dubius		Dhole	*			
Cuon spp.		Dhole, jackal	*	*	*	*
Vulpes spp.		Fox	*	*		*
Nyctereutes spp.		Raccoon dog		*	*	*
Megantereon spp.		Sabretooth cat	*	*		
Felis spp.		Cat	*	*	*	*
Felis teilhardi		Cat	*			
Felis (Panthera) pardus		Cat	*			
Catopuma spp.		Asiatic golden cat			*	*
Pardofelis spp.		Marbled cat				*
Prionailurus spp.		Leopard cat				*
Cynailurus pleistocaenicus		hunting leopard	*			
Neofelis spp.		Clouded leopard		*	*	*
Acinonyx (Sivapanthera) sp.		Cheetah			*	
Lynx sp.		Lynx		*		
Mustela spp.		Polecat, ferret		*	*	*
Arctictis spp.		Bearcat			*	*
Martes spp.		Marten, sable		*		*
Herpestes spp.		Mongoose			*	*
Amblonyx spp.		Otter				*
Lutra spp.		River otter		*	*	*
Lutrogale spp.		Smooth otter				*
Meles spp.		Old World badger		*	*	*
Melogale spp.		Chinese ferret badger			*	*

(continued)

Table 7.2 (continued)

Order	Genus/species	Common name	Early Pleistocene	Middle Pleistocene	Late Pleistocene	Holocene
	Mydaus spp.	Stink badger				*
	Arctonyx collaris	Hog Badger		*		
	Arctonyx minor	Hog Badger	*			
	Arctonyx spp.	Hog Badger	*	*	*	*
	Paguma larvata	Masked palm civet	*	*	*	*
	Hemigalus spp.	Banded palm civet			*	*
	Paradoxurus spp.	Palm civet		*	*	*
	Viverra spp.	Civet	*	*	*	*
	Viverricula malaccensis	Lesser oriental civet		*		
	Viverricula malaccensis fossilis	Lesser oriental civet		*		
	Viverricula spp.	Lesser oriental civet	*	*	*	*
	Arctogalidia spp.	Palm civet				*
	Cynogale spp.	Otter-civet				*
	Diplogale spp.	Hose's palm civet				*
	Chrotogale spp.	Owston's palm civet				*
	Prionodon spp.	Linsang				*
PROBOSCIDEA						
	Elephas spp.	Indian elephant		*	*	*
	Palaeoloxodon namadicus	Elephant		*	*	
	Stegodon sp.	Stegodont	*	*	*	
	Stegodon preorientalis	Stegodont	*			
	Stegodon orientalis	Stegodont		*		
	Gomphotherium serridentoides	Mastodont	*			
PERISSODACTYLA						
	Dicerorhinus spp.	Two horned rhinoceros		*	*	*
	Rhinoceros chiai	Rhinoceros	*			
	Rhinoceros sinensis	Rhinoceros		**		
	Rhinoceros spp.	Rhinoceros		*	*	*
	Nestoritherium sp.	Rhinoceros	*			
	Nestoritherium praesinensis	Rhinoceros	*			
	Dicoryphochoerus ultimus	Hippopotamus	*	*		
	Equus spp.	Horse		*	*	
	Equus yunnanensis	Horse	*			
	Megatapirus augustus	tapir		*		
	Megatapirus sp.	Giant tapir		*	*	
	Tapirus peii	tapir	*			
	Tapirus spp.	Tapir	*	*	*	*
ARTIODACTYLA						
	Axis spp.	Spotted deer	*	*	*	*
	Caprinae gen. et sp. indet.	Deer	*			
	Cervocerus (Cervavitus) fenqii	Deer	*			
	Cervocerus sp.	Deer	*			
	Cervus sp. indet.	Deer		*		
	Cervus spp.	Red deer	*	*	*	*
	Dicoryphochoerus ultimus	Deer	*	*		
	Elaphodus spp.	Tufted deer		*		*
	Elaphurus davidianus	David's deer		*		
	Hydropotes inermis	Chinese water deer		*		*
	Megaloceros pachyosteus	Giant antlered deer		*		
	Moschus sp.	Musk deer				*
	Muntiacus lacustris	Muntjac	*			
	Muntiacus spp.	Barking deer	*	*	*	*
	Paracervulus attenuatus	Deer	*			
	Procapreolus stenos	Roe deer	*			
	Pseudaxis grayi	Deer		*	*	
	Rusa yunnanensis	Rusa deer	*			

(continued)

Table 7.2 (continued)

Order	Genus/species	Common name	Early Pleistocene	Middle Pleistocene	Late Pleistocene	Holocene
Rusa spp.		Rusa deer	*	*	*	
Tragulus spp.		Asiatic mouse deer			*	*
Metacervulus capreolinus		Muntjak	*			
Caprinae gen. et sp. indet.		Serow		*		
Capricornis sumatraensis		Serow		*	*	*
Capricornis cf. sumatraensis		Serow		*		
Gazella spp.		Gazelle	*			
Megalovis guangxiensis		Giant goat	*	*		
Naemorhedus spp.		Antelope goral		*	*	*
Ovis spp.		Sheep	*	*	*	
Suidae		Suid	*			
Sus australis		Pig	*			
Sus bijiashanensis		Pig		**		
Sus liuchengensis		Pig	*			
Sus peii		Pig	*			
Sus scrofa		Pig		*		
Sus xiaozhu		Pig	*			
Sus spp.		Wild boar/Pig		**	*	*
Potamochoerus nodosarius		Boar	*			
Dorcabune liuchengensis		Dear	*			
Bovinae gen. et sp. indet.		Bison		*		
Bibos spp.		Wild ox	*	*		
Bison spp.		Bison		*	*	
Bos spp.		Auroch	*	*	*	*
Bubalus spp.		Water buffalo		*	*	*
Budorcas spp.		Takin			*	*
Hexaprotodon sp.		Pygmy hippo		*		

[a]Data were gathered from different published sources. In order to avoid slight discrepancies between species identifications, when two or more members of the same genus appeared they were collapsed into one category and given a "spp." designation

[b]*Homo erectus* may be present during the Late Pleistocene in this region (Swisher et al. 1996)

[c]Probable presence of many early *Homo sapiens* in SE Asia during the late Middle Pleistocene (Wu and Poirier 1995)

in deposits in central-east China south of the Yangtze River (Wei et al. 2006). Typical Oriental genera (e.g., *Macaca*, *Ailuropoda*, *Stegodon*) are also found in the same deposits as Palearctic taxa (e.g., *Ursus arctos*) in the Middle Pleistocene sites of Huludong (Tangshan), Longtandong (Hexian), and Yiyuan (Table 7.3). Further vertebrate paleontological investigations in this region will lead to better descriptions and modeling of the nature of the Quaternary central-east China faunal migration corridor. For instance, despite the presence of an abundance of limestone deposits, containing a diversity of late Neogene-Quaternary faunal materials, the regions around Huainan (northern Anhui Province) and the Yiyuan *Homo erectus* site (central Shandong Province) have yet to be systematically surveyed.

Discussion

Substantial variation exists between the Palearctic and Oriental biogeographic zones (Tables 7.1 and 7.2). Even though earlier researchers (e.g., Luchterhand 1978; Pope 1984, 1988) have suggested that SE Asia was never open enough for open-steppe dwelling species to survive there during the Pleistocene, recent paleontological research indicates that during glacial periods open-steppe taxa were able to migrate southward into SE Asia and survive for extended periods of time. For instance, *Equus sanmeniensis*, a typical Palearctic species, is found in Early Pleistocene deposits from at least three sites (Renzidong, Longgupo, Liucheng) that are south of the Yangtze River and fall in the Oriental biogeographic region (Dong et al. 2000; Jin et al. 2000). This suggests that the colder climate during the Early Pleistocene prompted certain open-steppe dwelling taxa to migrate south into the Oriental region, where they were able to survive successfully for extended periods of time. In addition, *Equus yunnanensis* appears in Members 3 and 4 of the Yuanmou Fauna (Early Pleistocene) also located in the Oriental biogeographic zone (Deng and Xue 1999; Dong et al. 2000). *Equus hemonius* fossils were identified in Late Pleistocene cave deposits from Yunnan and Guangxi Provinces, suggesting a temporarily more open environment (Wang and Mo 2004; Wang et al. 2007).

Even though the *Ailuropoda-Stegodon* faunal complex is considered to have been similar throughout the Plio-Pleistocene,

Table 7.3 Faunal list for important Early and Middle Pleistocene sites in China (Wu and Poirier 1995; Dong et al. 2000; Jin et al. 2000; Wang et al. 2007; Rink et al. 2008). Data from sites not discussed in the text are presented here for comparative purposes (site locations on Fig. 7.1)

Early Pleistocene										Middle Pleistocene							
Taxon	Mohui Cave (Baise)	Renzidong (Fanchang)	Longgupo (Wushan)	Liucheng	Danawu (Yuanmou)	ZKD Loc. 18	Nihewan	ZKD Loc. 12	Gongwangling (Lantian)	Chenjiawo (Lantian)	ZKD Loc. 1	Xuetangliangzi (Yunxian)	Huludong (Tangshan)	Longtandong (Hexian)	Yiyuan	Daxin	Wuming
Primates																	
Homo erectus					*				*	*	*	*	*	*	*		
Pongo sp.				*												*	
Macaca robusta											*			*	*		
Macaca sp.	*															*	*
Rhinopithecus sp.																*	
Rhinopithecus lantianensis									*			*					
Procynocephalus sp.		*	*					*									
Hylobates sp.																*	
Hominoidea gen. et sp. indet	*		*														
Gigantopithecus blacki	*		*	*												*	*
Insectivora																	
Scaptochirus primitivus											*						
Scaptochirus cf. *P. primitivus*		*															
Scaptochirus moschatus									*								
? *Scaptochirus* sp.														*			
Neomys bohlini											*						
Neomys sinensis											*						
Crocidura sp.											*						
Erinaceus olgai											*						
Sorex sp.		*					*				*						
Blarinella quadraticauda														*			
Blarinella cf. *B. quadraticauda*		*	*														
Beremandia fissidens		*															
Chodsigoa youngi														*			
Anourosorex squamipes														*			
Anourosorex quadratidens		*	*														
Soriculus cf. *S. parva*		*	*														
Soriculus sp.		*	*														
Chimarrogale cf. *C. himalayica*		*	*														
Talpa sp.		*	*														
Chiroptera																	
Rhinolophus pleistocanicus											*						
Rhinolophus cf. *R. pleistocancus*														*			
Rhinolophus cf. *R. ferrumequinum*		*	*														
Miniopteris schrebersi														*			
Miniopteris cf. *M. schrebersi*											*						

(continued)

Table 7.3 (continued)

Taxon	Early Pleistocene: Mohui Cave (Baise)	Renzidong (Fanchang)	Longgupo (Wushan)	Liucheng	Danawu (Yuanmou)	ZKD Loc. 18	Nihewan	ZKD Loc. 12	Gongwangling (Lantian)	Middle Pleistocene: Chenjiawo (Lantian)	ZKD Loc. 1	Xuetangliangzi (Yunxian)	Huludong (Tangshan)	Longtandong (Hexian)	Yiyuan	Daxin	Wuming
La io											*						
? *Pipistrelus* sp.											*						
Hipposideros sp.		*	*											*			
? *Myotis* sp.		*						*						*			
Murina cf. *M. aurata*		*	*														
Rodentia																	
? *Castor* sp.											*						
? *Eothenomys* sp.											*						
? *Phaiomys* sp.											*						
Alticola sp.											*						
Allocricetus ehiki		*	*			*											
Apodemus agrarius														*			
Apodemus cf. *A. sylvaticus*		*	*			*				*			*				
Apodemus dominans		*	*														
Apodemus sp.									*								
Arvicola sp.					*												
Arvicola terrae-rubrae									*								
Atherurus sp.				*												*	
Bahomys hyposodonta									*	*							
Borsodia sp.		*					*										
Brachyrhizomys cf. *B. shansius*		*															
Chiropodomys sp.		*	*														
Citellus cf. *C. mongolicus*											*						
Clethrionomys rofucanus											*						
Cricetinus cf. *C. griseus*											*						
Cricetinus cf. *C. obscurus*											*						
Cricetinus varians								*			*		*	*			
Cricetulus cf. *C. griseus*									*								
Cricetulus sp.		*						*	*								
Cricetulus varians									*					*			
Cromeromys cf. *C. gansunicus*		*															
Gerbillus roborowskii											*						
Gerbillus sp.								*	*								
Hapalomys sp.	*	*	*														
Hystrix kiangsenensis	*																
Hystrix magna	*	*	*	*			*										
Hystrix subcristata				*												*	
Hystrix cf. *H. subcristata*									*		*						*
Hystrix sp.										*							

Table 7.3 (continued)

	Early Pleistocene									Middle Pleistocene							
Taxon	Mohui Cave (Baise)	Renzidong (Fanchang)	Longgupo (Wushan)	Liucheng	Danawu (Yuanmou)	ZKD Loc. 18	Nihewan	ZKD Loc. 12	Gongwangling (Lantian)	Chenjiawo (Lantian)	ZKD Loc. 1	Xuetangliangzi (Yunxian)	Huludong (Tangshan)	Longtandong (Hexian)	Yiyuan	Daxin	Wuming
Hystrix subcristata					*							*					
Kowalskia sp.		*															
Leopoldamys edwardsioides		*	*														
Leopoldamys sp.	*																
Marmota bobak											*						
Marmota complicidens											*						
Marmota sp.											*						
Micromys cf. *M. mimutus*		*									*						
Microtus brandtioides											*			*			
Microtus epiratticeps									*		*		*				
Microtus sp.					*												
Mimomys cf. *M. peii*		*	*				*										
Mus musculus											*						
Mus sylvaticus											*						
Myospalax epitingi											*						
Myospalax fontanieri									*	*							
Myospalax sp.									*		*						
Myospalax tingi									*	*							
Myospalax wongi											*						
Niviventer preconfuciannus		*	*														
? *Niviventer fulvescens*		*	*														
Niviventer sp.	*																
Petaurista brachyodous											*						
Petaurista cf. *P. brachyodous*		*															
Petaurista sp.									*								
Pitymys simplicidens											*						
Rattus cf. *R. bowersi*		*	*														
Rattus edwardsi														*			*
Rattus norvegicus													*	*			
Rattus rattus								*			*			*		*	
Rattus sp.		*	*														
Rhizomys sp.					*												*
Sciurotamias cf. *S. praecox*		*	*			*											
Tamias wimani											*						
Trogontherium cuvieri											*						
Trogontherium sp.															*		
Trogopterus sp.		*	*														

(continued)

Table 7.3 (continued)

Early Pleistocene										Middle Pleistocene							
Taxon	Mohui Cave (Baise)	Renzidong (Fanchang)	Longgupo (Wushan)	Liucheng	Danawu (Yuanmou)	ZKD Loc. 18	Nihewan	ZKD Loc. 12	Gongwangling (Lantian)	Chenjiawo (Lantian)	ZKD Loc. 1	Xuetangliangzi (Yunxian)	Huludong (Tangshan)	Longtandong (Hexian)	Yiyuan	Daxin	Wuming
Typhlomys cf. *T. intermedius*		*	*														
Typhlomys cinerus	*																
Wushanomys cf. *W. hypsodontus*		*	*														
Vernaya sp.		*	*														
Sciurus sp.								*									
Siphnenus sp.								*									
Lagomorpha																	
Ochotona koslowi											*						
Ochotona cf. *O. thibetana*									*								
Ochotona sp. A											*						
Ochotona sp. B											*						
Ochotonoides complicidens					*				*	*							
Lepus wongi										*							
Lepus cf. *L. wongi*											*						
Lepus sp. A											*						
Lepus sp. B											*						
Hypolagus brachypus		*				*											
Carnivora																	
Viverricula malaccensis																*	
Vivirricula malaccensis fossilis					*											*	
Canis lupus											*						
Canis lupus variabilis									*		*						
Canis cyonoides											*						
Canis sp.								*						*			
Nyctereutes sinensis											*		*				
Nyctereutes cf. *N. sinensis*		*	*				*										
Nyctereutes sp.		*															
Cuon antiquus											*						
Cuon javanicus												*				*	
Cuon alpinus										*			*	*			
Cuon dubius	*			*													
Arctonyx collaris	*											*	*			*	
Arctonyx minor				*													
Viverra sp.				*													
Vulpes cf. *V. vulpes*											*			*			
Vulpes cf. *V. corsac*											*						
Vulpes sp.													*	*			
Canidae gen. et sp. indet.											*						
Canis variabilis															*		

Table 7.3 (continued)

Early Pleistocene										Middle Pleistocene							
Taxon	Mohui Cave (Baise)	Renzidong (Fanchang)	Longgupo (Wushan)	Liucheng	Danawu (Yuanmou)	ZKD Loc. 18	Nihewan	ZKD Loc. 12	Gongwangling (Lantian)	Chenjiawo (Lantian)	ZKD Loc. 1	Xuetangliangzi (Yunxian)	Huludong (Tangshan)	Longtandong (Hexian)	Yiyuan	Daxin	Wuming
Ursus thibetanus	*										*	*	*	*	*	*	
Ursus cf. thibetanus				*													
Ursus arctos											*		*	*	*		
Ursus cf. *U. angustidens*								*									
Ursus cf. *U. spelaeus*											*						
Ursus cf. *U. etruscus*									*								
Ursus cf. *U. microdontus*		*	*	*		*	*										
Ursus sp.																	*
Ailuropoda melanoleuca									*			*				*	*
Ailuropoda sp.											?			*			
Ailuropoda microta	*	*	*	*													
Martes pachygnatha		*				*	*										
Martes anderssoni		*				*	*										
Mustela cf. *M. altaica*		*				*											
Meles chiai		*	*	*		*	*										
Meles cf. *M. leucurus*									*	*	*	*					
Lutra melina											*						
Lutra sp.														*			*
Gulo sp.											*						
Mustela cf. *M. sibirica*											*						
Mustela sp.								*			*						
Pachycrocuta brevirostris									*		*		*	*			
Pachycrocuta licenti	*	*			*							*					
? *Pachycrocuta licenti*			*			*	*										
Crocuta crocuta ultima											*						
Hyaena brevirostris licenti				*													
Hyaena sp.															*		*
Paramachairodus sp.		*															
Megantereon inexpectatus											*						
Megantereon lantianensis									*								
Megantereon cf. *M. nihowanensis*					*												
Megantereon sp.		*		*			*	*						*			
Homotherium crenatidens		*	*														
Homotherium ultimus											*						
Homotherium sp.												*					
Paguma larvata				*												*	
Panthera youngi											*						
Panthera pardus												*	*	*			

(continued)

Table 7.3 (continued)

	Early Pleistocene									Middle Pleistocene							
Taxon	Mohui Cave (Baise)	Renzidong (Fanchang)	Longgupo (Wushan)	Liucheng	Danawu (Yuanmou)	ZKD Loc. 18	Nihewan	ZKD Loc. 12	Gongwangling (Lantian)	Chenjiawo (Lantian)	ZKD Loc. 1	Xuetangliangzi (Yunxian)	Huludong (Tangshan)	Longtandong (Hexian)	Yiyuan	Daxin	Wuming
Panthera cf. *P. pardus*								*	*		*						
Panthera tigris												*	*		*		
Panthera cf. *P. tigris*									*	*	*						
Panthera sp.	*																
Felis teilhardi	*			*					*		*						
Felis peii												*					
Felis chinensis														*			
Felis cf. *F. lynx*								*									
Felis (Panthera) pardus				*													
Felis sp.				*													
Felis sp. A											*						
Felis sp. B											*						
Felis cf. *F. microtis*		*	*	*							*						
Cynailulus pleistocaenicus				*					*								
Cynailulus sp.											*						
Lynx sp.		*				*	*										
Proboscidea																	
Gomphotherium serridentoides				*													
Palaeoloxodon cf. *P. namadicus*											*						
Stegodon orientalis									*			*		*		*	
Stegodon preorientalis	*			*													
Stegodon elephantoides					*												
Stegodon sp.					*								*				*
Elephas sp.										*							
Sinomastodon intermedius		*	*	*													
Sinomastodon sp.	*																
Perissodactyla																	
Dicerorhinus choukoutiensis											*						
Dicerorhinus lantianensis									*								
Dicerorhinus kirchbergensis									*				*				
Dicerorhinus mercki															*		
Dicerorhinus sp.		*	*				*							*			
Coelodonta antiquitatis											*						
Rhinoceros chiai				*													
Rhinoceros sinensis	*	*	*		*		*					*				*	*
Rhinoceros sp.								*									
Tapirus peii				*													
Tapirus sinensis									*			*		*			

Table 7.3 (continued)

Early Pleistocene										Middle Pleistocene							
Taxon	Mohui Cave (Baise)	Renzidong (Fanchang)	Longgupo (Wushan)	Liucheng	Danawu (Yuanmou)	ZKD Loc. 18	Nihewan	ZKD Loc. 12	Gongwangling (Lantian)	Chenjiawo (Lantian)	ZKD Loc. 1	Xuetangliangzi (Yunxian)	Huludong (Tangshan)	Longtandong (Hexian)	Yiyuan	Daxin	Wuming
Tapirus sanyuanensis	*	*	*	*													
Megatapirus augustus									*							*	
Megatapirus sp.														*			
Equus sanmeniensis		*	*	*		*	*		*		*	*			*		
Equus yunnanensis				*	*							?					
Equus sp.													*	*			
Nestoritherium praesinensis				*													
Nestoritherium sinense									*								
Nestoritherium sp.		*	*	*	*												
Hesperotherium sp.	*																
Artiodactyla																	
Suidae				*													
Sus australis				*													
Sus bijiashanensis																*	*
Sus liuchengensis				*													
Sus lydekkeri									*	*	*	*	*	*	*		
Sus scrofa												*					*
Sus xiaozhu	*			*								*					
Sus cf. *S. xiaozhu*														*			
Sus peii	*	*	*	*			*										
Sus sp.					*			*								*	*
Paracamelus gigas											*						
Camelidae gen. et sp. indet											*						
Moschus moschiferus											*						
Moschus sp.		*	*	*		*	*										
Eostyloceros longchunanensis					*												
Dicoryphochoerus ultimus				*												*	
Dorcabune liuchengense	*			*													
Muntiacus lacustris				*	*												
Muntiacus nanus		*	*	*		*	*										
Muntiacus sp.	*											*				*	
Metacervulus capreolinus		*	*		*		*					*					
Paracervulus attenuatus					*												
Potamochoerus nodosarius				*													
Cervavitus ultimus		*	*	*	*												
Hydropotes inermis														*			
? *Hydropotes* sp.											*		*				
Elaphodus cephalophus									*				*				

(continued)

Table 7.3 (continued)

Taxon	Early Pleistocene: Mohui Cave (Baise)	Renzidong (Fanchang)	Longgupo (Wushan)	Liucheng	Danawu (Yuanmou)	ZKD Loc. 18	Nihewan	ZKD Loc. 12	Gongwangling (Lantian)	Middle Pleistocene: Chenjiawo (Lantian)	ZKD Loc. 1	Xuetangliangzi (Yunxian)	Huludong (Tangshan)	Longtandong (Hexian)	Yiyuan	Daxin	Wuming
Axis shansius					*												
Axis cf. *A. rugosus*					*												
Caprinae gen. et sp. indet.				*												*	
Cervus cf. *C. philisi*		*	*			*	*										
Cervus cf. *C. fengqii*	*			*													
Cervus (Sika) *grayi*									*	*	*		*	*			
Cervus (Rusa) *elegans*												*					
Cervus (Rusa) *yunnanensis*	*			*	*							*					
Cervus (Rusa) *stehlini*					*												
Cervus (Rusa) sp.	*				*												
Cervus (Rusa) *unicolor*		*	*				*										
Cervus sp. indet.																*	
Megaloceros pachyosteus											*		*	*	*		
Megaloceros konwanlinensis									*								
Megaloceros sp.										*		*					
Elaphurus davidianus														*			
Cervus sp.					*						*				*		
Procapreolus stenos					*												
Capreolus sp.											*		*				
Capricornis sumatraensis									*								
Capricornis cf. *sumatraensis*																	*
Gazella sp.					*						*						
Spirocerus peii											*						
Spirocerus cf. *S. wongi*											*						
Ovis cf. *O. ammon*											*						
Ovis sp.											*						
Ovibovinae gen. et sp. Indet											*						
Leptobos brevicornis									*			*					
Leptobos sp.									*								
Bubalus teilhardi											*		*				
Bubalus sp.												*					
Bison sp.											*			*			
? *Naemorhedus* sp.											*						
Bos (*Bibos*) sp.		*	*	*	*											*	
Bos sp.				*	*												
Megalovis guangxiensis	*															*	
Cervicornia (or Bovidae)								*									
Bovinae gen. et sp. indet	*										*						*

variation is present (Rink et al. 2008). For instance, Wang et al. (2007) found that Middle and Late Pleistocene faunal complexes in South China are different, with some open-steppe taxa (e.g., *Equus hemonius*) present in the latter sub epoch. Although it has traditionally been thought that *Ailuropoda* and *Stegodon* are always found in association, Wang et al. (2007) also identified *Ailuropoda* in Lower Pubu, but instead of *Stegodon*, they found *Elephas*.

During interglacial periods, warm-humid adapted taxa (e.g., *Ailuropoda, Bubalus, Macaca*) were able to expand their ranges and survive north of the Qinling Mountain Range for long time periods. Consequently, the Middle and Late Pleistocene can probably be better characterized by relatively extreme climatic conditions that at times facilitated the movement and survival of warm, humid faunal groups as far north as Zhoukoudian and the Korean Peninsula, but at other times were cool enough to allow open-steppe dwelling species to migrate south of the Qinling Mountain Range and into the Oriental biogeographic region where they currently cannot survive.

Evidence for climatic variation within each of these sub epochs is growing. For instance, a recent stable carbon isotope study was conducted by Gaboardi et al. (2005) on herbivore teeth from Zhoukoudian Locality 1. In their analysis, they noted that after 520 ka, the intensity of the winter monsoon increased, resulting in a transition from a C_3/C_4 to a C_3 dominated environment. Additional stable carbon isotope analyses of tooth enamel from late Neogene-Quaternary sites in China are increasing (e.g., Deng et al. 2002; Deng and Li 2005; Wang et al. 2006, 2007). However, these studies concentrate on southwestern or northern China. Future stable carbon isotope research will be conducted on material from the central-east China region. In particular it will be important to document the variation between C_3 and C_4 plants through spatial–temporal facies. This would facilitate reconstructions of the complexity of the faunal migration corridor(s) in the region.

Based on this brief review of the Chinese biogeography, the late Neogene-Quaternary environment may be characterized as follows:

1. During the Pliocene, the environment was warm and humid. Beginning in the Late Pliocene, there was a significant environmental change related to the increase in intensity of the East Asian monsoon system, the Northern Hemisphere glaciations, and increased orbital cycling ("Milankovitch cycles") (Liu and Ding 1998; An 2000; Zhang et al. 2007).
2. During the Early Pleistocene the climate was cooler and drier. Open grassland fauna (e.g., *Equus sanmeniensis*) and cold adapted taxa (e.g., *Beremandia, Villanyia fanchangensis, Mimomys* cf. *M. peii, Cromeromys* cf. *C. gansunicus*) appear in the Oriental biogeographic zone (Deng and Xue 1999; Wang et al. 2007; Jin et al. 2009; Zhang et al. 2008).
3. During the Middle Pleistocene the climate was sometimes warmer and more humid. Warm and humid adapted plant and animal taxa (*Bambusa, Bubalus, Macaca*) from the SE Asian region were able to successfully penetrate the NE Asian region (Jablonski et al. 2000; Norton 2000). Environmental variation between warm/humid and cool/dry climates increased.
4. During the Late Pleistocene the climate was cooler and drier (Xie et al. 2004; Wang et al. 2007). Warm and humid adapted plant and animal taxa retreated back south of the Qinling Mountain Range. Increase in Gramineae in the north indicates a cooler, more open environment (Whyte 1984).
5. During the Holocene the climate stabilized to within 1–3°C of the present day. There was a warming trend during the Early-Middle Holocene that coincides, and may be a direct, or indirect, cause for the advent of plant and animal domestication in East Asia.

Conclusions

Review of the Quaternary Palearctic-Oriental biogeographic boundary in China indicates that it fluctuated significantly throughout the course of the past 2.6 Ma. Because of the significant fluctuation we draw the conclusion that a strict boundary between the two biogeographic zones is not valid. Nevertheless, by examining the variation in taxonomic diversity across space and time, we note that several broader ranging patterns in China can be described (as discussed above). It is evident that more detailed studies that examine floral and faunal life histories across time and space are critical to understanding the nature of environmental change in China during the Quaternary. The work of Jablonski et al. (2000) in examining the diversification of non-human primate taxa and of Deng and Xue (1999) for equids in Quaternary China are excellent starts in this direction.

For paleoanthropologists, it is important to be able to link these changes in taxonomic diversity and paleoenvironment to the human evolutionary record. For instance, a few Early Pleistocene sites are known in northern China (e.g., Majuangou, Xiaochangliang, Donggutuo, Gongwangling, Xihoudu). The dates from these sites have been used to develop a southward dispersal model proposed by Zhu et al. (2003). That is, they suggested hominins moved from Majuangou (1.71–1.66 Ma) and Xiaochangliang (1.36 Ma) in the Nihewan Basin southward to the southern part of the Chinese Loess Plateau at Xihoudu (1.27 Ma) and Gongwangling (1.15 Ma). Indirect support for the Zhu et al. model may be gleaned from the loess-soil record. In their study, Heslop et al. (2002) found a major coarsening in unit L_{15} (loess level 15), which corresponded with the time period around 1.2 Ma. They interpret this to indicate a much colder climate in northern China and associate this major peak with the effect of an intensive winter monsoon in the region during this time interval. This would have prompted a faunal and floral migration to the south. However, new magnetostratigraphic studies are constantly changing the assumed age of these Early Pleistocene sites. For instance, recent paleomagnetic analyses suggest Xiaochangliang may be 1.26 Ma (Li et al. 2008) and not 1.36 Ma as suggested by Zhu et al. (2003). In light of these ever changing chronologies, we must be careful how these data are applied to the human evolutionary record, particularly because paleomagnetic studies are in reality a relative dating technique. Cross-checking these relative dating approaches with chronometric dating methodologies is critical.

Even though the Zhu et al. (2003) model may be correct, several questions can be raised. For instance, since there was a general cooling trend during the Early Pleistocene, with a southward migration of flora and fauna, do we see similar patterning in the micro- and macromammal migratory patterns? In addition, regions that have not been systematically surveyed (e.g., central-east China) need to be investigated in order to verify whether this southward migration was a broader regional pattern or only applicable to the four sites discussed by Zhu et al. (2003). Identifying additional traces of this southward dispersal during the Early Pleistocene from northern to central China should be the work of future research.

Since the dynamic Plio-Pleistocene environment likely heavily influenced hominin dispersal and behavioral patterning (Potts 1996, 1998; Dennell and Roebroeks 2005; Dennell 2010), understanding the unique environmental settings throughout the Old World in which hominin traces appear is critical to developing a synthetic view of human prehistoric lifeways. Only by comparing and synthesizing the biogeographic data with loess-soil, deep sea oxygen isotope, bathymetric fluctuation, and pollen analysis datasets, will paleoanthropologists be able to fully understand the dynamics involved with hominin decision-making during the Quaternary in East Asia. Only then will paleoanthropologists be able to confidently identify the position of early hominins in the biotic community and understand how this position changed as our behavior evolved from the Early Pleistocene to the modern day.

Acknowledgements The writing of this manuscript was supported by the Major Basic Research Projects of MST of China Grant (2006CB806400). We thank Eric Delson, Nina Jablonski, Geoff Pope, and Guoqin Qi for comments on an earlier draft of this manuscript. We take full responsibility for the ideas presented here and for any errors that may exist.

References

Aigner, J. S. (1981). *Archaeological Remains in Pleistocene China.* Munchen, Germany: Verlag C.H. Beck.

An, Z. (2000). The history and variability of the East Asian paleomonsoon climate. *Quaternary Science Reviews, 19*, 171–187.

An, Z. S., Kutzbach, J. E., Prell, W. L., & Porter, S. C. (2001). Evolution of Asian monsoons and phased uplift of the Himalaya-Tibetan plateau since Late Miocene times. *Nature, 411*, 62–66.

Anton, S. C. (2007). Climatic influences on the evolution of early Homo? *Folia Primatol, 78*, 365–388.

Anton, S. C., & Swisher, C. C., III. (2004). Early dispersals of Homo from Africa. *Annual Review of Anthropology, 33*, 271–296.

Belmaker, M. (2010). On the road to China: the environmental landscape of the early Pleistocene in western Eurasia and its implication for the dispersal of *Homo*. In C.J. Norton, D. Braun (Eds.), *Asian paleoanthropology: from Africa to China and beyond. Vertebrate paleobiology and paleoanthropology series.* Dordrecht, The Netherlands: Springer (this volume).

Ciochon, R., Olsen, J., & James, J. (1990). *Other origins: the search for the giant ape in human prehistory*. New York: Bantam.

Deng, T., & Xue, X. X. (1999). *Chinese fossil horses of Equus and their environment.* Beijing: China Ocean Press.

Deng, T., & Li, Y. M. (2005). Vegetational ecotype of the Gyirong Basin in Tibet, China and its response in stable carbon isotopes of mammal tooth enamel. *Chinese Science Bulletin, 50*, 1225–1229.

Deng, T., Dong, J. S., & Wang, Y. (2002). Variation of terrestrial ecosystem recorded by stable carbon isotopes of fossils in northern China during the Quaternary. *Chinese Science Bulletin, 47*, 76–78.

Dennell, R. (2010). The colonization of "Savannahstan": issues of timing(s) and patterns of dispersal across Asia in the Late Pliocene and Early Pleistocene. In C. J. Norton, D. Braun (Eds.), *Asian paleoanthropology: from Africa to China and beyond. Vertebrate paleobiology and paleoanthropology series.* Dordrecht, The Netherlands: Springer (this volume).

Dennell, R., & Roebroeks, W. (2005). An Asian perspective on early human dispersal from Africa. *Nature, 438*, 1099–1104.

Derbyshire, E., Keen, D. H., Kemp, R. A., Rolph, T. A., Shaw, J., & Meng, X. (1995). Loess-palaeosol sequences as recorders of palaeoclimatic variations during the last glacial-interglacial cycle: some problems of correlation in North-Central China. *Quaternary Proceedings, 4*, 7–18.

Ding, Z., Rutter, N. W., & Liu, T. S. (1997). The onset of extensive loess deposition around the G/M boundary in China and its palaeoclimatic implications. *Quaternary International, 40*, 53–60.

Dong, W. (1997). From Megaloceros pachyosteus to Procapreolus jinensis: a review on the study of fossil cervids in China. In Y.S. Tong et al. (Eds.), *Evidence for evolution – essays in honor of Prof. Chungchien Young on the hundredth anniversary of his birth* (pp. 165–175). Beijing: China Ocean Press.

Dong, W., Jin, C. Z., Xu, Q. Q., Liu, J. Y., Tong, H. W., Zheng, L. T. (2000). A comparative analysis on the mammalian faunas associated with Homo erectus in China. *Acta Anthropologica Sinica, 19*(Supplement), 246–256.

Fort, M. (1996). Late Cenozoic environmental changes and uplift on the northern side of the central Himalaya: a reappraisal from field data. *Palaeogeography, Palaeoclimatology, Palaeoecology, 120*, 123–145.

Gaboardi, M., Deng, T., & Wang, Y. (2005). Middle Pleistocene climate and habitat change at Zhoukoudian, China, from the carbon and oxygen isotopic record from herbivore tooth enamel. *Quaternary Research, 63*, 329–338.

Guo, Z. T., Ruddiman, W. F., Hao, Q. Z., Wu, H. B., Qiao, Y. S., Zhu, R. X., et al. (2002). Onset of Asian desertification by 22 Myr ago inferred from loess deposits in China. *Nature, 416*, 159–163.

Han, D. F., & Xu, C. H. (1985). Pleistocene mammalian faunas of China. In R. K. Wu & J. W. Olsen (Eds.), *Palaeoanthropology and palaeolithic archaeology in the People's Republic of China* (pp. 267–286). New York: Academic.

Heller, F., & Liu, T. S. (1982). Magnetostratigraphical dating of loess deposits in China. *Nature, 300*, 431–439.

Heslop, D., Dekkers, M. J., & Langereis, C. G. (2002). Timing and structure of the Mid-Pleistocene transition: records from the loess deposits of northern China. *Palaeogeography, Palaeoclimatology, Palaeoecology, 185*, 133–143.

Hooijer, D. A. (1951). Two new deer from the Pleistocene of Wanhsien, Szechwan, China. *American Museum Novitates, 1495*, 1–18.

Jablonski, N. G. (1997). The relevance of environmental change and primate life histories to the problem of hominid evolution in East Asia. In N. G. Jablonski (Ed.), *The changing face of East Asia during the tertiary and quaternary* (pp. 462–475). University of Hong Kong: Centre of Asian Studies.

Jablonski, N. G., Whitfort, M. J., Roberts-Smith, N., & Xu, Q. Q. (2000). The influence of life history and diet on the distribution of catarrhine primates during the Pleistocene in eastern Asia. *Journal of Human Evolution, 39*, 131–157.

Ji, H. X. (1980). The classification of Quaternary mammalian fauna in Lantian area. *Paleovertebrata and Paleoanthropologica, Sinica, 18*, 220–228.

Jin, C. Z., Kawamura, Y., & Taruno, H. (1999). Pliocene and Early Pleistocene insectivore and rodent faunas from Dajushan, Qipanshan and Haimao in North China and the reconstruction of the faunal succession from the Late Miocene to Middle Pleistocene. *Journal of Geosciences, Osaka City University, 42*, 1–19.

Jin, C. Z., Dong, W., Liu, J. Y., Wei, G. B., Xu, Q. Q., Zheng, J. J., Zheng, L. T., Han, L. G., & Wang, F. Z. (2000). A preliminary study on the Early Pleistocene deposits and the mammalian fauna from the Renzi Cave, Fanchang, Anhui, China. *Acta Anthropologica Sinica, 19*(Suppl.), 235–245.

Jin, C. Z., Zhang, Y. Q., Sun, C. K., & Zheng, L. T. (2009). First discovery of the large shrew, Beremendia (Insectivora, Soricidae), from the lower Pleistocene of South China and its paleoclimatic implications. *Vertebrata PalAsiatica, 47*, 153–163.

Li, H. M., Yang, X. Q., Heller, F., & Li, H. T. (2008). High resolution magnetostratigraphy and deposition cycles in the Nihewan Basin (North China) and their significance for stone artifact dating. *Quaternary Research, 69*, 250–262.

Liu, T. S. (1985). *Loess and the environment*. Beijing: China Ocean Press.

Liu, T. S., & Ding, M. L. (1984). The characteristics and evolution of the palaeoenvironment of China since the Late Tertiary. In R. O. Whyte (Ed.), *The evolution of the East Asian environment* (Vol. I, pp. 11–40). University of Hong Kong: Centre of Asian Studies.

Liu, T., & Ding, Z. (1998). Chinese loess and the paleomonsoon. *Annual Review of Earth and Planetary Sciences, 26*, 111–145.

Louys, J., Curnoe, D., & Tong, H. (2007). Characteristics of Pleistocene megafauna extinctions in Southeast Asia. *Palaeogeography, Palaeoclimatology, Palaeoecology, 243*, 152–173.

Luchterhand, K. (1978). Late Cenozoic climate, mammalian evolutionary patterns and Middle Pleistocene human adaptation in Eastern Asia. In L. G. Freeman (Ed.), *Views of the past* (pp. 363–421). The Hague: Mouton.

Luchterhand, K. (1984). Mammalian endemism and diversity and Middle Pleistocene hominid distribution and adaptation in Eastern Asia. In R. O. Whyte et al. (Eds.), *The evolution of the East Asian environment* (pp. 848–863). Hong Kong: University of Hong Kong Press.

Norton, C. J. (2000). The current state of Korean paleoanthropology. *Journal of Human Evolution, 38*, 803–825.

Pei, W. C. (1957). The zoogeographical divisions of Quaternary mammalian fauna in China. *Vertebrata Palasiatica, 1*(1), 9–24.

Pope, G. G. (1984). The antiquity and paleoenvironment of the Asian Hominidae. In R. O. Whyte et al. (Eds.), *The evolution of the East Asian environment* (pp. 822–847). Hong Kong: University of Hong Kong Press.

Pope, G. G. (1988). Recent advances in Far Eastern paleoanthropology. *Annual Review in Anthropology, 17*, 43–77.

Pope, G. G., & Keates, S. G. (1994). The evolution of human cognition and cultural capacity. In R. S. Corruccini & R. L. Ciochon (Eds.), *Integrative paths to the past* (pp. 531–567). Englewood Cliffs, NJ: Prentice Hall.

Potts, R. (1996). *Humanity's descent: the consequences of ecological instability*. New York: William Morrow.

Potts, R. (1998). Variability selection in hominid evolution. *Evolutionary Anthropology, 7*, 81–96.

Qi, G. Q. (1989). The Pleistocene human environment of North China. *Acta Anthropologica Sinica, 9*, 340–349.

Qi, G. Q. (1990). Quaternary mammalian faunas and environment of fossil humans in North China. In: R.K. Wu, X.Z. Wu, S.S. Zhang (Eds.), *Early Humankind in China* (pp. 277–337). Beijing: Science Press.

Rink, W. J., Wei, W., Bekken, D., & Jones, H. L. (2008). Geochronology of Ailuropoda-Stegodon fauna and Gigantopithecus in Guangxi Province, southern China. *Quaternary Research, 69*, 377–387.

Spicer, R. A., Harris, N. B., Widdowson, M., Herman, A. B., Guo, S., Valdes, P. J., et al. (2003). Constant elevation of southern Tibet over the past 15 million years. *Nature, 421*, 622–624.

Sun, X. J., & Wu, Y. S. (1988). Recent progress in Late Cenozoic palaeobotany and palynology in China: a brief review. In P. Whyte et al. (Eds.), *The Palaeoenvironment of East Asia from the Mid-Tertiary* (Vol. I, pp. 503–519). University of Hong Kong: Centre of Asian Studies.

Swisher, C. C., III, Rink, W. J., Anton, S. C., Schwarcz, H. P., Curtis, G. H., Suprijo, A., et al. (1996). Latest Homo erectus of Java: potential contemporaneity with Homo sapiens in Southeast Asia. *Science, 274*, 1870–1874.

Teilhard de Chardin, P. (1941). *Early man in China*. Beijing: Institute de Geo-Biologie Publication No.7.

Tougard, C. (2001). Biogeography and migration routes of large mammal faunas in South-East Asia during the Late Middle Pleistocene: focus on the fossil and extant faunas from Thailand. *Palaeogeography, Palaeoclimatology, Palaeoecology, 168*, 337–358.

Tougard, C., & Montuire, S. (2006). Pleistocene paleoenvironmental reconstructions and mammalian evolution in South-East Asia: focus on fossil faunas from Thailand. *Quaternary Science Reviews, 25*, 126–141.

Wang, W., & Mo, J. Y. (2004). Human fossil teeth newly discovered in Nanshan cave of Fusui, Guangxi. *Acta Anthropologica Sinica, 23*, 130–137.

Wang, Y., & Deng, T. (2005). A 25 m.y. isotopic record of paleodiet and environmental change from fossil mammals and paleosols from the NE margin of the Tibetan Plateau. *Earth and Planetary Science Letters, 236*, 322–338.

Wang, H., Ambrose, S. H., Liu, C. L. J., & Follmer, L. R. (1997). Paleosol stable isotope evidence for early hominid occupation of East Asian temperate environments. *Quaternary Research, 48*, 228–238.

Wang, Y., Deng, T., & Biasatti, D. (2006). Ancient diets indicate significant uplift of southern Tibet after ca. 7 Ma. *Geology, 34*, 309–312.

Wang, W., Potts, R., Yuan, B. Y., Huang, W. W., Cheng, H., Richards, R. L., et al. (2007). Sequence of mammalian fossils, including hominoid teeth, from the Bubing Basin caves, South China. *Journal of Human Evolution, 52*, 370–379.

Wei, Q. (1978). New discoveries from the Nihewan formation and other aspects regarding its study. In IVPP (Eds.), *Treatises on Paleoanthropology* (pp. 136–150). Beijing: Science Publishing Society.

Wei, Q. (1988). The stratigraphic, geochronological, and biostratigraphic background of the earliest known Paleolithic sites in China. *L'Anthropologie, 92*, 931–938.

Wei, G. B., Taruno, H., Kawamura, Y., & Jin, C. Z. (2006). Pliocene and Early Pleistocene primitive mammoths of northern China: their revised taxonomy, biostratigraphy and evolution. *Journal of Geosciences, Osaka City University, 49*, 59–101.

Whyte, R. O. (1984). The Gramineae in the palaeoenvironment of East Asia. In R. O. Whyte (Ed.), *The evolution of the East Asian environment* (Vol. I, pp. 622–650). University of Hong Kong: Centre of Asian Studies.

Wu, X. Z., & Poirier, F. E. (1995). *Human Evolution in China*. Oxford: Oxford University Press.

Wu, R. K., & Li, X. X. (Eds.). (2001). *Homo erectus from Nanjing*. Nanjing: Jiangsu Science and Technology Publishing House.

Xie, Y., Mackinnon, J., & Li, D. (2004). Study on biogeographical divisions of China. *Biodiversity Conservation, 13*, 1391–1417.

Xu, Q. Q. (1997). Homo erectus lantianensis at Chenjiawo, the first appearance of Peking Man in China. In N. G. Jablonski (Ed.), *The changing face of East Asia during the tertiary and quaternary* (pp. 446–461). University of Hong Kong: Centre of Asian Studies.

Xu, Q. Q., Mu, X. N., Xu, H. K., Zhong, S. L., & Mu, D. C. (1993). Discovery of Middle Pleistocene mammal fauna from Tangshan Karst Cave, Nanjing and its significance. *Chinese Science Bulletin, 38*, 1742–1746.

Xue, X., & Zhang, Y. (1991). Quaternary mammalian fossils and fossil human being. In Z. H. Zhang & S. X. Shao (Eds.), *The Quaternary of China* (pp. 307–374). Beijing: China Ocean Press.

Zhang, Z. S., Wang, H. J., Guo, Z. T., & Jiang, D. B. (2007). Impacts of tectonic changes on the reorganization of the Cenozoic paleoclimatic patterns in China. *Earth and Planetary Science Letters, 257*, 622–634.

Zhang, Y. Q., Kawamura, Y., & Jin, C. Z. (2008). A new species of the extinct vole *Villanyia* from Renzidong Cave, Anhui, East China, with discussion on related species from China and Transbaikalia. *Quaternary International, 179*(1), 163–170.

Zheng, H., Powell, C. M., An, Z., Zhou, J., & Dong, G. (2000). Pliocene uplift of the northern Tibetan Plateau. *Geology, 28*, 715–718.

Zheng, L. T., Huang, W. P., & IVPP (Eds.). (2001). *Hexian Hominid Site*. Beijing: Chonghua Book Bureau Publisher.

Zhu, R. X., An, Z. S., Potts, R., & Hoffman, K. A. (2003). Magnetostratigraphic dating of early humans in China. *Earth-Science Reviews, 61*, 341–359.

Chapter 8
The History of Hominin Occupation of Central Asia in Review

Michelle M. Glantz

Abstract The timing of hominin dispersals during the early Pleistocene, specifically into East Asia, is well established. The pattern of migration across inner Asia and the subsequent duration/intensity of hominin colonization of these areas, however, are still poorly resolved. The large territory of Central Asia defines a clear path within Eurasia through which hominin dispersals farther east may have occurred. The purpose of the present study is to examine the degree to which an autochthonous evolutionary trajectory is supported in Central Asia and how the potential connections between this and neighboring regions may be characterized during the Pleistocene. Archaeological and human paleontological evidence from the region is reviewed and compared to that from the Near East, the Altai, and China. This review informs a more detailed analysis of the Central Asian Middle Paleolithic record. Prevailing theoretical models suggest that Central Asia was inhabited by Neandertals migrating from the west to seek refuge from expanding modern human populations during the Middle Paleolithic. Morphological analyses of the newly discovered Obi-Rakhmat hominin and a re-evaluation of the Teshik-Tash child, both from sites in Uzbekistan, provide a test of this model. Results indicate that evidence of the morphological pattern that typically describes European Neandertals is equivocal in Central Asia. Although both Obi-Rakhmat and Teshik-Tash express some Neandertal features, their morphologies also suggest some admixture with local populations and/or those migrating into Central Asia from the North and East.

Keywords Anghilak Cave • Obi-Rakhmat Grotto • Uzbekistan • Central Asia • Early Pleistocene • Middle Paleolithic variability • Initial Upper Paleolithic • Hominin migration • Colonization

M.M. Glantz (✉)
Department of Anthropology, Colorado State University, Fort Collins, CO 80523, USA
e-mail: Mica.Glantz@colostate.edu

Introduction

As the preeminent hominin fossil from Central Asia, the Teshik-Tash child has encouraged scholarly interest primarily in Late Pleistocene hominin habitation of the region (Gremyatskii 1949; Glantz et al. 2003; Derevianko et al. 2004). Teshik-Tash 1 is significant in part because it represents the only relatively complete cranium from its developmental cohort and the Middle Paleolithic that is associated with well represented lithic and faunal assemblages from east of the Aral Sea to the Pacific Ocean. However, a full understanding of hypothesized Neandertal dispersals into the area during Oxygen Isotope Stages ("OIS") 5 and 4 requires an investigation of the context of Early and Middle Pleistocene hominin occupation. Whether evidence of long-term, relatively uninterrupted periods of hominin occupation existed prior to the Late Pleistocene has a direct bearing on how Neandertal presence there is modeled as well as on the role the region played in hominin dispersals farther east.

The purpose of this paper is to review the evidence of the earliest hominin occupation of Asia's heart in order to address two related questions. First, is Central Asia a corridor through which hominins dispersed during the Early and early Middle Pleistocene? And second, is there evidence of an autochthonous evolutionary trajectory in the region that laid the foundations for comparatively high Middle Paleolithic site densities? The research reviewed here has been summarized in the English literature by Ranov and Davis (1979), Davis and Ranov (1999) and more recently and thoroughly by Leonid Vishnyatsky (1999). These reviews, particularly Vishnyatsky's, provide a wealth of detailed information concerning the Paleolithic of Central Asia and how it compares to neighboring areas. My purpose here is not to expand upon the descriptions of the sites and their assemblages presented in those contributions. Instead, I will present more current information on paleoclimatic reconstructions of the region, describe newly discovered sites, and incorporate a more detailed analysis of Central Asian hominin fossils in order to address the questions posed above. In this regard, hypothesized population dynamics during the Pleistocene may be more fully addressed.

C.J. Norton and D.R. Braun (eds.), *Asian Paleoanthropology: From Africa to China and Beyond*, Vertebrate Paleobiology and Paleoanthropology, DOI 10.1007/978-90-481-9094-2_8, © Springer Science+Business Media B.V. 2010

Defining Central Asia: How Big Is It?

Defining the extent of the territory identified as Central Asia is not a straightforward matter, as the size of the region is delineated from often contradictory political, cultural, historical, and physio-geographical criteria. The former Soviet republics of Kazakhstan, Uzbekistan, Turkmenistan, Tajikistan, and Kyrgyzstan comprise the region that most westerners identify as Central Asia. Referred to as "Asia's heart", this politically circumscribed area covers approximately four million square kilometers. The former Soviet republics are also geographically linked by drainages into the Aral and Caspian Seas.

Ethno-historical accounts of Central Asia, however, expand the region to include Xinjiang Autonomous Region in NW China, Mongolia, Iran, Afghanistan, and Babur's Mogul India (Hopkirk 1994). Contemporary ecological studies also place Mongolia in Central Asia, while Soviet geographical tradition dictated identifying Kazakhstan as separate from the rest of the region based on that country's ecological similarity to southern Siberia. In the present study, Central Asia is defined as the territory that encompasses the former Soviet republics as well as northern Afghanistan. As more is known about the Paleolithic occupation of Xinjiang Autonomous Region in northwestern China, I would argue that this region should be included in discussions of Paleolithic Central Asia in the future.

The Paleoclimate of Central Asia

Central Asia is vast and ecologically diverse, characterized by desert basins, steppes, and mountain landscapes. These contemporary physio-geographical zones and their corresponding plant and animal communities are the product of the activity of the Caspian Sea in the western corner of this region and the Tien Shan and Pamir high mountains zones situated in the east and south, respectively, during the Cenozoic (Velichko and Nechaev 2005). The area can be divided into two paleoclimate zones that are identifiable from the Plio-Pleistocene boundary; first, the intermontane depressions and piedmonts of the Tien Shan and Pamirs of south Central Asia and second, the plains of Kazakhstan, Turkmenistan, and Uzbekistan (Akhmetyev et al. 2005). The paleoclimate of the second region is more difficult to reconstruct because of incomplete geological records, a lack of preserved organic remains, and a complete absence of radiometric dates. In the plains region, Quaternary divisions are based on geomorphology; terrace cycles are correlated with the stratigraphic scheme of the Caspian Sea (Akhmetyev et al. 2005). Because of the lack of any supporting chronology, I will not provide details concerning paleoclimatic reconstructions of the plains area. Overall, the ancient plains of Central Asia experience the same aridization as the mountainous zone described below; this process is more extreme in the plains and contributed to the formation of the contemporary deserts of the region during the Late Pleistocene.

The mountainous regions of south Central Asia are filled with relatively thick Late Cenozoic deposits that preserve rich faunal sequences, some palynological data, and a limited chronological record. A close study of the loess-soil series of the Tajik depression, the Issyk-Kul depression of Kyrgyzstan, and the Fergana depression of Uzbekistan provide a window onto climatic fluctuations over the last 2 million years. Basic stratigraphic subdivisions in the area follow the Late Pliocene (3.5–1.8 Ma), the Eopleistocene (1.8–0.8 Ma), and the Pleistocene (0.8–0.12 Ma) (Akhmetyev et al. 2005). During the Late Pliocene, biostratigraphic and paleoecological data support the reconstruction of heavily forested mountainous areas giving way to open areas of savanna and semi-savanna on the foothills and adjacent plains; some swampy locations also existed. In general, the Pleistocene in Central Asia is characterized by increasing aridization and continentality. This trend toward aridity caused a reduction in floral genera, with the Late Pleistocene characterized by the fewest number of plant species (Vishnyatsky 1999)

The changes in representative mammal fauna over the Pleistocene also provide corroboration of reconstructed climate trends as well as indicate dispersal patterns. The Kuruksai locality from the southern part of the Tajik depression is situated in the reversed interval of the lower part of the Matuyama epoch, lower than the Olduvai episode, and provides a faunal sequence that can be correlated to the Villafranchian (Dodonov et al. 1991). From the Kuruksai complex and other faunal assemblages from stratigraphically analogous sites, it is possible to reconstruct the late Pliocene environment as dominated by dry open areas (shrubby savannas) with patchily distributed swampy biotopes. On the whole, Mediterranean faunal elements are widely represented during this period. In contrast, younger biostratigraphic levels from the south of Tajikistan at the Lakhuti-2 locality and Tepke-1 in Kyrgyzstan are dominated by Palearctic elements and indicate mixed landscapes combining mountain slopes and coniferous-platyphillous forests with forest-steppe and steppe vegetation (Akhmetyev et al. 2005). Mammalian populations from this time period (strata with reverse magnetization, situated several meters below the Matuyama-Brunhes inversion) were more directly influenced by fauna from the European-Siberian subregion (Vangengeim and Pevzner 1991).

The youngest biostratigraphic levels from the area are also from the Tepke section in Kyrgyzstan. The bone-bearing horizon at Tepke-2 is situated in the normally magnetized interval above the Matuyama-Brunhes reversal. Spore-pollen analyses indicate a relatively dry climate with greater afforestation than the contemporary environment. Starting earlier and continuing into the Late Pleistocene, Kahlke (1994) has found that taxa originating in Asia were central to the formation of the large mammal fauna during the western and southern European Late Glacial. In contrast, the likelihood of European taxa extending their range of distribution eastward was significantly lower (Kahlke 1994: 52; Keates 2004). Keates (2004) uses these observations to infer an analogous dispersal pattern among Pleistocene hominins.

Tectonic activity and dust from the deserts surrounding the mountains of Central Asia encouraged loess formation that accelerated during the late part of the Middle Pleistocene and into the terminal phases of the Late Pleistocene; roughly 45 paleosols have been identified in the loess sequence from the Tajik depression, ten of which are from the Middle to Late Pleistocene (Akhmetyev et al. 2005). A curve reflecting short period paleoclimatic events recorded in the Central Asian loess-soil sections correspond to paleoclimate conditions reconstructed from the loess sections of China (Liu 1985; Ding et al. 2002), the glacial-interglacial sequence of Europe (Molodkov and Bolikhovskaya 2006; Kukla 1978), and the oxygen-isotope curve. More precise correlation of Central Asian paleoclimatic fluctuations to those of other parts of the Old World are hampered by the absence of geochronological data. However, over the past 0.75 Ma, the number of dry-cold, wet-warm intervals probably reached 20, with the warm intervals consisting of several optima. The geochemistry and palynospectra of the paleosols clearly indicates their formation during periods of relatively humid and mild conditions (Akhmetyev et al. 2005). Although the paleoclimatic implications related to the timing and trajectory of hominin dispersals will be discussed in more detail below, it is interesting to note that artifact-bearing localities are almost exclusively associated with paleosols.

Earliest Evidence of Hominin Occupation of Central Asia

The Lower Paleolithic of Central Asia is comprised of four technological complexes, each of which is identified by key tool types such as Acheulian-like bifaces, pebbles, cores and flakes, and 'small' artifacts. The chronological framework of these industries is poor, as only a handful of sites have been reliably dated. The oldest stone tool localities come from paleosols 12 and 11 of the loess-soil formation in Tajikistan (Fig. 8.1). Situated below the Brunhes/Matuyama boundary and above the Jaramillo subchron, the Kul'dara locality represents the oldest occupation of Central Asia known thus far, at 800–900 ka (Ranov et al. 1995). This observation is reinforced by the absence of hominin material and/or evidence

Fig. 8.1 Topographic map of the mountainous zones of southern Central Asia. *Triangle* indicates the site of Karatau/Lakhuti and the *cross* indicates the site of Kul'dara. Both sites are in Tajikistan and are part of the loess formation of the Tajik depression (map adapted from www.untj.org/files/map/overview_CA)

Table 8.1 Central Asian sites discussed in the text from chronologically oldest to youngest[a]

Site	Location	Radiometric date	Hominin remains	Toolkit	Significance
Kul'dara	Tajikistan	800–900 ka Paleomagnetism	n	Pebble	Oldest site in region
Karatau	Tajikistan	500–600 ka Paleomagnetism	n	Pebble	Additional early Middle Pleistocene site
Lakhuti	Tajikistan	500–600 ka Paleomagnetism	n	Pebble	Additional early Middle Pleistocene site
Barikazgan, Tanirkazgan, Akkol	Kazakhstan	n	n	Core/flake	Defines Lower Paleolithic of region
Koshkurgan I, II	Kazakhstan	500–170 ka ESR	n	'Micro'	Only site with this industrial complex
Yangadja	Kazakhstan	n	n	Bifacial	Surface scatter of Acheulian tools
Sel'ungur	Uzbekistan	126 ka uranium–thorium	n	Pebble	
Obi-Rakhmat	Uzbekistan	90–40 ka radiocarbon, U-series, ESR, final results pending	y	MP/IUP[b]	Long stratigraphic sequence, putative appearance of IUP and hominin specimen with mixed morphology
Kul'bulak	Uzbekistan	n	n	MP	Long stratigraphic sequence, foliate points
Teshik-Tash	Uzbekistan	n	y	MP	Relatively complete hominin specimen, discovery anchored easternmost aspect of Neandertal range in CA
Khudji	Tajikistan	~40 radiocarbon	y	MP	
Anghilak	Uzbekistan	~27–40 ka AMS	y	MP	Late occurrence of typical MP toolkit
Darra-i-Kur	Afghanistan	n	y	MP	hominin interpreted as modern human
Samarkandskaya	Uzbekistan	n	y	UP/M[c]	Material is probably Mesolithic

[a]With the exception of Anghilak Cave (Glantz et al. 2003; Glantz et al. 2006) and Darra-i-Kur (Angel 1972), the sites listed above and discussed in the text represent a subsection of those described by Davis and Ranov 1999 and Vishnyatsky 1999

[b]MP/IUP (Middle Paleolithic/Initial Upper Paleolithic)

[c]UP/M (Upper Paleolithic/Mesolithic)

of their presence in the form of stone tools or modified animal bones among several Central Asian Plio-Pleistocene and Early Pleistocene faunal series (Sotnikova et al. 1997). The long loess and pedocomplex sequence from the Chasmanigar profile in Tajikistan also lacks hominin fossil material and evidence of their activities (Davis and Ranov 1999).

At Kul'dara, 40 artifacts made mostly from quartzite pebbles were recovered (Table 8.1). Although Ranov has attributed this material to the Karatau pebble culture, most workers find the assemblage typologically difficult to define (Vishnyatsky 1999). Aside from a few isolated finds in paleosols 11 and 6, the next noteworthy assemblages associated with this formation are Karatau 1 (PC 6) and Lakhuti (PC 5). Thermoluminescence determinations originally placed these localities at 200 and 130 ka, respectively. Better correlations with the oxygen-isotope record, however, suggest dates much older than previously thought, placing the sites at approximately 600–500 ka. Karatau 1 has yielded more than 600 artifacts made mostly from metamorphic pebbles; prepared cores and bifaces are completely absent. At Lakhuti, another 500 artifacts were recovered. This industry is very similar to the one at Karatau with the exception of a few prepared cores and blades (Ranov and Schafer 2000).

During two seasons of foot surveys (1958–1959), Alpysbaev collected over 6,000 lithic pieces from the Karatau Mountains of southern Kazakhstan. His description of this material (Alpysbaev 1979), primarily from the sites of Borikazgan, Tanirkazgan, and Akkol, defines the Lower Paleolithic of this region, specifically the Karatau mountain zone, as characterized by irregular orthogonal cores, 'citron' spalls, massive *racloir*-like tools, and large cutting tools. Although Shunkov (2005) and others posit that the Alpysbaev material is similar to that from Kul'dara, it is undated and primarily derived from deflated surfaces. In a re-analysis of the Alpysbaev collection, Glantz et al. (2009b) found that the material departs somewhat from previous descriptions (i.e., Vishnyatsky 1999; Shunkov 2005, and others) in that it is predominately characterized by discoidal and fan cores and modified flakes. It also contains a significant number of tool types more reliably placed within the Middle Paleolithic.

The only other Lower Paleolithic localities in the region with chronometric dates are from Koshkurgan I and II, stratified sites with a 'small tools' industrial complex (Derevianko et al. 2003, Derevianko 2006). The Koshkurgan sites are also located in the Karatau Mountains, within the area of Alpysbaev's original surveys. A series of ESR dates were generated from travertine deposits; they range from ~500–170 ka (Derevianko, 2003).

In addition to the sealed open-air sites in Tajikistan and Kazakhstan and the surface collections in Kazakhstan of core/flake tools, bifacially worked tools have also been recovered. Provocative surface finds from the desert plateaus between the Aral and Caspian Seas of Kazakhstan, Uzbekistan, and Turkmenistan appear similar to the Acheulian, particularly bifaces from the site of Yangadja on the Krasnovodsk Peninsula of Kazakhstan (Vishnyatsky 1999). Because the localities are without a stratigraphic context, working out the chronology seems presently impossible.

In general, core and flake industries predominate in Central Asia during the late Early and Middle Pleistocene and appear to be differentially distributed in the foothill zones of the Karatau, Pamir, and Tien Shan Mountains. Examples of Acheulian-like bifaces are relatively rare and exclusively from surface deposits. Although isolated bifaces were recovered from the deserts of Turkmenistan and Uzbekistan, the highest densities of this tool type are from western and northern Kazakhstan. Based on the distribution of these various industries, Vishnyatsky (1999) suggested that the pebble and core/flake industries of the foothill zones may represent an autochthonous development, while the Acheulian seems to be relegated to the steppe areas of western and northern Central Asia. The impact of chronology, raw material availability, climate, and/or site function on the pattern of Lower Paleolithic variability in Central Asia is currently unknown.

Given that the new dates from the Nihewan basin in northern China place hominin occupation of this region at roughly 1.6 Ma (Zhu et al. 2004) the relatively younger dates from Central Asia are surprising, as the region represents a possible corridor to the east. The observation that paleoclimatic reconstructions support a general similarity between the savanna conditions of Africa, the Levant and southern Eurasia (i.e., Dmanisi) with those of the foothill regions of Central Asia makes the relative absence of Early Pleistocene sites in the latter area even more unexpected. It is only later during the Middle and Late Pleistocene that delta temperature changes between the warmest and coldest months were relatively large and the climate was overall cooler and drier in Central Asia (Akhmetyev et al. 2005). Although sampling bias may represent the ultimate reason behind the lack of evidence of early hominin colonization, this explanation is unsatisfactory. It is also difficult to invoke the pressures of an inhospitable environment to explain this absence of evidence. Ironically, it seems that hominin exploitation of the region intensified as the climate deteriorated. Also, the extent to which hominin habitation only occurred during warm and wet intervals during the broader trend toward cool and xeric environments cannot be completely resolved because of the lack of chronological control in the region.

The hypothesis that hominins did not explore the vast expanse of Central Asia until after their initial migration to East Asia serves as an alternative explanation to their absence in the region during the Early Pleistocene. It has been suggested that the lithic complex from Kul'dara shares some characteristics with Early Pleistocene assemblages from sites from the Nihewan basin of north China, unlike the surface lithic scatters from the western half of Central Asia. In addition, paleoclimatic studies of the Xinjiang region point to a relatively mild wet climate during the Early Pleistocene (Rolland 1992). This evidence taken in conjunction with the work of Kahlke (1994) suggests that Central Asia may have been colonized by hominins moving in an east-west trajectory, perhaps sometime after the initial colonization of China.

Sel'ungur Cave, located in the Ferghana region of Uzbekistan, represents the next oldest assemblage from the region (Islamov et al. 1988). A date of 126 ka was derived from a travertine sample overlying the uppermost cultural layer. Sel'ungur has a relatively large but poorly preserved faunal sample as well as over 1,500 stone artifacts. The original excavators interpreted the industry as Acheulian, although Vishnyatsky (1999) and others suggest it is more likely that the Sel'ungur material is attributable to the Lower Paleolithic pebble industries of the region. Original workers also identified six teeth and some long bone fragments as hominin, calling the material a local variant of *Homo erectus* (Islamov et al. 1988). However, Glantz et al. (2004) determined that the teeth are a mixture of cave bear and possibly ungulate, while the long bone is probably a juvenile hominin humerus, species indeterminate (Viola, pers.comm.).

The Late Pleistocene of Central Asia

The Middle Paleolithic record of Central Asia is much richer than that of earlier periods. This period is represented by a mixture of cave and intact open-air sites from each of the former Soviet republics and Afghanistan. Chronology is still a major obstacle to interpreting the dynamics of hominin adaptations to the region. Only eight of roughly 50 excavated Late Pleistocene sites have radiometric determinations and these determinations are not necessarily reliable. Consequently, it is difficult to adequately characterize Middle Paleolithic variability in the region. For example, Ranov (1971); Ranov and Davis (1979) and others (e.g., Suleimanov (1972) have maintained that the Middle Paleolithic of Central Asia is represented by a number of different facies or variants, while Vishnyatsky (1999) interprets the material as less mosaic than other regions, with similar inventories across cave and open-air sites. However, some recent finds of bifacial/foliate implements at three Middle Paleolithic sites in the Kyzylkum desert of Uzbekistan (Sayfullaev and Cauche 2004) have been interpreted as similar to 'anomalous' leaf points from Kul'bulak, Uzbekistan, as well as to surface occurrences from Turkmenistan. Bifacial implements and foliate points have also been reported from a site in Central Iran, Qaleh Bozi (Biglari et al. 2009) and are rather common and associated with Levallois technology in the Siberian Altai at the sites of Denisova, Ust'Kansk, and Okladnikov (Derevianko and Shunkov 2002). It is difficult to conclude whether this techno-typological theme is due to cultural diffusion or discrete innovative events. Regardless of the etiology of this theme, the presence of bifacial implements and foliate points and their distribution certainly increases the degree of variability commonly associated with the Middle Paleolithic of Central Asia.

Because Lower Paleolithic sites are scarce in Central Asia, the geographic origin of the Middle Paleolithic of the region has often been placed in the Levant and/or Taurus-Zagros region (see Davis 1990 and Vishnyatsky 1999). However, this question remains unresolved; if the basic technological feature of the Middle Paleolithic, prepared-core flaking, developed out of bifacial technology (see Schick 1994; Lycett 2007) then a number of regions as possible antecedents for the Middle Paleolithic of Central Asia may be identified. Lithic techno-typology as well as hominin morphology (see discussion below) point to Central Asia as a zone of interaction during the Middle Paleolithic, with contacts to the north, east, and west.

Another important debate has centered on the putative identification of the Initial Upper Paleolithic in the region at the cave site of Obi-Rakhmat in Uzbekistan (see Derevianko et al. 2004). This site, along with the open-air site of Kul'bulak in Uzbekistan, has the longest stratigraphic sequences in Central Asia. Much of the 10 m of continuous stratigraphy at Obi-Rakhmat has been dated with a variety of techniques and results are pending. Conservatively, it appears that the sequence at Obi-Rakhmat can be dated to roughly 90–40 ka (Blackwell et al. 2006).

The Upper Paleolithic, like the Lower Paleolithic, also is not well represented in Central Asia. Sites like Samarkanskaya in Uzbekistan that can be identified as Upper Paleolithic appear to date to closer to the Mesolithic boundary (Ginzburg and Gokhman 1974). Ranov and Davis (1979); Davis and Ranov (1999) have suggested that the local environmental conditions became too harsh to support hominin occupation during the last Ice Age. Or, alternatively, local Upper Paleolithic traditions retained a Middle Paleolithic tool inventory (i.e., Anghilak Cave, Uzbekistan), but are unidentifiable as such without some radiometric information (Glantz et al. 2006; Derevianko et al. 2003). In any event, there is some evidence of the continuous occupation of Central Asia during the Late Pleistocene and perhaps across the Middle to Upper Paleolithic divide. A continuous occupation of the Siberian Altai also has been suggested and seems to be supported at sites like Kara Bom (Derevianko et al. 2000; Derevianko and Markin 1992).

A handful of ecological/biogeographic models have been proposed to explain the relatively patchy spatial and temporal distribution of Middle and Upper Paleolithic sites in Central Asia. These models (Davis 1990; Davis and Ranov 1999; Schafer et al. 1998; Mangerud et al. 2004; Nat 1971) rely heavily on climatic reconstructions, most of which underscore the extreme differences between the hospitality of the foothill zones, the desert regions, and the dry grassy plains of the territory during the interglacial and glacial periods. Some models focus on specific foothill regions being the preferred hominin habitat during interglacials (see Davis 1990 and Nat 1971 for alternative views) and also suggest abandonment of the area during glacial periods, certainly during the Last Ice Age. While other models suggest that the inner deserts and grassy plains of western Central Asia would represent oases during glacial periods (Mangerud et al. 2004 and Vishnyatsky 1999). With the exception of models posed by Davis (1990) and Ranov and Davis (1979), other scenarios posit that Central Asia was never completely abandoned during the Late Pleistocene.

Who Inhabited Central Asia During the Pleistocene? The Hominin Fossil Record

Presently, it is not possible to comment on the affiliation of hominins living in Central Asia during the Early and Middle Pleistocene because no fossil remains from those periods have been found. It may be possible, however, to identify a hominin group responsible for the Middle Paleolithic assemblages in Central Asia. Paleoanthropological dogma suggests that this region was inhabited by Neandertals during the Middle Paleolithic as they retreated east during periods of climatic deterioration in Europe or to escape dispersing modern humans from Africa and the Levant (Vishnyatsky 1999; Stringer and Gamble 1993). The interpretation of the Teshik-Tash child as a Neandertal is central to the 'Central Asia as a periphery' paradigm. The fossil's association with a Middle Paleolithic industry has also been the implicit basis for linking all Mousterian-like assemblages from Central Asia to this hominin group. In this regard, the Neandertal status of the Teshik-Tash child provides taxonomic corroboration of cultural links made between Middle Paleolithic industries in Central Asia and those from the Taurus-Zagros region such as Shanidar Cave. A closer look at the hominin fossil evidence from the region is necessary to shed light on the validity of this model, in addition to examining evidence of continuities vs. discontinuities in the Middle to Upper Paleolithic archaeological record.

The hominin fossil record of Central Asia is chronologically contained between OIS-5b and OIS-2 (Fig. 8.2). Although Angel (1972) interpreted the broken right temporal bone from a Mousterian level of Darra-i-Kur cave,

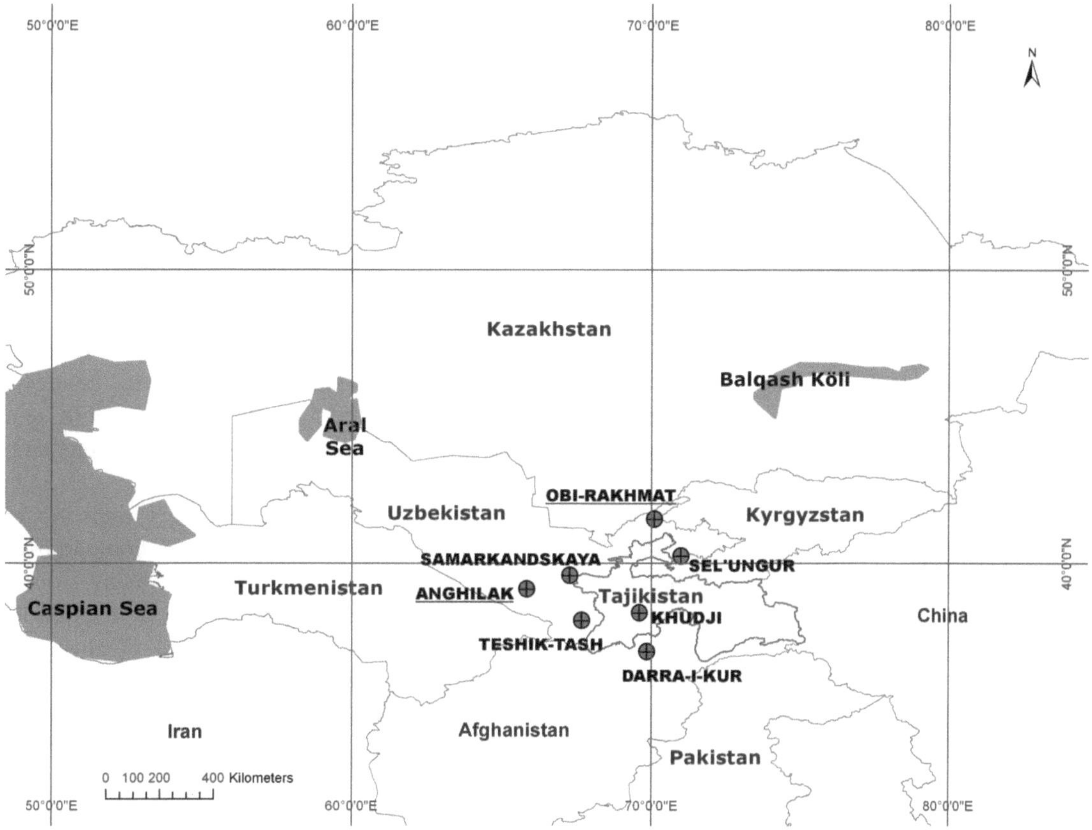

Fig. 8.2 Map showing the distribution of Central Asian fossil hominin sites, adapted from Davis and Ranov (1999). The Anghilak and Obi-Rakhmat fossils are recent additions to this record, 2002 and 2003, respectively

Afghanistan as being morphologically similar to modern humans, the attribution of this fossil and its archaeological association has had virtually no impact on the prevailing characterization of Central Asia as a Neandertal zone. Aside from Darra-i-Kur, other lesser known hominin material includes a deciduous lower lateral incisor from Khudji, Tajikistan (Trinkaus et al. 2000) and a fifth metatarsal from Anghilak Cave in Uzbekistan (Glantz et al. 2006, 2008). Other hominin material has been found at the late Upper Paleolithic site of Samarkandskaya in Uzbekistan, but it is not well provenienced and some interpret that archaeology as Mesolithic (pers. comm. R. Suleimanov).

Darra-i-Kur has a minimum age of 30 ka (Trinkaus 2005; Wolpoff 1999), the deciduous lateral incisor from Khudji, Tajikistan is estimated to be 40 ka (Trinkaus et al. 2000), and the fifth metatarsal from Anghilak Cave, Uzbekistan is roughly 32 ka (Glantz et al. 2006). Each of these sites provides substantial associated archaeological materials both in the form of lithic and faunal assemblages that can be characterized as typically Mousterian in many ways (Vishnyatsky 1999). In fact, the relatively recent dates associated with the lithic assemblages from Anghilak Cave (Glantz et al. 2006) point to a relatively late persistence of Mousterian-like assemblages in Central Asia.

The most complete hominin fossils from Central Asia are the Teshik-Tash child and a newly discovered individual from Obi-Rakhmat Grotto, Uzbekistan. Both of these specimens are juveniles; Teshik-Tash is more complete than Obi-Rakhmat 1, which consists of six permanent maxillary teeth and 150 cranial fragments (Glantz et al. 2008; Bailey et al. 2008). No radiometric determinations have been generated for the Teshik-Tash Cave stratigraphy or hominin fossil material, although the archaeology places it within the Middle Paleolithic. The Obi-Rakhmat individual may be as old as 90 ka or as young as 70 ka (Derevianko et al. 2004). Both specimens have been recently analyzed in studies that question the hypothesis that European Neandertals moved into *vacant* Central Asia during the Middle Paleolithic and that this region was that group's easternmost periphery (Glantz et al. 2008, 2009a; but see Krause et al. 2007).

In a reanalysis of the Neandertal status of the Teshik-Tash child, the affinity of this specimen with respect to other Late Pleistocene sub-adult hominins was examined (Ritzman et al. 2006; Glantz et al. 2009a). We found that linear variables, taken from areas of the cranium and mandible of Teshik-Tash that were not heavily reconstructed, more closely aligned this fossil with an Upper Paleolithic modern human sample rather than that of Neandertal juveniles.

The Obi-Rakhmat specimen (OR-1) is estimated to be roughly 9 years of age at death. The six permanent maxillary teeth are exceptionally large and preserve morphologies that are more commonly associated with the Neandertals. The lateral upper incisor is shovel shaped and the first molar expresses a skewed occlusal profile and a relatively large hypocone in comparison to the metacone (Bailey et al. 2008). Also, a relatively anomalous third or possible second molar was recovered.

Interpretations of the morphological affinities of the OR-1 cranium, however, are complicated by its age and extreme fragmentation. Although the relative dimensions of the bony labyrinth are more closely aligned to the Neandertals (Viola et al. 2004), other discrete traits present on the temporal bone that are commonly associated with this group are absent. The most pronounced departure from Neandertal morphology is observed in the reconstruction of the left parietal of Obi-Rakhmat. The parietal of OR-1 is long, beyond the range of variation present in similarly aged as well as adult Neandertals (Glantz et al. 2004, 2008). *Homo erectus* and *H. heidelbergensis* specimens have even shorter parietals than the Neandertals (Fig. 8.3). This observation suggests that a long parietal, like in OR-1, is a condition that distinguishes modern humans from the other hominins examined in this sample.

Given current models of Late Pleistocene hominin systematics that are typological by definition, we are left with identifying Teshik-Tash and Obi-Rakhmat as either Neandertal or modern human. In the case of these specimens, either attribution is problematic. Moreover, the categories of modern human and Neandertal obscure the spatio-temporal variability that existed among these respective groups. Does the morphology of Teshik-Tash and other hominin material from the region point to a geographic variant of the Neandertals – a Central Asian Neandertal group? It is well established that Neandertals exhibit spatio-temporal variability (Smith 1981; Smith et al. 1989a). In this regard, Neandertal variability in some regions and during specific times may, in part, be related to gene flow between Neandertals, other archaics and, modern humans and/or changing environmental conditions (Arensburg and Belfer-Cohen 1998).

The presence of a Neandertal variant in Central Asia is very difficult to assess when most comparative frameworks lack Late Pleistocene fossil specimens from eastern Asia (e.g., fossil material summarized in Wu and Poirier 1995). This criticism would also be true of any genetic analyses potentially used to support the presence of a Neandertal variant in Central Asia (Pennisi 2007; Krause et al. 2007). To that end, it would not be possible to describe Neandertal clinal variation that would satisfactorily exclude all other Late Pleistocene hominin morphologies and geneologies (see Rosenberg et al. 2006).

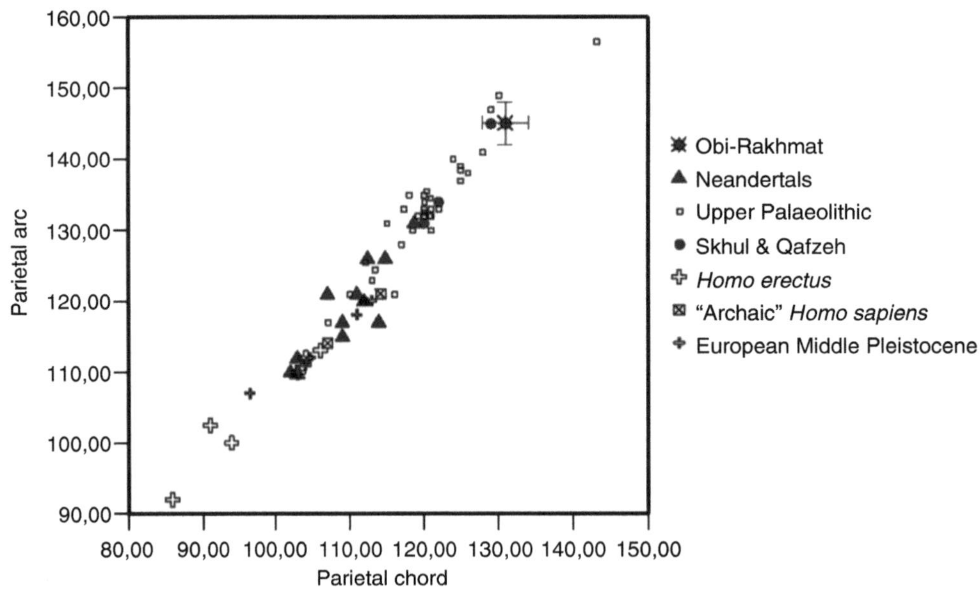

Fig. 8.3 Scatterplot illustrating parietal length with 3 mm error bars associated with OR-1. The OR-1 parietal is clearly positioned among the Skhul and Qafzeh group as well as Upper Paleolithic modern humans. Specimens in comparative groups include; *Homo erectus* – Zhoukoudian II, III, X, XI, XII, Hexian; Middle Pleistocene *Homo* – Xujiayao 10, Maba, Sima de los Huesos 2, 3, 4, 5, 6; Skhūl and Qafzeh – Qafzeh 9, Skhūl 4, 5, 9; Neandertals – La Ferrassie 1, 3, Shanidar 1, 2, Tabūn 1, Amud 1, La Chapelle aux Saints, Spy 1, 2, Neanderthal, Circeo 1, La Quina 5, Teshik-Tash; Upper Paleolithic modern humans – Sungir' 3, 4, Předmostí 1, 3, 4, 7, 9, 10, Dolni Vestonice 3, 11, 13, 14, 15, 16, Pavlov 1, Combe Capelle, Cro-Magnon 1, 2, 3, Grotte des Enfants 4, Barma Grande 1, 2, 5, 6, Chancelade, Oberkassel 1, 2, Upper Cave 101, 102, 103, Ziyang 1, Chuandong 1, Lijiang 1. All measurements taken from the literature (Wu and Poirier 1995, Arsuaga et al. 1997; Minugh-Purvis 1988; Trinkaus 1983; Vandermeersch 1981; Sladek et al. 2000)

According to Trinkaus (2005:222) "early modern humans prior to 25,000 years B.P. and outside of east Africa/southwest Asia exhibit complex and varying mosaics of early modern, late archaic, and regional anatomical features." However, as Trinkaus (2005, 2006) makes clear, no hominin fossils with a relatively complete suite of modern human derived morphologies resided in Eurasia during the Middle Paleolithic. Although the hominin material from Central Asia (e.g., Teshik-Tash, Obi-Rakhmat, and Darra-i-Kur), fits within this general description of early modern humans, the lack of associated radiometric determinations for Teshik-Tash and Darra-i-Kur, the observation that OR-1 may be as old as 90 ka, and the presence of many archaic/Neandertal features argues against re-attributing these remains to early modern humans. A taxonomic reattribution, in this regard, still obscures the morphological variability expressed in these Central Asian remains, nor does it effectively treat Late Pleistocene hominin variation in Eurasia on an analytical or theoretical level. Instead, the results of the previously described studies as well as other evidence suggest that the characterization of Central Asia as the eastern periphery of the European Neandertal range is an oversimplification.

The hypothesized demarcation of the Neandertal range is a product of the presumption of reproductive isolation. Evidence that supports potential isolation/separation of Central Asian hominins from northern and eastern Asian populations is not easily delineated. Geographic barriers like the high mountain zones of the Tien Shan and Pamirs did not seem to impede faunal dispersals across Eurasia (see Keates 2004). In this regard, the notion that Neandertals moved east from their core area in Europe and stopped in Central Asia because of being geographically hemmed in is unsupported, certainly in light of Middle Paleolithic evidence from the Siberian Altai and other sites in China. Pope (1992), Wolpoff (1999), and Wu et al. (2005) also suggest that the Chinese fossil record (i.e., Jinniushan and Maba) provide morphological support of interaction between eastern Asian and other Eurasian populations via sustained migrations (Keates 2004). In support of this Sohn and Wolpoff (1993) have interpreted the morphology of Zuttiyeh, a Middle Pleistocene fossil from the Levant, as being influenced by gene flow from the east.

In addition, the suggestion that Neandertals retreated to Central Asia to escape a deteriorating European environment is too general a statement to be testable. As previously discussed, the Central Asian environment was becoming increasingly arid and continental during the Late Pleistocene, to the extent that Ranov and Davis (1979), Davis and Ranov 1999) suggested that the region was uninhabitable during the Last Glacial Maximum. The chronology at Anghilak Cave overturns this general hypothesis (Glantz et al. 2006); it seems hominin groups lived in Central Asia during the coldest and driest periods of the Late Pleistocene. And finally, if we adopt the most rigid view of the Out of Africa/replacement hypothesis that stipulates dispersing modern humans from Africa/the Levant were forcing Neandertals into the peripheries of their range, sites with early modern human skeletal remains in Eurasia that can be securely dated to the Middle Paleolithic should exist; this evidence presently is lacking.

The hominin fossil record from Central Asia is chronologically limited and fragmentary. The hominin fossils that preserve key morphological areas express features common to Neandertals as well as modern humans – who are provocatively absent from Eurasia during the Middle Paleolithic. Because the mechanism of reproductive isolation is difficult to identify in this region [no geographic barriers and some general similarities between toolkits from Central and East Asia during the Middle Paleolithic, especially those made from low quality raw materials (but see Gao and Norton 2002; Norton et al. 2006)], it seems most parsimonious to interpret Eurasia as a large reticulating cline (Wolpoff et al. 2000). Neandertal morphology appears east of Europe, even east of Central Asia, as genes are passed around the Old World. Neandertal-like morphology in Central Asia may also be a reflection of a generalized cold adaptation, useful in the continental environments of Central, North and parts of East Asia (Rosenberg et al. 2006).

Concluding Remarks

The interpretation of Central Asia as a core zone of hominin interaction during the Late Pleistocene would be bolstered by evidence of long-term, continuous hominin occupation of the region during the Early and Middle Pleistocene. Currently, this evidence does not exist. Hominin groups appear to have colonized East Asia long before arriving in Central Asia. While the Kul'dara locality in Tajikistan places hominins in Central Asia at least by the late Early Pleistocene, this occupation seems to have been sporadic during the Middle Pleistocene. The archaeological record during this period, however, is diverse as evidence of Acheulian, pebble, core/flake, and 'small tool' industries are present, perhaps indicating a zone of interaction between east and west. Better evidence for continuous occupation of the region can be observed during the Late Pleistocene, although data suggest that assemblages across the Middle-Upper Paleolithic boundary retain a typically Mousterian toolkit.

While new hominin fossil remains and reinterpretations of previously discovered material are provocative, because of their relative incompleteness and developmental age, it is difficult to conclusively state that Neandertals, in the European sense, are absent from Central Asia. It is equally important to note, however, that characterizing the ways by which Central Asian hominins compare to (both morphologically and genetically) their East Asian contemporaries during the Middle Paleolithic is difficult, as this sample is often not included in most current analyses of the European, African, or Central Asian hominin record (e.g., Krause et al. 2007). Although it seems obvious that hominin population movement during the Pleistocene across Asia was not solely in a west to east, or south to north trajectory, until archaeological and paleontological material from East Asia is rigorously and regularly included into our existing models of population dynamics, this dogma will remain unchanged.

Acknowledgements I would like to thank the editors Christopher Norton and David Braun for inviting me to present this paper at the 2007 annual American Association of Physical Anthropologists meeting in Philadelphia and for providing editorial support during the preparation of the manuscript for publication. I would also like to thank Eric Delson and Allison Brooks for their constructive comments as discussants at this symposium. Nicolas Rolland provided exceptionally useful information, references, and opinions in his review of this article that I have tried to faithfully incorporate into the manuscript. His comments allowed me to greatly improve the manuscript.

References

Akhmetyev, M., Dodoniv, A., Somikova, M., Spasskaya, I., Kremenetsky, K., Klimanov, V. (2005). Kazakhstan and Central Asia (Plains and Foothills). In A. Velichko, V. Nechaev (Eds.), *Cenozoic climatic and environmental changes in Russia. Special paper 382*. Boulder Colorado: The Geological Society of America.

Alpysbaev, Kh A. (1979). *Lower Paleolithic monuments of southern Kazakhstan (on the oldest settling of Kazakhstan by primitive man)* (pp. 1–208). Almata: Nauka.

Angel, J. L. (1972). A Middle Palaeolithic temporal bone from Darra-i-Kur, Afghanistan. Prehistoric research in Afghanistan (1959–1966). *Transactions of the American Philosophical Society*, *62*, Part 4, 54–56.

Arensburg, B., & Belfer-Cohen, A. (1998). Sapiens and neandertals. In T. Akazawa, K. Aoki, & O. Bar-Yosef (Eds.), *Neandertals and Modern Humans in Western Asia* (pp. 311–322). New York: Plenum.

Arsuaga, J. L., Martinez, I., Gracia, A., & Lorenzo, C. (1997). The Sima de los Huesos crania (Sierra de Atapuerca, Spain). A comparative study. *Journal of Human Evolution, 33*, 219–281.

Bailey, S., Glantz, M., Weaver, T., & Viola, B. (2008). The affinity of the dental remains from Obi-Rakhmat Grotto, Uzbekistan. *Journal of Human Evolution, 55*, 238–248.

Biglari, F., Javeri, M., Mashkour, M., Yazdi, Y., Shidrang, S., Tengberg, M., Taheri, K., & Darvish, J. (2009). Test Excavations at the Middle Paleolithic sites of Qaleh Bozi, southwest of central Iran: a preliminary report, in M. Otte, F. Biglari & J. Jaubert (Eds.) *Iran Palaeolithic. Proceedings of the XV World Congress UISPP, Lisbon, 4-9 September 2006* (British Archaeological Reports International Series 1968): 29-38. Oxford: Archaeopress

Blackwell, B., Mian, A., Baboumian, S., Blickstein, J., Skinner, A., Wrinn, P., Krivoshapkin, A., Derevi'anko, A., Wagner, J., and Patchett, P. (2006). Settling the age dispute for the Late Middle Paleolithic at the Obi-Rakhmat hominid site, Uzbekistan. Presentation at the Paleoanthropology Society Meetings in San Juan, Puerto Rico.

Davis, R. (1990). Pleistocene climates and migration into Asia: evidence from loess. Unpublished paper, read at the International Symposium on the Chronostratigraphy of Paleolithic in the North, Central, and Eastern Asia and America, Novosibirsk, pp. 1–16.

Davis, R., & Ranov, V. (1999). Recent work on the Paleolithic of Central Asia. *Evolutionary Anthropology, 8*(5), 186–193.

Derevianko, A. P. (Ed.) (2003). *The Stone Age of Kazakhstan: archaeological studies carried out by the joint Russian-Kazakhstan expedition in Kazakhstan, 1998–2001* (pp. 1–183). Novosibirsk: Institute of Archaeology and Ethnography SB RAS press.

Derevianko, A. P. (2006). The Lower Paleolithic small tool industry in Eurasia: migration or convergent evolution? Archaeology. *Ethnology and Anthropology of Eurasia, 1*(25), 2.

Derevianko, A. P., & Markin, S. (1992).The Mousterian of the Gorny Altai. Novosibirsk: Nauka (in Russian).

Derevianko, A. P., & Shunkov, M. V. (2002). Middle Paleolithic industries with foliate bifaces in Gorny Altai. *Archaelogy, Ethnology, and Anthropology of Eurasia*, *1*(9), 16–42.

Derevianko, A. P., Petrin, V. T., & Rybin, E. P. (2000). The Kara-Bom site and the characteristics of the Middle-Upper Paleolithic transition in the Altai. *Archaeology, Ethnology, and Anthropology of Eurasia, 2*, 33–52.

Derevianko, A. P., Shunkov, M. V., Adadjanian, A. K., Baryshnikov, G. F., Malaeva, V. A., Ulianov, V. A., Kulik, N. A., Postnov, A. V., Anoikin, A. A. (2003). Prirodnaiia Sreda I Chelovek v Paleolite Gornogo Altaiia. Novosibirisk: Izdatel'stvo Instituta Archaeologii Etnografii S.O.RAN.

Derevianko, A. P., Krivoshapkin, A. I., Anoikin, A. A., Wrinn, P. J., Islamov, U. I. (2004). The lithic industry of Obi-Rakhmat Grotto. In A. P. Derevianko (Ed.) *Grot Obi Rakhmat Institute of Archaeology and Ethnography, Siberian Branch* (pp. 5–29). Novosibirsk: Russian Academy of Sciences.

Ding, Z. L., Ranov, V., Yang, S. L., Finaev, A., Han, J. M. & Wang, G. A. (2002). The loess record in southern Tajikistan and correlation with Chinese loess. *Earth and Planetary Science Letters,* 200, 387–400.

Dodonov, A., Zhegallo,V., Penkov, A., Sotnikova, M. (1991). Stratigraphy of sites of the Late Pliocene vertebrates in southern Tadjikistan: USSR Academy of Sciences. Izvestiya: Ser. Geoll., v. 5, pp. 12–21 (in Russian).

Gao, X., & Norton, C. J. (2002). A critique of the Chinese 'Middle Palaeolithic'. *Antiquity, 76*, 397–412.

Ginzburg V. V., & Gokhman I.I. (1974). Kostnye ostatki cheloveka iz Samarkandskoi paleoliticheskoi stoyanki. In Problemy etnicheskoy antropologii i morfologii cheloveka. Nauka: Leningrad, pp 5–12.

Glantz, M. M., Suleimanov, R., Hughes, P., Schauber, A. (2003). Anghilak cave, Uzbekistan: documenting Neandertal occupation at the periphery. *Antiquity*, *77*, 295.

Glantz, M. M., Viola, B., & Chikisheva, T. (2004). New hominid remains from Obi-Rakhmat Grotto. In A. P. Derevianko (Ed.), *Grot Obi-Rakhmat* (pp. 77–93). Novosibirsk: Institute of Archaeology and Ethnography, Siberian Branch, Russian Academy of Sciences.

Glantz, M., Tostevin, G., Suleimanov, R., Ritzman, T., Adams, J., Derr, K. (2006). The chronological context of Middle Paleolithic deposits at Anghilak Cave, Uzbekistan. Paper presented at the SAA meetings in San Juan, Puerto Rico.

Glantz, M., Viola, B., Wrinn, P. J., Chikisheva, T., Derevianko, A., Krivoshapkin, A. I., et al. (2008). New hominin remains from Uzbekistan. *Journal of Human Evolution, 55*, 223–237.

Glantz, M., Athreya, S., Ritzman, T. (2009a). Is Central Asia the eastern outpost of the Neandertal range? A reassessment of the Teshik-Tash Child. *American Journal of Physical Anthropology*, *138*, 45–61.

Glantz, M., Galm, J., Suleimanov, R., Taimagembetov, Zh. (2009b). Another look at the Lower Paleolithic of Kazakhstan: a preliminary description of Alpysbaev's lithic collections. Abstracts of the PaleoAnthropology Society 2009 Meetings. *PaleoAnthropology, 2009*, A1–A40.

Gremyatskii, M. A. (1949). Skull of the Neandertal child from Teshik-Tash Cave, Southern Uzbekistan. In M. A. Gremyatskii (Ed.), *Teshik-Tash: paleolithic man* (pp. 137–182). Moscow: Moscow State University.

Hopkirk, P. (1994). *The great game: the struggle for empire in Central Asia*. New York: Kodansha America.

Islamov, U. I., Zubov, A. A., & Kharitonov, V. M. (1988). Paleoliticheskaya stoyanka Sel Ungur v Ferganskoi doline. *Voprosy antropologii, 80*, 38–49.

Kahlke, R.-D. (1994). Die Enystehung-, Entwicklungs- und Verbreitungsgeschichte des oberpleistozanen Mammuthus-Coelodonta-Faunenekomplexes in Eurasien (GroBsauger). Abhandlungen der Senckenbergischen Naturforschenden Gesellschaft 546:Frankfurt: Verlag Waldemar Kramer.

Keates, S. G. (2004). Home range size in Middle Pleistocene China and human dispersal patterns in Eastern and Central Asia. *Asian Perspectives, 43*(2), 227–247.

Krause, J., Orlando, L., Serre, D., Viola, B., Prufer, K., Richards, M., et al. (2007). Neandertals in central Asia and Siberia. *Nature, 449*, 902–904.

Kukla, G. (1978). The classical European glacial stages: correlation with deep-sea Sediments: Nebraska Academy of Sciences. *Transactions, 6*, 57–93.

Liu, T. S. (1985). *Loess and the environment*. Beijing: China Ocean Press.

Lycett, S. (2007). Is the Soanian techno-complex a Mode 1 or Mode 3 phenomenon? A morphometric assessment. *Journal of Archaeological Science*, 34, 1434–1440.

Mangerud, J., et al. (2004). Ice-dammed lakes and rerouting of the drainage of northern Eurasia during the Last Glaciation. *Quaternary Science Review, 23*, 133–1332.

Minugh-Purvis, N. (1988). Patterns of Craniofacial Growth and Development in Upper Pleistocene Hominids, PhD Dissertation, University of Pennsylvania.

Molodkov, A., & Bolikhovskaya, N. (2006). Long-term palaeoenvironmental changes recorded in palynologically studied loess-palaeosol and ESR-dated marine deposits of Northern Eurasia: implications for sea–land correlation. *Quaternary International, 152–153*, 37–47.

Nat, D. (1971). Elements de Prehistoire et d'Archeologie Nord-Siberiennes. Fascicule 1. Contributions du Centre d'Etudes Arctiques et Finno-Scandinaves, no. 9.

Norton, C. J., Bae, K. D., Harris, J. W. K., & Lee, H. Y. (2006). Middle Pleistocene handaxes from the Korean Peninsula. *Journal of Human Evolution, 51*, 527–536.

Pennisi, E. (2007). No sex please, we're Neandertals. *Science, 316*, 967.

Pope, G. (1992). Craniofacial evidence for the origin of modern humans in China. *YPA, 35*, 243–298.

Ranov. V. (1971). On the study of the Mousterian culture in Central Asia. *Materialy i issledovania po arkheologii, 173*, 209–232.

Ranov, V., & Davis, R. (1979). Toward a new understanding of the Soviet Central Asian Paleolithic. *Current Anthropology, 20*, 249–270.

Ranov, V., & Schafer, J. (2000). The Palaeolithic of the late Middle Pleistocene in Central Asia, 400–100 Ka ago. In A. Ronen & M. Weinstein-Evron (Eds.), *Toward Modern Humans: Yabrudian and Micoquian, 400–50 k-years ago, BAR International Series 850* (pp. 77–94). Oxford: Archaeopress.

Ranov, V. A., Carbonell, E., & Rodriguez, X. P. (1995). Kul'dara: earliest human occupation in Central Asia in its Afro-Asian context. *Current Anthropology, 36*(2), 337–346.

Ritzman, T., Glantz, M., Athreya, S. (2006). Opening the stone: a multivariate reassessment of the Neandertal status of the Teshik-Tash child. *Paleoanthropology, 4*(appendix), 80.

Rolland, N. (1992). The Paleolithic colonization of Europe: an archaeological and biogeographical perspective. *Trabajos de Prehistoria, 49*, 69–111.

Rosenberg, K., Zune, L., & Ruff, C. (2006). Body size, body proportions, and encephalization in a Middle Pleistocene archaic human from northern China. *Proceedings of the National Academy of Sciences of the United States of America, 103*(10), 3552–3556.

Sayfullaev, B. C., & Cauche, D. (2004). Nouvelles decouvertes paleolithiques dans la partie nord-est du Kizilkum Central (Ouzbekistan). *L'Anthropologie, 108*, 55–67.

Schafer, J., Ranov, V., & Sosin, P. (1998). The cultural evoution of man and the chronostratigraphical background of changing environments in the loess palaeosoil sequences of Obi-Mazar and Khonako (Tadzjikistan). *Anthropologie, 36*(1), 121–135.

Schick, K. (1994). The Movius Line reconsidered. In R. S. Corruccini & R. L. Ciochon (Eds.), *Integrative paths to the past* (pp. 569–596). Englewood Cliffs, NJ: Prentice Hall.

Shunkov, M. V. (2005). The characteristics of the Altai (Russia) Middle Paleolithic in regional context. *Indo-Pacific Prehistory Association Bulletin, 25(Taipei papers, vol. 3)*, 69–77.

Sladek, V., Trinkaus, E., Hillson, S., & Holliday, T. W. (2000). *The People of the Pavlovian: Skeletal Catalogue and Osteometrics of the Gravettian Fossil Hominids from Dolni Vestonice and Pavlov*. Academy of Sciences of the Czech Republic. Institute of Archaeology, Brno.

Smith, F. H. (1981). Upper Pleistocene hominid evolution in South-Central Europe: a review of the evidence and trends. *Current Anthropology, 23*, 667–703.

Smith, F. H., Falsetti, A. B., Donnelly, S. M. (1989a). Modern human origins. *Yearbook of Physical Anthropology, 32*, 35–68.

Smith, F. H., Simek, J. F., Harrill, M. S. (1989b). Geographic variation in supraorbital torus reduction during the later Pleistocene (c. 80,000–15,000 BP). In P. Mellars, C. Stringer (Eds.), *The human revolution* (pp.172–193). New Jersey: Princeton University Press.

Sohn, S., & Wolpoff, M. (1993). Zuttiyeh face: a view from the East. *Amercian Journal of Physical Anthropology, 91*, 325–347.

Sotnikova, M. V., Dodonov, A. E., & Penkov, A. V. (1997). Upper Cenozoic bio-magnetic stratigraphy of Central Asian mammalian localities. *Palaeogeography, Palaeoclimatology, Palaeoecology, 133*, 243–258.

Stringer, C. B., & Gamble, C. (1993). *In search of the Neanderthals*. New York: Thames & Hudson.

Suleimanov, R. (1972). *Statisticheskoe Izuchenie Kultury Grotta Obi-Rakhmat*. Fan,Tashkent.

Trinkaus, E. (1983). *The Shanidar Neanderthals*. Academic Press, New York.

Trinkaus, E. (2005). Early modern humans. *Annual Review of Anthropology, 34*, 207–230.

Trinkaus, E. (2006). Modern human vs Neandertal evolutionary distinctiveness. *Current Anthropology, 47*(4), 597–620.

Trinkaus, E., Ranov, V. A., & Laukhin, S. (2000). Middle Paleolithic human deciduous incisor from Khudji, Tajikistan. *Journal of Human Evolution, 38*, 575–583.

Vangengeim, E., & Pevzner, M. (1991). Villafranchian of the USSR: bio and magnetostratigraphy. In E. Vangengeim (Ed.), *Paleogeography and biostratigraphy of the Pliocene and Anthrogene* (pp. 124–145). Moscow: USSR Academy of Sciences, Geological Institute (in Russian).

Vandermeersch, B. (1981). *Les Hommes Fossiles de Qafzeh (Israel)*. Éditions du CNRS, Paris.

Velichko, A., & Nechaev, V. (2005). *Cenozoic climatic and environmental changes in Russia. Special paper 382.* Boulder Colorado: The Geological Society of America.

Viola, B., Seidler, H., zur Nedden, D. (2004). Computer tomographic investigations of the OR-1 petrosals. In A.P. Derevianko (Ed.), *Grot Obi-Rakhmat* (pp. 100–105). Institute of Archaeology and Ethnography, Siberian Branch, Russian Academy of Sciences, Novosibirsk.

Vishnyatsky, L. (1999). The Paleolithic of Central Asia. *Journal of World Prehistory, 13*, 69–122.

Wolpoff, M. H. (1999). *Paleoanthropology*. Boston: McGraw-Hill.

Wolpoff, M. H., Hawks, J., & Caspari, R. (2000). Multiregional, not multiple origins. *American Journal of Physical Anthropology, 112*, 129–136.

Wu, X., & Poirier, F. (1995). *Human evolution in China: a metric description of the fossils and a review of the sites*. New York: Oxford University Press.

Wu, L., Zhang, Z., & Wu, X. (2005). Middle Pleistocene human crania from Tangshan (Nanjing), southeast China: a new reconstruction and comparisons with *Homo erectus* from Eurasia and Africa. *AJPA, 127*, 253–262.

Zhu, R., Potts, R., Xie, F., Hoffman, K. A., Deng, C. L., Shi, C. D., Pan, Y. Z., Wang, Y. C., Shi, G. H., & Wu, N. Q. (2004). New Evidence on the earliest human presence at high northern latitudes in northeast Asia. *Nature, 431*, 559–562.

Chapter 9
Core-and-Flake Assemblages of Central and Peninsular India

Parth R. Chauhan

Abstract As in most regions of the Old World, Lower Paleolithic assemblages in the Indian subcontinent have been traditionally divided into biface (Acheulean) or non-biface (core-and-flake) categories. Non-biface or Mode 1 assemblages have been reported from throughout the region, including the sub-Himalayan region or the Siwalik Hills as well as from various other eco-zones of India. This evidence has important implications for the understanding of diverse ecological and functional adaptations by South Asian hominin populations throughout the Pleistocene. This record comprises stratified and surface occurrences that include cores, core-tools, flakes and flake-tools, irregular bifacial knapping but a conspicuous absence of classic Acheulean bifaces and early Middle Paleolithic (diminutive) handaxes. Virtually all of these non-Acheulean assemblages currently remain undated, although a few occurrences in the Narmada Basin may date to the Mid-Pleistocene. From general geological observations and comparative typology, however, most appear to be comparatively younger in relative age and often mixed with different technological industries in surface contexts. This paper provides a broad regional perspective of these core-and-flake assemblages known from north-central and peninsular India.

Keywords India • Paleolithic • Mode 1 • Mode 2 • Core-and-flake assemblages

Introduction

Mode 1 assemblages[1] are generally represented by simple core-and-flake industries and form the earliest archaeological evidence in the Paleolithic record (Schick and Toth 2006). This conservative lithic technology is present throughout prehistory as a singular technology as well as in combination with other lithic traditions on a global scale, highlighting its functional efficiency. The initial emergence of Mode 1 assemblages have been acknowledged by some (e.g., Semaw et al. 1997) as an evolutionary phenomenon that highlights the cognitive capabilities of early hominins. However, the cultural capacities of later hominins (various species of *Homo*) primarily producing core-and-flake assemblages (e.g., Zhoukoudian, Atapuerca, Middle Awash) has been seen by some as very close to that of modern-primates (Foley and Lahr 2003), in spite of the ability to create standardized forms. While most reported African Mode 1 assemblages are generally Late Pliocene and Early Pleistocene in age (see Clark et al. 1994 for a young exception), similar assemblages in eastern Asia continue to appear well into the Late Pleistocene (Pope and Keates 1994). In Europe, they also occur in Middle Pleistocene contexts (Tieu 1991; Dennell and Roebroeks 1996; Roebroeks 2001) with some older exceptions (e.g, Carbonell et al. 1995; Oms et al. 2000). This implies that the dichotomy between Mode 1 and Mode 2 technologies may vary significantly from region to region. Possible explanations for this phenomenon include a method of technological dispersal or differential modes of technological development in these respective regions (e.g., Larick and Ciochon 1996; Carbonell et al. 1999; Burdukiewicz and Ronen 2003).

In Lower Paleolithic contexts, most exclusively Mode 1 assemblages are distinguished by the *absence* of Acheulean types (see Movius 1969) such as bifacially flaked handaxes and cleavers. Most core-and-flake assemblages consist of clasts or nodules of stone that are marginally flaked. The core forms from this industry are generally asymmetrical or amorphous in planform (particularly the heavy-duty implements).

[1]The utilization of the *Mode* system was first proposed by Clark (1968, 1977). These modes were based on the manner in which the basic flake-core relationship occurred and are considered to be reflective of raw material availability, functional differentiation, and manifestations of hominin technological strategies (Foley and Lahr 2003). Mode 1 is represented by pre- or non-Acheulian lithic technologies (Lower Paleolithic); Mode 2 includes Acheulian or biface technology (Lower Paleolithic); Mode 3 comprises Middle Paleolithic flake-based technologies; Mode 4 is represented by Upper Paleolithic blade technology; and Mode 5 is generally characterized by Mesolithic or Late Stone Age assemblages.

P. R. Chauhan (✉)
226 Centennial Avenue, Cranford, NJ 07016, USA
e-mail: parthrchauhan@gmail.com

C.J. Norton and D.R. Braun (eds.), *Asian Paleoanthropology: From Africa to China and Beyond*, Vertebrate Paleobiology and Paleoanthropology, DOI 10.1007/978-90-481-9094-2_9, © Springer Science+Business Media B.V. 2010

Variable degrees of retouch and differential modes of flake removal can be attributed to several factors: age of the assemblage, type or form of raw material or the intended function of the tools. Despite the amorphous nature of these assemblages in general, they still reflect certain foresight and mental template(s) (Pope and Keates 1994) and regional and typo-technological variability (Gao and Norton 2002).

Pleistocene hominin behavioral patterns and their adaptations to various paleo-landscapes of the Indian subcontinent are indirectly reflected through site and artifact densities, raw material exploitation and site locations in specific ecological contexts. South Asian Paleolithic sites are found in a variety of geological contexts and represent a diversity of typo-technological modes (Misra 1987). The Lower Paleolithic record of the region has traditionally been divided into Mode 1 (non-Acheulean) and Mode 2 (Acheulean) industries (Misra 1987; 1994; Petraglia 1998; Gaillard and Mishra 2001; Chauhan 2009a). Clark's Mode system has never been formally utilized by South Asian archaeologists. These industries often occur in overlapping geographic, geomorphologic, and stratigraphic contexts although this is not always the case (e.g., de Terra and Paterson 1939; see Sankalia 1974; Jayaswal 1982).

Traditionally, archaeologists in India have often relied on relative dating methods (e.g., terrace sequences, biochronology, typology, regional stratigraphy). In recent years, the application of absolute dating methods, primarily uranium-thorium, have become increasingly important for identifying the ages of Lower Paleolithic sites and site clusters (Mishra 1992, 1995; Agrawal and Yadava 1995; Badam and Sathe 1995; Singhvi et al. 1998; Kusumgar and Yadava 2002). The array of absolute dates has served to extend the lower age limit for the earliest Acheulean evidence in South Asia to the early Middle Pleistocene and possibly beyond. Since the beginning of Paleolithic studies in the subcontinent, most research has focused on the more easily-recognized Mode 2 assemblages. As a result, an accurate understanding of the Mode 1 evidence (regardless of age) is currently lacking. The chronological framework for Mode 2 assemblages from South Asia does not necessarily imply a similar framework for the core-and-flake assemblages in the region, although considerable temporal overlap may be evident in several parts of the subcontinent (Jayaswal 1978).

Currently, the only well-studied pre-Acheulean evidence in South Asia comes from northern Pakistan and includes the c. 2.0 Ma site at Riwat (Rendell et al. 1987, 1989) and the 2.2–0.9 Ma old Mode 1 assemblages from the nearby Pabbi Hills (Hurcombe 2004). However, these early lithic occurrences do not necessarily suggest a southern dispersal into peninsular India at that time and still require supporting evidence (Dennell 2003, 2007). Additionally there are reports of Late Pliocene and Early Pleistocene palaeoanthropological evidence from northern India (e.g., Varma 1975, 1991; Sharma 1977; Singh 2003) however these have yet to be substantiated (Chauhan 2009b). Recently, Paddayya et al. (2002) have reported an ESR age of c. 1.2 Ma for the Early Acheulean evidence from Isampur in the Hunsgi Basin in southern India. Unfortunately, this chronostratigraphic attribution remains tentative and requires more extensive study (Chauhan 2004), especially when considering the possibility of geological reworking at the site (A. Skinner, personal communication, 2006) and current problems with ESR on Indian faunal specimens in specific depositional environments (e.g., Blackwell et al. 2007; also see Chauhan 2010). These problems indicate that evidence for a pre-Acheulean or pre-Middle Pleistocene occupation of the Indian subcontinent is yet to be clearly established.

In addition to the Soanian pebble/cobble based lithic assemblages from the Siwalik Hills in the northern part of the Indian sub-continent (e.g. de Terra and Paterson 1939; Karir 1985; Corvinus 2002; Chauhan 2007a), comparable evidence has been reported from central and peninsular India to the south (e.g., Jayaswal 1982). With some possible exceptions (discussed below), current stratigraphic and geomorphic data suggest that most of the core-and-flake assemblages are either contemporary with or post-date Acheulean assemblages (Ghosh 1985; Pappu 2001; Corvinus 2002; Chauhan 2003, 2007a). These assemblages in peninsular India occur in variable topographic settings and are found in surface as well as in situ contexts. Although these assemblages remain undated, they have direct implications on our understanding of diverse Pleistocene ecological adaptations, technological continuity, tool functions and subsistence strategies within a regional Asian context. These assemblages were often assigned regional cultural names on the basis of differences in tool-type frequencies, tool-size, degrees of retouch, and so forth (Jayaswal 1982). However, considering often small data samples and single-site contexts, it is difficult to view some of these assemblages as representing individual cultural/industrial identities.

No reviews of core-and-flake assemblages from India have been published since the works of Jayaswal (1978, 1982) although numerous sites and lithic scatters have since been reported. Dividing India into broad arbitrary geographic zones, the purpose of this paper is to review the current state of archaeological research on this region's core-and-flake assemblages (Table 9.1 and Fig. 9.1).

Northern India

The Siwalik Hills or the Siwalik Foreland Basin consist of fluvial sediments deposited by Himalayan and sub-Himalayan rivers flowing south and southwestwards (Gill 1983) from the Lesser and Greater Himalayas, during a time when the region south of these mountains was a vast depression

Table 9.1 Key core-and-flake assemblages discussed in the text

	Site	Specimens	Context	Age	Reference
PAKISTAN	Riwat	< 31	In conglomerate horizon	ca. 2 Ma	Rendell et al. (1989)
	Pabbi Hills	607	Surface scatters	ca. 2.2 - 0.9 Ma	Hurcombe (2004)
INDIA					
Uttar Pradesh	Kalpi	65	Stratified in fluvial sediments	ca. 45 Ka	Tewari et al. (2002)
	Lahchura	345	Surface occurrences	n/a	Pant (1982)
	Bangawan	69	Surface occurrences	n/a	Pant (1982)
Madhya Pradesh	Durkadi	>650	In and above gravels	Mid-Pleistocene?	Armand (1983)
	Mahadeo-Piparia	>1215	In and above gravels	Mid-Pleistocene?	Khatri (1963); Supekar (1985)
	Murghakhera	n/a	In and above gravels	n/a	Khatri (1963); Supekar (1985)
Andhra Pradesh	Rushikonda	45	In conglomerate horizon	n/a	Reddy et al. (1995)
Goa (Malvan Area)	Kuperichi Ghati	33	Surface	n/a	Guzder (1980)
	Salel	24	Surface	n/a	Guzder (1980)
	Chowke Nullah	87	Surface	n/a	Guzder (1980)
	Haddi	65	Surface	n/a	Guzder (1980)
Karnataka	Nittur	31	In conglomerate horizon	n/a	Ansari (1970)
Meghalaya	Rongram	n/a	Within alluvium & colluvium	n/a	Sharma (2002)
	Nangwalbibra-A	27	Surface	n/a	Sharma and Roy (1985)
	Nangwalbibra-B	57	In conglomerate horizon	n/a	Sharma and Roy (1985)

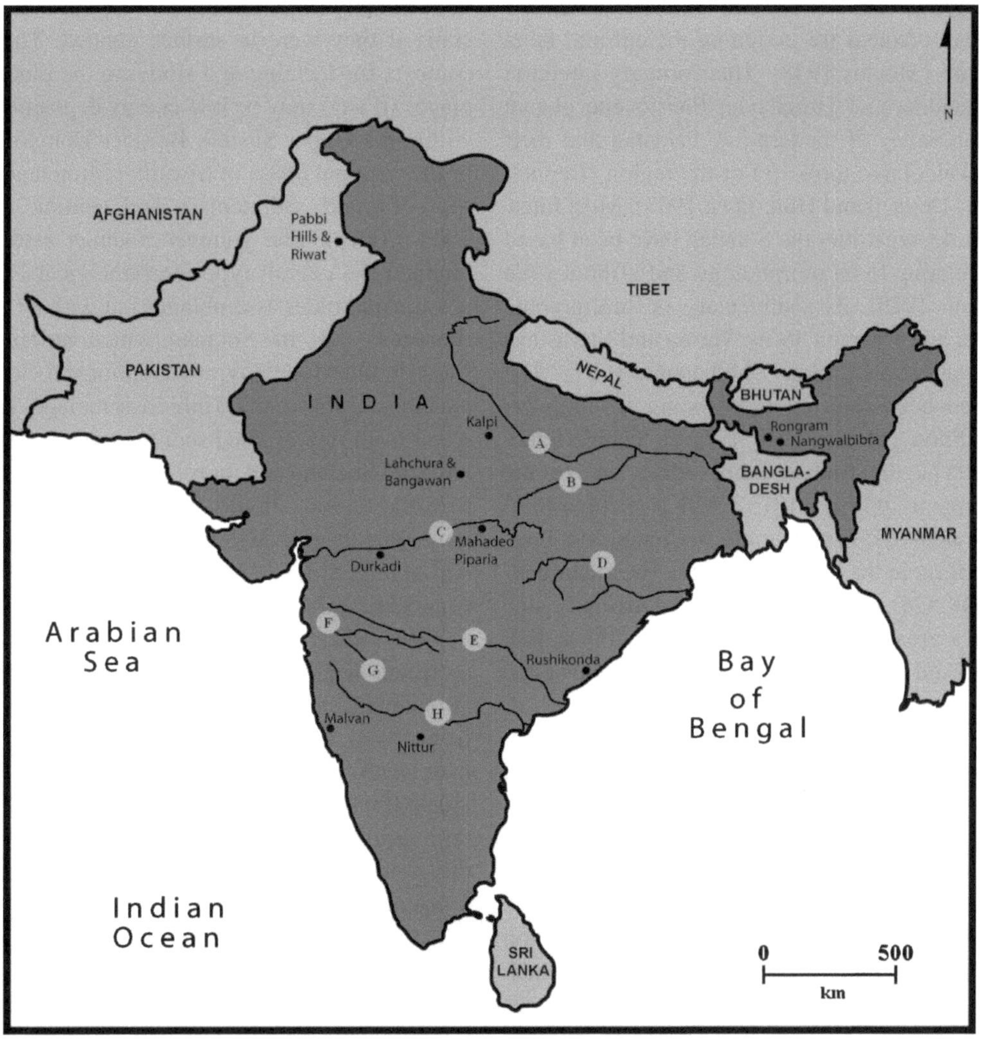

Fig. 9.1 Distribution of key core-and-flake assemblages in South Asia. Rivers: A: Ganges, B: Son, C: Narmada, D: Mahanadi, E: Godavari, F: Manjra, G: Ghod, H: Krishna

(often referred to as the foredeep) (Brozović and Burbank 2000). The Hills span from the western side of the Indus (northern Pakistan to the west) to the Bay of Bengal (Sikkim/Assam region to the east), covering a total length of approximately 2,400 km. The topography of the Siwalik Hills became a prominent feature on the landscape and reached its present elevation during the Middle Pleistocene (Kumar et al. 1994). The range is less than 13 km wide in places (average of 24 km), and it reaches an elevation between 900 and 1,200 m. Quartzite pebbles and cobbles was the primary raw-material exploited by the hominin occupants of this eco-zone at multiple temporal intervals (Chauhan 2007a,c; Dennell 2007). In addition to being located within the Boulder Conglomerate Formation of the Upper Siwalik Subgroup (Johnson et al. 1982), these localized quartzite clasts also occur in stream beds, on Siwalik land surfaces of varying ages, and in the terrace sections of intermontane valleys (Chauhan 2007c; Dennell 2007).

Paleolithic sites in the Siwaliks have been traditionally divided into two types: Acheulean and Soanian. De Terra and Paterson (1939) are credited for assigning the cultural label 'Soan' or 'Soanian' (Movius 1948). Their primary intention was to identify evidence of Himalayan Pleistocene glacial cycles in the Soan valley of modern-day Pakistan and their influence on the Paleolithic toolkit(s) of the region (Dennell and Rendell 1991; Dennell and Hurcombe 1993). Most interpretative frameworks regarding the Soanian have been based on assumptions relating to its morphology and affinities (de Terra and Paterson 1939), its sedimentary or stratigraphic contexts (de Terra and Paterson 1939; Verma and Srivastava 1984), and geographical locations (Mohapatra 1981). The Soanian has often been referred to as being a 'chopper-chopping tool tradition' (Movius 1948, 1957), a 'pebble-flake' tradition (Sen 1957), the 'Indian Abbevellian set' of the 'Pebble-Core element' (Ghosh 1975), and a 'pebble-tool' complex (Jayaswal 1982). For example, following de Terra and Paterson's efforts in the early part of the twentieth century, Movius (1948:376), regarded the Soan culture as "... one manifestation of a great complex of chopper-chopping-tool found in Southern and Eastern Asia." More recently however, Dennell (1995) states that the Soanian as a lithic industry or tradition, is too inadequately defined to use as a typological category. Some core-and-flake assemblages (though not labeled as Soanian) have also been reported further north from the Himalayas (e.g., Nautyal et al. 1982; Agrawal et al. 1991), signifying a possible adaptation to higher elevations and colder climates (Moran 1982).

Since the initial investigations by de Terra and Paterson (1939), researchers in India have broadly divided such assemblages on a regional basis depending upon the types of Soanian artifacts recovered (e.g., *Early Soan* in Beas Valley, *Late Soan* in Sirsa Valley) and identified Soanian sites based on the preponderance of choppers and absence of Acheulean bifaces (Lal 1956; Sen 1957; Karir 1985). However, the Soanian industry in general does not resemble conventional Early Paleolithic Mode 1 assemblages such as the Oldowan Industrial Complex (Leakey 1971; Schick and Toth 2006) or the Zhoukoudian industry (i.e., non-bifacial technology) (Zhang 1985). The majority of Soanian assemblages imply a Mode 3 affinity or a prominent Levallois component in both its flake and core specimens (Paterson and Drummond 1962; Gaillard and Mishra 2001; Corvinus 2002; Lycett 2007; Chauhan 2003, 2007a). Indeed, Gaillard (2006) has argued for the provisional separation of all Soanian assemblages as simply Early or Late categories, possibly equivalent to Lower and Middle Paleolithic distinctions respectively.

It is possible that the oldest Soanian assemblages may be at least from the Middle Pleistocene and thus, classifiable as an exclusively Mode 1 phenomenon prior to the development/dispersal of prepared-core or Levallois technology in the Siwalik Hills. Nevertheless, such older assemblages may be extremely difficult to assign to a particular industry especially if they were in surface context. The most promising contexts for locating and studying the older Soanian assemblages (if any) may be low-energy depositional environments within the Upper Siwalik Boulder Conglomerate Formation or the terminal phase of Siwalik sedimentation or prior to the 'post-Siwalik' sedimentary and tectonic regime (Chauhan 2003, 2007c). The younger Soanian assemblages possibly retained the overall typo-morphological composition of the presumably older assemblages but with an increased Mode 3 character – i.e., the Soanian, which variably comprises both Mode 1 and Mode 3 typo-technological elements at different locations. Until stratified lithic assemblages are radiometrically dated from fine-grained sedimentary contexts in the Siwalik Hills, all Soanian and morphologically similar evidence in the Siwalik region should be provisionally interpreted as a non-Acheulean post-Siwalik Paleolithic occupation in the region (e.g. Stiles 1978; Corvinus 2002; Soni and Soni 2005; Chauhan 2007a,c).

One recent example of a lithic assemblage typo-morphologically similar to the Soanian and south of the sub-Himalayan region is from Kalpi, a Middle Paleolithic site from the Ganga Valley of northern India dated to approximately 45 ka using thermoluminescence. Kalpi is located about 400 km south of the Himalayan foothills and probably represents the first absolute-date for such an assemblage in this zone of the subcontinent. Here, Tewari et al. (2002) report a prominent 'pebble tool' component (on quartzite), along with flakes, bone tools and vertebrate fossil remains. Virtually all of the 12 choppers range in size from 1 cm to 4.5 cm in length and about 2 cm in width. One explanation for their diminutive size may be a constraint in raw material size such as the virtual absence of larger clasts in the Ganga Basin in general. However, the unusually small size of the quartzite pebble-based tools warrants a study on their functional significance and even a re-consideration of their integrity as hominin-produced specimens. In contrast to the Middle Paleolithic classification of Kalpi, the contemporaneous assemblage from Site 55 in northern Pakistan (Rendell and Dennell 1987; Rendell et al. 1989) has been classified as 'initial Upper Paleolithic' or 'late Early Paleolithic' because of a prominent blade component in an Asian context (Dennell et al. 1992:27). Such a classification for this site, where blades were being systematically produced at c. 45 ka, suggests that the South Asian Paleolithic sequence is in need of a revised taxonomy (e.g., James and Petraglia 2005) as much of it is regionally distinct from its European, African and East Asian counterparts (see Misra 1962).

Central India

In the past, numerous prehistorians (e.g., Sankalia 1974; Jayaswal 1982) have implied that central India was the 'meeting ground' for the 'pebble tool' culture in the north (Soanian) and the 'handaxe culture' (Acheulean) from peninsular India. However, such observations have been largely abandoned since the geographic distribution of both types of assemblages is now known to overlap significantly. The central region of the subcontinent is dominated by the Narmada Valley from where several key Mode 1 occurrences are known. Paleomagnetic studies by Agrawal et al. (1988) suggest that the Narmada deposits fall within the Brunhes Chron (<0.73 Ma), while studies by the Geological Survey of India (Rao et al. 1997) suggest Early Pleistocene ages. However, their respective investigations focused on different stratigraphic sections and thus, the earliest Pleistocene contexts in the basin remain undated.

In this vast region, Jayaswal (1982) has separated all core-and-flake assemblages as those occurring exclusively without conventional Acheulean bifaces. One of these occurrences is the Mahadevian industry named after the site of Mahadeo-Piparia, reported by Khatri (1963, 1966, 1975) who suggested an Oldowan affiliation for this assemblage. Although there is a strong Mode 1 typo-morphological component at the site, later excavations and stratigraphic observations by Supekar (1968) refuted Khatri's claim that the Mahadevian industry was a predecessor to the Indian Acheulean. For example, bifacial flaking in the form of 'proto-handaxes' and handaxes have been reported at the site as well as Middle Paleolithic occurrences within the conglomerate (Supekar 1985). A similar claim as that of Khatri's was also made by Armand (1979) who defined the Durkadian industry at the site of Durkadi from excavated contexts 2 km south of the lower Narmada channel (Table 9.2).

Table 9.2 Fresh and rolled tool types from Durkadi (From Armand 1983)

Type	Fresh Specimens	Rolled Specimens
HEAVY DUTY TOOLS:	**48**	**28**
Choppers:	37	23
Discoide	10	2
Side	19	19
Angular (on angle)	4	1
End	4	0
Proto-cleavers	1	0
Proto-handaxes	6	0
Handaxes	1	0
Heavy duty complex tools	4	2
Heavy hollow side-scrapers	0	4
LIGHT DUTY TOOLS:	**63**	**35**
Scrapers:	46	28
Round	5	4
Double side	2	1
Single side	33	19
Hollow side	1	2
Single end	5	2
Borers:	6	2
Angular (on angle)	5	2
End	1	0
Burins	1	0
Light duty complex tools	10	5
Other types of heavy or light duty tools	8	18
TOTAL TOOLS:	**119**	**82**
Prepared cores:	22	11
On-pebble	17	4
On-flake	5	7
Unprepared cores:	20	6
On-pebble	16	3
On-flake	4	3
Other types of cores	9	7
TOTAL CORES:	51	24
Prepared -platform flakes:	59	83
Plain	50	67
Faceted	9	16
Unprepared-platform flakes	63	49
Vertical	50	46
Horizontal	13	3
Other types of unworked flakes:	9	27
TOTAL UNWORKED FLAKES:	**131**	**158**
Percuter	0	2
Anvil	0	1
WASTE:	**62**	**17**
TOTAL DEBITAGE:	**244**	**202**
Non-classified artifacts	0	3
TOTAL ARTIFACTS	**363**	**287**

At both Mahadeo-Piparia and Durkadi, a high frequency of unusually large non-biface artifacts, including cores, choppers, flakes and other formal tool types (Figs. 9.2 and 9.3) were recovered in stratified contexts. While Mahadeo-Piparia has yielded Acheulean bifaces, Durkadi continues to be enigmatic, despite Armand's (1983, 1985) report of 1

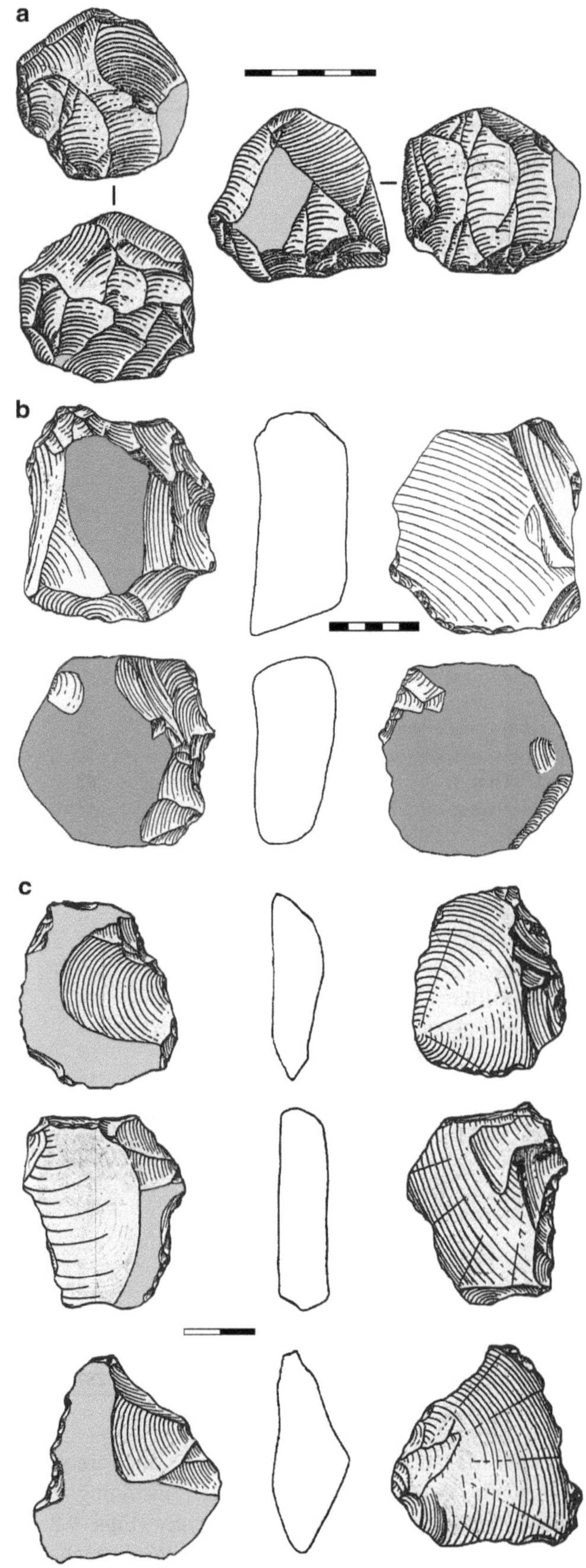

Fig. 9.2 (**a**) Polyhedron from Durkadi. Scale is in centimeters (Modified after Armand 1983). (**b**) Core and chopper from Durkadi. Scale is in centimeters (Modified after Armand 1983). (**c**) Flake specimens from Durkadi. Scale is in centimeters (Modified after Armand 1983)

'proto-cleaver', 6 'proto-handaxes' and 1 'Abbevillian' or evolved Durkadian handaxe. These eight specimens do not conform to the current typo-morphological definition of Acheulean bifaces as they lack bilateral and planform symmetry and adequate bifacial reduction (Fig. 9.4). They also do not appear to resemble typical early developmental stages of the Acheulean as known from, for example, Olduvai Gorge, Konso-Gardula, Peninj and 'Ubeidiya (see Clark 1998). In sum, neither Mahadeo-Piparia and Durkadi, nor any other site in the Indian subcontinent, shows any convincing stratigraphic evidence for a technological transition from an Oldowan-type into the more sophisticated Acheulean technology (Jayaswal 1982).

The Mahadeo-Piparia and Durkadi assemblages have been recovered from within and over-lying the high-energy gravels of the Narmada River and many artifacts at both sites are in relatively fresh condition. This signifies the use of the conglomerate surface through multiple visits for clast acquisition and stone tool production prior to the surface's burial by fine-grained sediments. This is a key geoarchaeological context at many Paleolithic sites in the subcontinent (Chauhan 2009b). From a comparison of the applied methodologies, associated interpretations, and published literature, the evidence from Durkadi appears to be much more archaeologically and stratigraphically robust than that from Mahadeo-Piparia. While Mahadeo-Piparia was assigned an early Middle Pleistocene age (Khatri 1963), Durkadi was interpreted to be about 1 Ma in age (Armand 1983). Considering the geological contexts of these sites, it is plausible that the preponderance of Mode 1 tool-types is linked to the predominant availability of size delimited river-worn clasts (i.e., virtually no boulders) over tabular material. Large tabular blocks are more suitable for detaching the flake blanks necessary for the production of bifaces, specimens which most core-and-flake sites are lacking.

Currently, the most convincing evidence for a possible Early Pleistocene archaeological occurrence in India comes from the central Narmada Basin. At Dhansi, Patnaik et al. (2009) have recently recovered in situ Paleolithic artifacts from a stratum below the Brunhes-Matuyama boundary (magnetostratigraphy reported by Rao et al. 1997). If these paleomagnetic observations prove to be correct, Dhansi may represent the first *unequivocal* evidence of an Early Pleistocene human presence in India and also extend the maximum age of the Narmada River. Though the lithic assemblage (a few cores and flakes) are currently un-diagnostic, a revision of the chronostratigraphic context of this site and additional excavations are currently underway. Close to the Narmada valley, core-and-flake assemblages have been again reported to occur stratigraphically below Acheulean horizons, implying the development of the Acheulean from non-biface tool-types (see Jayaswal 1978, 1982). This observation has been made during excavations of some Bhimbetka rock-shelters in the Raisen District of Madhya Pradesh. These sandstone and quartzite

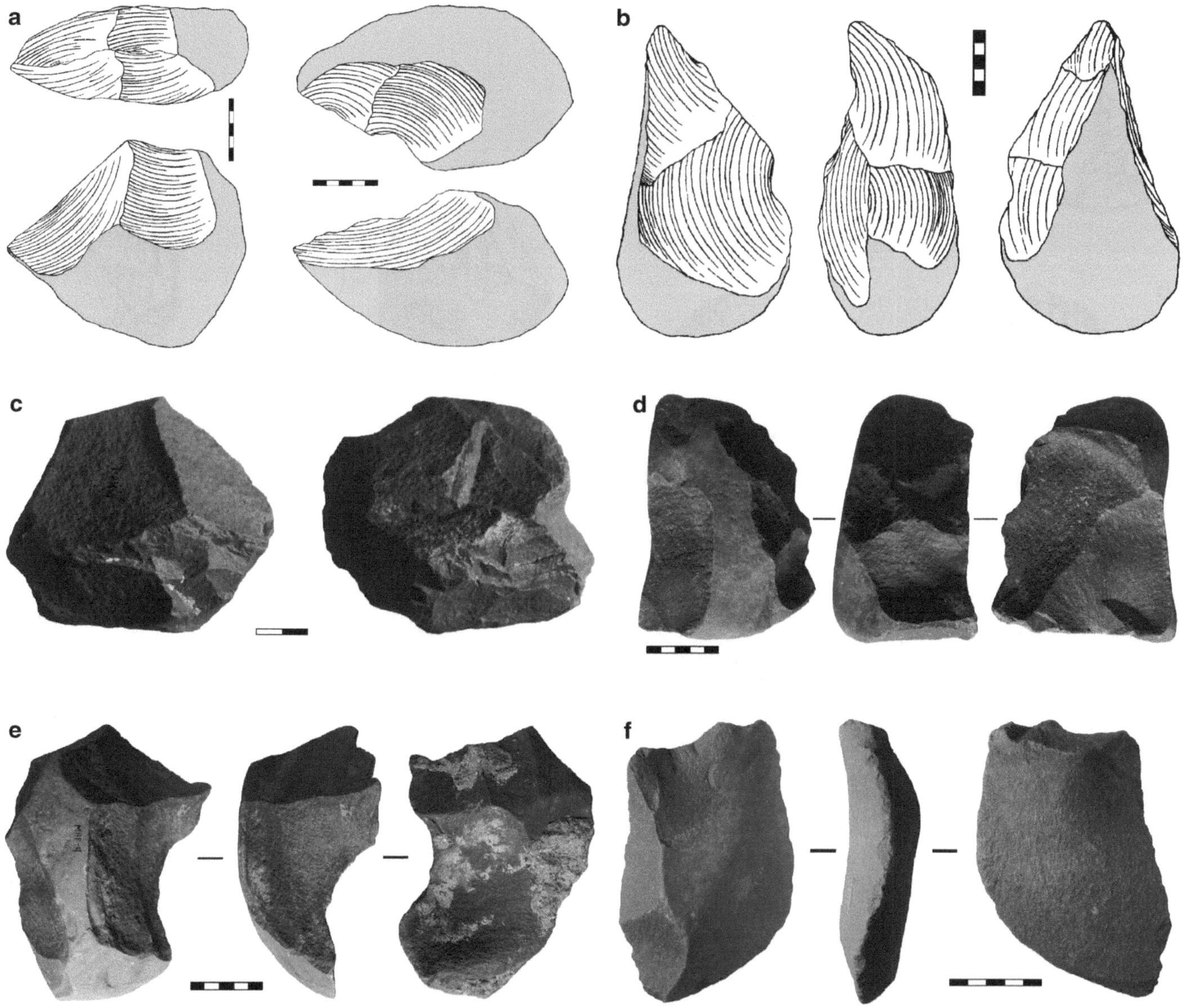

Fig. 9.3 (**a**) Cores/choppers from Mahadeo Piparia. Scale is in centimeters (modified after Khatri 1963). (**b**) Pointed core/chopper from Mahadeo Pipari. Scale is in centimeters (modified after Khatri 1963). (**c**) Polyhedron from Mahadeo Piparia. Scale is in centimeters. (**d**) Core/core-scraper from Mahadeo Piparia. Scale is in centimeters. (**e**) Core from Mahadeo Piparia. Scale is in centimeters. (**f**) Flake scraper from Mahadeo Piparia. Scale is in centimeters

hills, also in Madhya Pradesh, contain several hundred caves and rockshelters in association with perennial springs and seasonal streams. Excavations have been conducted at one cave (III-F-24) and two rock-shelters (III-A-29, III-F-23) by different researchers. Three trenches in the cave indicated a cultural sequence with assemblages assigned to the Lower and Middle Paleolithic. Although Wakanker (1973, 1985) recorded pebble tools below the Acheulean levels at Bhimbetka, Barkheda, Jawad, Rampura, Punjabpura, Agra, and Ramgad-Khediaghat, later excavations by Misra (1985) at Bhimbetka (Shelter III F-23) could not confirm this (Jayaswal 1978, 1982). Rather, the earliest cultural levels in the excavated rock-shelters are Late Acheulean and preclude an exclusive presence of Mode 1 assemblages in the Bhimbetka sequence.

In this central region, Lower Paleolithic assemblages dominated by choppers rather than bifaces have also been observed in the Chattisgarh area (the upper Mahanadi River) (see Joshi in Mohapatra 1985). From Uttar Pradesh, the main evidence comes from the Lahchura locality in Uttar Pradesh, on the banks of the river Dhasan. Here, over 64% of about 345 specimens were choppers, the rest being scrapers, flakes, cores, split pebbles, and debitage (Fig. 9.5 and Table 9.3) and is designated as the Lahchura industry from Lahchura-1 (Jayaswal 1982:59; see also Pant 1982). Another exclusively Mode 1 assemblage comes from the Betwa River in the same zone, and is designated as the Bangawan industry (Fig. 9.6 and Table 9.4).

Unfortunately, there are several problems with the contextual and behavioral interpretations of these occurrences.

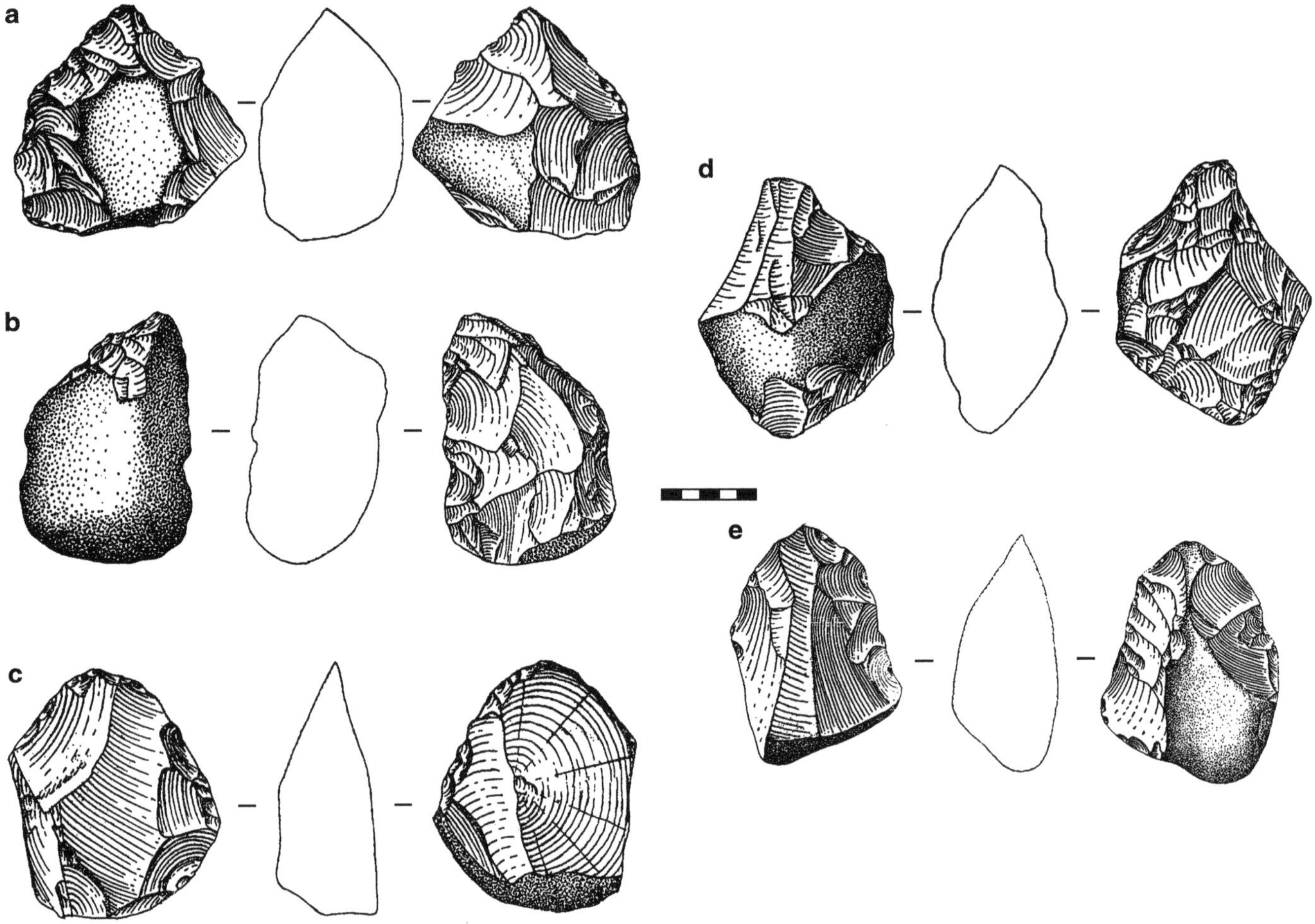

Fig. 9.4 Some specimens from Durkadi classified as 'proto-handaxes' (**a–d**) and 'Abbevillian' handaxe (**e**) by Armand (1983). Scale is in centimeters

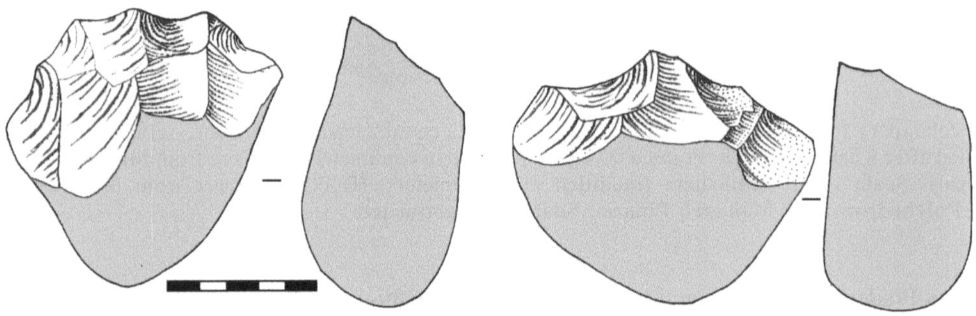

Fig. 9.5 Choppers from Lahchura-1. Scale is in centimeters (Modified after Pant 1982)

Table 9.3 The assemblage composition of Lahchura (From Jayaswal 1982)

Type	Number	%
Chopper	190	55.07
Chopping tools	31	8.98
Scraper	4	1.15
Flake	67	19.42
Core	35	10.14
Split Pebble	13	3.76
Chip	5	1.44
TOTAL:	345	99.96

As elsewhere in the subcontinent, there is a high amount of typological variability in the Paleolithic assemblages in this area. In addition, Pant (1982) classified and chronologically organized some of the material based on patination. However many of these sites are found in high-energy fluvial sediments (i.e., gravels) as well as surface occurrences on nearby hill slopes. At another locality, Lahchura-2, the investigator reports the surface occurrence of Mode 1 evidence with an increase in flakes and flake-tools including Levallois and blade types. Such interpretations appear to

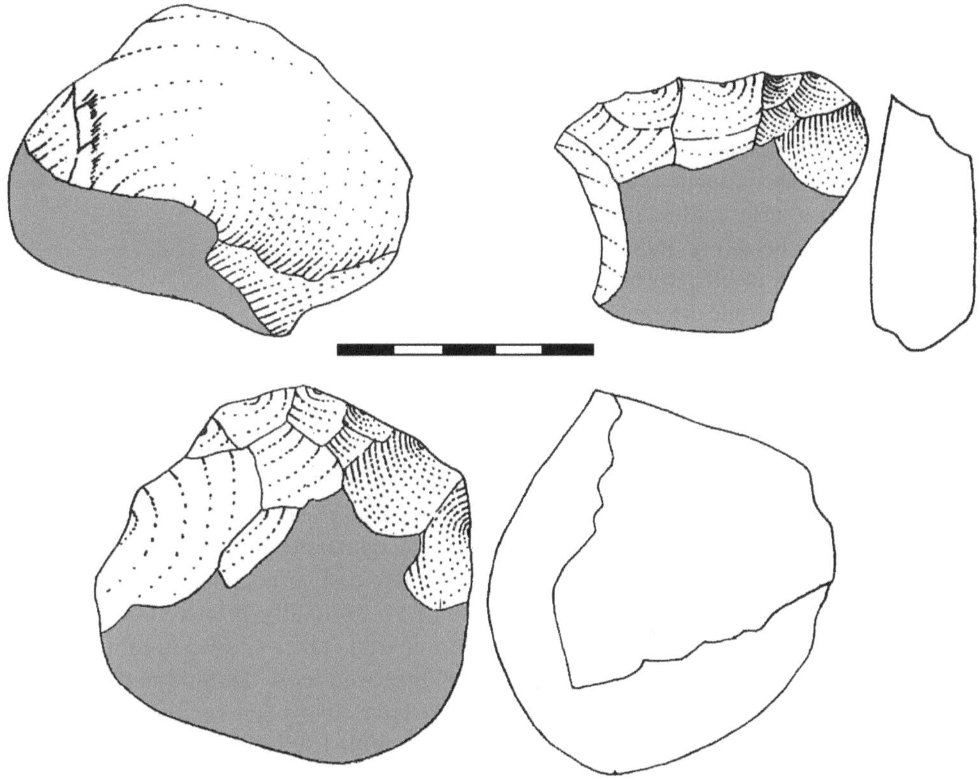

Fig. 9.6 Choppers from Bangawan. Scale is in centimeters (Modified after Pant 1982)

Table 9.4 The assemblage composition of Bangawan (From Jayaswal 1982)

Type	Number	%
Chopper	38	55.07
Chopping tools	9	13.04
Split Pebble	2	2.89
Flake	13	18.84
Blade	1	1.44
Core	6	8.69
TOTAL:	69	99.97

have erroneously relied on disturbed surface occurrences and secondary stratified occurrences (in fluvial gravels), thus warranting a critical re-evaluation and revision of this evidence in relation to mixed industries, assemblage variability, and the seemingly suspect age of these respective occurrences.

Western India

In the western region of India, the states of Maharashtra, Goa, and Gujarat have yielded core-and-flake assemblages produced on shale, dolerite, and quartzite. In Gujarat, D.B. Chitale has reported both unimarginal and bimarginal choppers (see Chauhan 2007a for chopper typology) near Unai on the Ambika River as well as the Dangs area (see Sali's comments in Mohapatra 1985). In Maharashtra, such evidence has been found near the Waiganga, Mula, and Krishna Rivers, but technological details are not available (Jayaswal 1982). In some localities such as Songaon, only single or a few choppers have been reported (Pappu 1974). In the Konkan region of southwestern Maharashtra, the landscape and climate is tropical and marked by a coastal ecology, where the Mode 1 evidence is particularly abundant and often in association with secondary laterite deposits (Guzder 1980; also see Sali's comments in Mohapatra 1985). Many artifacts from the Malvan area in Ratnagiri and Kolaba Districts were recovered from a pebble-bed and on terrace deposits of the Savitri and Kajvi Rivers. The assemblages include choppers, split-pebbles, convex-edged and denticulate specimens, unretouched flakes, proto-handaxes, and a possible unifacial handaxe. Both unmarginal and bimarginal choppers (Chauhan 2007a) are found in the area, and most artifacts are referred to as non-standardized tool-types. Guzder (1980) mentions the recovery of mostly fresh artifacts from laterite plateaus as well as rolled specimens in streambeds, 15 km inland from the coast. In addition to these presumably Lower Paleolithic assemblages, a large number of Middle Paleolithic and Mesolithic assemblages were also recovered from the same general region. Guzder distinguishes the pebble-tool evidence from coastal Maharashtra as forming a "more or less distinct group." and "On the other hand these chopper assemblages may be late variants of the

Lower Paleolithic, possibly contemporary with Middle Palaeolithic cultures in other parts of India and representing some form of relatively isolated development" (Guzder 1980:79). With rare exceptions (e.g., Marathe 2006), the Acheulean is not as prominent in western and southwestern India as in other parts of the subcontinent (Pappu 2001; Petraglia 2006). This observation also applies in a general sense for northeastern India, collectively indicating that Acheulean technology did not penetrate all parts of the Indian subcontinent. Existing Lower Paleolithic Mode 1 technology, by substituting the Acheulean, appears to have continued as a part of subsequent Middle and Upper Paleolithic industries in those regions.

Eastern India

In eastern India, early Quaternary deposits are represented by ferruginous conglomerates with occasional Paleolithic material. Reasons for the dearth of early archaeological evidence in east and northeast India, particularly Acheulean industries, may be due to marginal hominin presence during the Early and Middle Pleistocene. It is also possible that the high-relief terrain, thick-vegetation and heavy rainfall has hindered Paleolithic surveys. This region has yielded a number of non-biface assemblages, most of which may be culturally related to younger lithic industries from Southeast Asia such as the Hoabinian (Sharma 2002). Most of the core-and-flake assemblages in eastern India come from Bihar and West Bengal (Ghosh 1970), Orissa (Ota 1982–1983; Behera 1992), Meghalaya (Sharma and Roy 1985; Sharma 2002), and Manipur (Singh and Singh 1990; Singh 1997). In Meghalaya, Sharma and Roy (1985) recovered choppers, flakes, scrapers, points, and cores of dolerite and chert at the confluence of Simsang and Rongu rivers at Nangwalbibra in Meghalaya (also see Sharma 1991; Singh 1991). Using geological and stratigraphic data, Sharma (2002) demonstrates that several types of different assemblages appear to be contemporaneous at Rongaram, including flake-blade specimens, pebble short axes of the Hoabinhian tradition and bifaces. This work supports the rather young age of many core-and-flake assemblages in this part of India and possibly elsewhere on the subcontinent. An example of this is provided through an interesting pattern of chopper frequency from the Singhbhum region of Bihar. Ghosh (1970) notes the initial occurrence of choppers as components of the Acheulean industry in the region. This chopper component is virtually absent in the subsequent flake industry but reappears in the (youngest) flake-blade industry of the region, perhaps implying a functional need for heavy-duty tools at varying times. A similar example is found in the Nguomian flake industry of Vietnam where choppers re-appear in greater frequency in the upper layers of a cave habitation (Van Tan 1997). However, such observations need to be verified through proper excavations and stratigraphic control at buried sites rather than relying on interpretations of disturbed surface scatters.

Southern India

Most Paleolithic sites in Karnataka are situated along the Malaprabha and Ghataprabha Rivers. The best known core-and-flake locality in this zone comes from Nittur (Ansari 1970, 1985), where unimarginal and bimarginal choppers were recovered, with a few miscellaneous flakes. The 31 specimens were produced on basalt/dolerite and quartzite pebbles, and fossils of *Bos namadicus* were also recovered. The assemblage comprises choppers (65.2%), bifacially-prepared cores (19.2%), and massive unprepared flakes (7.6%) combined with the presence of the prepared core technique (Jayaswal 1978). Pebble-tools have also been recovered from Kerala (Rajendran 1977, 1998–1999) where the environment and terrain are similar to that of Goa or the Konkan region. This has been regarded as a possible seasonal phenomenon by Guzder (1980:79). She cites the work of Kleindienst (1961:44–46) who highlighted differences in Acheulean tool-type frequencies and their respective contexts at Olorgesailie and Isimila in Africa (see Paddayya 1982 for Acheulean example of possible seasonal adaptations in southern India). Unfortunately, not enough paleoenvironmental and geochronological data is available to propose such an explanation for the geographic differences between the Indian core-and-flake evidence and the Acheulean occurrences.

Another occurrence of a 'pebble tool assemblage' from southern India comes from the Vishakapatnam coast of Andhra Pradesh. At the site of Rushikonda (where Middle and Upper Paleolithic, Mesolithic, and Neolithic artifacts were also found) excavations were carried out on Terraces I and II which revealed artifacts at the interface between a "pebbly and bouldery gravel" (Reddy et al. 1995:113). From a collection of 45 artifacts (Table 9.5) made on medium-to-coarse-grained quartzite pebbles, the investigators highlight the uniqueness of the technology from other assemblages in the region. The pebbles are reported to have been split both vertically as well as horizontally and a broken anvil shows marks of battering. Owing to the absence of flake-based specimens, the investigators assign the assemblage to the Lower Paleolithic. Interestingly, almost all the reported flakes (n = 20) are split and they interpret illustrated choppers as 'pre-Acheulian' (Reddy et al. 1995:116). They propose this because the Mode 1 assemblage derives from a buried

Table 9.5 The artifact composition at Rushikonda on the Vishakapatnam Coast, Andhra Pradesh (From Reddy et al. 1995)

Artifact type	Number
Unifacial choppers	10
Worked split flakes	2
Used split flakes	4
Split pebbles	14
Split flakes	14
Anvil	1
Total	45

context while the Acheulean tool-types and younger technological industries were surface collected from a wave-cut rock terrace 0.5 km away. Because individual lithic assemblages can have different depositional and geomorphic histories even in one small region, this type of broad 'stratigraphic correlation' is highly problematic and unreliable. Given the current evidence, the Rushikonda core-and-flake assemblage is most probably very young in age and may not even be Lower Paleolithic. For example, some specimens appear to resemble casual cores or un-utilized choppers found in generally younger contexts in the subcontinent (see e.g., Reddy 1969; Sharma 2002; Tewari et al. 2002). In addition, a key problem is that several of the illustrated chopper specimens possess only two or less flake scars and may represent natural flaking. In light of the ambiguous nature of the assemblages and their context, the lithic occurrences at Rushikonda and others like it merit further investigations that emphasize rigorous geological and geochronological analyses.

Discussion

As mentioned earlier, pebbles and cobbles of various types of raw material (predominantly quartzite, sandstone, and basalt) are found as parts of fluvial/gravel deposits, paleochannel exposures, terrace sequences, and colluvial fan gravels. These raw materials are distributed throughout the Indian subcontinent and the frequencies and levels of clast sorting vary considerably, conditional upon the environment of deposition and associated post-depositional geomorphology. All types of Paleolithic sites are associated with this raw material type, often with a prominent heavy-duty tool component (e.g. choppers, core-scrapers, single-platform cores on split cobbles). Such sites are rarely found away from these raw material sources suggesting that pebbles, cobbles and associated tools were not transported over long distances (Chauhan 2007c, 2009c). With few exceptions such as the Malvan sites in Goa (Guzder 1980) and Lalitpur in the Bundelkhand region (see Jayaswal 1982), angular or tabular raw material has rarely been used for the manufacture of core-and-flake assemblages in South Asia – in contrast to the Oldowan where various types of clasts were exploited (e.g., Stiles 1998). Not surprisingly, the dimensional or morphological restriction of these rounded quartzites affected the assemblage composition of both Acheulean and core-and-flake assemblages. For example, the Soanian industry rarely contains polyhedrons or spheroids, primarily due to the limited lateral thickness of the often flat quartzite cobbles utilized. This suggests that many of these assemblages may reflect discrete raw material preference and selection by Pleistocene hominins in the region (e.g. Chauhan 2007a).

None of the aforementioned core-and-flake assemblages in peninsular India have been dated on an absolute scale and convincing stratigraphic associations of Pleistocene faunal and floral remains are unknown. With some exceptions, most of the reported evidence is generally represented by small sample sizes and thus are inadequate for developing a clear understanding of assemblage compositions and typo-technological affinities. Another major problem in interpreting the core-and-flake occurrences in India is that they are often found in surface contexts or in secondary gravel/conglomerate deposits. In addition, Paleolithic assemblage compositions have often been influenced by post-depositional site-formation mechanisms (Paddayya 1987; Petraglia 2002), a crucial factor which was not systematically investigated by most previous investigators. The effects of such geological processes on surface or buried sites may have resulted in the mixture of two or more lithic industries or technological traditions. Therefore, one needs to be highly cautious when interpreting the un-diagnostic core-and-flake assemblages from India as Lower Paleolithic or pre-Acheulean. Significantly more research is required to properly explain their variations (or similarities) throughout the subcontinent in time, space, technology and overall function.

The earliest Sri Lankan lithic evidence also comprises undated and ambiguous core-and-flake assemblages on quartz and quartzite which persist until the Holocene at some sites (Kennedy 2003). The lithic assemblages are not clearly diagnostic and most probably post-date the Late or terminal Acheulean evidence in India. In Nepal, in the Tui Valley, an industry of flakes and cores (bifaces are absent) was recovered from the basal alluvium of a quartzite cobble-boulder gravel, occurring below the stratified silts and clays of the Babai Formation at Brakhuti (Corvinus 2002). Similar specimens are found elsewhere in the Tui valley in high numbers, where the associated cobble-boulder gravel is exposed (above the bedrock and below the silt). The Brakhuti industry possibly represents a special requirement for heavy-duty tools in a forested habitat, implying considerable wood-work (Corvinus 1995). These assemblages from Nepal are of Late Pleistocene age (some may be even younger) and appear to share some typo-technological characters with similar industries from Southeast Asia (e.g. Van Tan 1997).

With the possible exceptions of Riwat, the Pabbi Hills, Durkadi and a *part* of Mahadeo-Piparia, the majority of South Asian core-and-flake assemblages, including a part of the Soanian industry, are probably not Lower Paleolithic and thus, have little relevance to a discussion of the Movius Line (Chauhan 2007a; Lycett 2007). In the Siwalik Hills region numerous Paleolithic investigations attest to a rich but discontinuous occupational history of both Acheulean and non-Acheulean traditions. Most of these non-Acheulean Siwalik assemblages have been classified as being a part of the *Soanian* tradition but do not appear to be related to the Late Pliocene-Early Pleistocene occurrences from northern Pakistan. Many occurrences within the Soanian industry often comprise a diagnostic Levallois component but there are also scatters in the Siwaliks that only contain non-diagnostic chopper, core, and flake specimens. Such ambiguous occurrences may be of variable age or belong to different Paleolithic traditions in the region. The conspicuous absence of pre-Acheulean or pre-Middle Pleistocene lithic assemblages in peninsular India may be due to a number of factors: either hominins did not occupy the region until the terminal Early Pleistocene or early Middle Pleistocene, or appropriate sediments are not well preserved/exposed (Chauhan 2006). An additional explanation is that possible Early Pleistocene evidence (if any exists) has been misinterpreted as being younger because it remains undated and/or because Oldowan or pre-Acheulean evidence is not expected in the South Asian archaeological record (Chauhan 2007b).

Similarly, the absence or low frequency of Mode 2 assemblages at specific locations, particularly in Kerala and northeastern India requires further discussion. Either existing Acheulean sites have not yet been discovered from these zones or perhaps simple core-and-flake technology was adequate for the exploitation of these areas (Schick 1994; Clark 1998; Petraglia 1998; Keates 2002; Norton et al. 2006; Lycett and von Cramon_Tabadel 2008). The presence and absence of Mode 2 technology at regional and sub-regional levels throughout the Old World, including Africa, probably reinforces this functional and behavioral trend. Another explanation is that, some if not most, Lower Paleolithic core-and-flake occurrences in India (and elsewhere in the Old World) may possibly represent Acheulean sites where bifaces were utilized (but *not* manufactured) at these locations and then transported elsewhere on the landscape. For example, the lack of biface roughouts, biface-thinning flakes and other associated debitage at the core-and-flake sites clearly eliminate them as Acheulean sites where bifaces are manufactured and then transported elsewhere. Whether or not there was a pre-Acheulean occupation in South Asia (evidence for which is equivocal), it also remains to be explained why and to what level non-biface toolkits continued to be produced once Acheulean technology appeared in the subcontinent. A comparative and systematic correlation of respective artifact types, associated raw materials and geomorphic contexts for both Acheulean and non-Acheulean occurrences in each region may clarify such issues.

Conclusions

Core-and-flake assemblages or Mode 1 technology in the Indian subcontinent are found in diverse geographical, ecological, and temporal contexts. Almost all of this evidence, from both the sub-Himalayan region and peninsular India, exhibits broad morphological similarities to other such assemblages in the Old World. They consist of the standard Mode 1 tool-types such as varieties of cores, discoids, choppers, core-scrapers, flakes, scrapers, notches, polyhedrons, sub-spheroids, unifaces, occasional *atypical* bifaces, debitage, and so forth and demonstrate a moderate diversity in knapping technique and tool-typology. The use of rounded quartzites reflects on hominin technological proficiency and associated cognitive levels which have major implications on their ability to reduce and shape unwieldy clasts or blanks in order to obtain suitable striking platforms. For example, to produce choppers, pebbles and cobbles were specifically selected with one flattish face, allowing easier primary-flake detachment. For the most part, artifact assemblages produced on pebbles-cobbles show minimal cortex removal. With the possible exception of Riwat and Pabbi Hills in northern Pakistan, there does not seem to be a diagnostic or *recognizable* pre-Acheulean Mode 1 industry in peninsular India. Most of the Mode 1 evidence appears to be restricted to Middle and Late Pleistocene stratigraphic or geomorphic contexts. Indeed, they may be very similar to Soanian assemblages where more than one technological mode is visible at different sites. The distribution, accessibility and morphology of rounded quartzite clasts appear to have been the principal factors in determining settlement location, inter and intra-regional mobility, and associated assemblage compositions and subsistence behaviors. Locations with a high density of quartzite clasts occasioned intensive exploitation of such sources, possibly represented by multiple visits for clast procurement. Core-and-flake assemblages that ultimately prove to be contemporary with but are spatially and ecologically separated from Acheulean assemblages in the subcontinent deserve a proper examination to explain this behavioral dichotomy.

Acknowledgements I would like to thank Christopher Norton, David Braun, and Jack Harris for the invitation to participate in the AAPA symposium in Philadelphia (2007), and for providing the opportunity to contribute to this volume. I am very grateful to Chris, David and Eric Delson for their patience and understanding during the preparation and revision of this manuscript. A part of the work discussed and related to the Soanian evidence and the Narmada Valley evidence was respectively funded by the Australian Research Council, the National Geographic Society, the Wenner-Gren Foundation for Anthropological Research and the Fulbright Scholar Program. Comments, suggestions and key insights by the editors and the three reviewers, including Shanti Pappu, helped improve and refine the manuscript significantly, including both specific statements and general observations. I am thankful to Nicholas Toth and Kathy Schick for their support of my research in India and for the postdoctoral research position at the Stone Age Institute & CRAFT Research Center, without which this research would not be possible.

References

Agrawal D. P., & Yadava, M. G. (1995). *Dating the human past*. Pune: Indian Society for Prehistoric and Quaternary Studies (Deccan College).

Agrawal, D. P., Kotlia, B. S., & Kusumgar, S. (1988). Chronology and significance of the Narmada Formations. *Proceedings of the Indian National Science Academy, 54*(3), 418–424.

Agrawal, D. P., Bhatt, D., Malayia, S., & Kharakwal, J. (1991). Archaeology of Kumaon: problems and prospects. *Man and Environment, 16*(1), 59–63.

Ansari, Z. D. (1970). Pebble tools from Nittur. *Indian Antiquary, IV*(1–4), 1–7.

Ansari, Z. D. (1985). Pebble tools from Nittur (Mysore State). In S. B. Deo & M. K. Dhavalikar (Eds.), *Studies in Indian Archaeology* (pp. 1–7). Bombay: Popular Prakashan.

Armand, J. (1979). The Middle Pleistocene pebble tool site of Durkadi in Central India. *Paleorient, 5*, 105–144.

Armand, J. (1983). *Archaeological Excavations in the Durkadi Nala-An Early Palaeolithic Pebble-Tool Workshop in Central India*. Delhi: Munshiram Manoharlal Publishers Pvt. Ltd.

Armand, J. (1985). The emergence of the handaxe tradition in Asia, with special reference to India. In V. N. Misra & P. Bellwood (Eds.), *Recent Advances in Indo-Pacific Prehistory* (pp. 3–8). New Delhi: Oxford/IBH.

Badam, G. L., & Sathe, V. (1995). Palaeontological research in India: Retrospect and prospect. In S. Wadia, R. Korisettar, & V. S. Kale (Eds.), *Quaternary Environments and Geoarchaeolgy of India* (pp. 473–495). Bangalore: Geological Society of India.

Behera, P. K. (1992). The pebble tool component of the Bonaigarh Neolithic Complex, Orissa. *Man and Environment, 17*(2), 57–63.

Blackwell, B. A. B., Montoya, A., Blickstein, J. I. B., Skinner, A. R., Pappu, S., Gunnell, Y., et al. (2007). ESR analyses for teeth from the open-air site at Attirampakkam, India: Clues to complex U uptake and paleoenvironmental change. *Radiation Measurements, 42*, 1243–1249.

Brozović, N., & Burbank, D. W. (2000). Dynamic fluvial systems and gravel progradation in the Himalayan foreland. *Geological Society of America Bulletin, 112*(3), 394–412.

Burdukiewicz, J. M., & Ronen, A. (Eds.) (2003). *Lower Palaeolithic small tools in Europe and the Levant. BAR international series 1115*. Oxford: Archaeopress.

Carbonell, E., Bermudez de Castro, J., Arsuaga, J., Diez, J., Rosas, A., Cuenca-Bescos, G., Sala, R., Mosquera, M., Rodriguez, X. (1995) Lower Pleistocene hominids and artifacts from Atapuerca-TD6 (Spain). *Science, 269*, 826–830.

Carbonell, E., Mosquera, M., Rodríguez, X. P., Sala, R., & van der Made, J. (1999). Out of Africa: The dispersal of the earliest technical systems reconsidered. *Journal of Anthropological Archaeology, 18*(2), 119–136.

Chauhan, P. R. (2003). An overview of the Siwalik Acheulian and reconsidering its chronological relationship with the Soanian – a theoretical perspective. Assemblage 7. http://www.shef.ac.uk/assem/issue7/chauhan.html.

Chauhan, P. R. (2004). A review of the Early Acheulian evidence from South Asia. Assemblage 8. (http://wwwshef.ac.uk/assem/issue8/chauhan.html)

Chauhan, P. R. (2006). Human origins studies in India: position, problems, and prospects. Assemblage 9. http://www.assemblage.group.shef.ac.uk/issue9/chauhan.html.

Chauhan, P. R. (2007a). Soanian cores and core-tools from Toka, Northern India: towards a new typo-technological organization. *Journal of Anthropological Archaeology, 26*, 412–441.

Chauhan, P. R. (2007b). The Indian subcontinent and 'Out of Africa I'. Proceedings of the Life Matters – Stony Brook Human Evolution Workshop Sept. 26–Oct. 1, 2005. J. Fleagle, J. Shea, R. Leakey (Eds.), *Paleobiology and Paleoanthropology series*. New York: Kluwer.

Chauhan, P. R. (2007c). Soanian lithic occurrences and raw material exploitation in the Siwalik Frontal zone, northern India: a geoarchaeological perspective. *Journal of Human Evolution, 54*, 591–614.

Chauhan, P. R. (2009a). The Lower Paleolithic of the Indian subcontinent. *Evolutionary Anthropology, 18*, 62–78.

Chauhan, P. R. (2009b). Was there an Oldowan occupation in the Indian Subcontinent?: A critical appraisal of the earliest paleoanthropological evidence. In K. Schick, N. Toth (Eds.), *The Cutting Edge: New Approaches to the Archaeology of Human Origins*. Stone Age Institute Press. pp. 35–54.

Chauhan, P. R. (2009c). Paleolithic exploitation of rounded and subangular quartzites in the Indian subcontinent. In B. Brookes, B. Adams (Eds.), *Lithic materials and Paleolithic societies* (pp. 97–115). Boston: Blackwell Publishing House.

Chauhan, P. R. (2010). Comment on 'Lower and Early Middle Pleistocene Acheulian in the Indian Sub-Continent' by Gaillard et al. 2009 (*Quaternary International*). *Quaternary International*. In press, available online: doi:10.1016/j.quaint.2010.01.021.

Clark, G. (1968). *World prehistory: a new outline*. Cambridge: Cambridge University Press.

Clark, G. (1977). *World Prehistory in New Perspective*. Cambridge: Cambridge University Press.

Clark, J. D. (1998). The Early Palaeolithic of the eastern region of the Old World in comparison to the West. In M. D. Petraglia & R. Korisettar (Eds.), *Early human behavior in global context: the rise and diversity of the Lower Palaeolithic record* (pp. 437–450). New York: Routledge.

Clark, J. D., de Heinzelin, J., Schick, K., Hart, W. K., White, T. D., WoldeGabriel, G., et al. (1994). African Homo erectus: old radiometric ages and young Oldowan assemblages in the Middle Awash Valley, Ethiopia. *Science, 264*, 1907–1909.

Corvinus, G. (1995). Quaternary stratigraphy of the intermontane Dun valleys of Dang-Deokhuri and associated prehistoric settlements in Western Nepal. In S. Wadia, R. Korisettar, V.S. Kale (Eds.), *Quaternary environments and geoarchaeology of India* (pp. 124–149). Bangalore: Geological Society of India.

Corvinus, G. (2002). Arjun 3, a Middle Palaeolithic Site, in the Deokhuri Valley, Western Nepal. *Man and Environment, 27*(2), 31–44.

de Terra, H., & Paterson, T.T. (1939). *Studies on the Ice Age in India and associated human cultures*. Washington D.C: Carnegie Institute Publication 493.

Dennell, R. W. (1995). The Early Stone Age of Pakistan: a methodological review. *Man and Environment, 20*(1), 21–28.

Dennell, R. W. (2003). Dispersal and colonisation, long and short chronologies: how continuous is the Early Pleistocene record for hominids outside East Africa? *Journal of Human Evolution, 45*(6), 421–440.

Dennell, R. (2007). "Resource-rich, stone-poor": early hominin land use in large river systems of northern India and Pakistan. In M.D. Petraglia, B. Allchin (Eds.), *The evolution and history of human populations in South Asia – interdisciplinary studies in archaeology, biological anthropology, linguistics and genetics. Vertebrate paleobiology and paleoanthropology series* (pp. 41–68). New York: Springer.

Dennell, R. W., & Rendell, H. M. (1991). De Terra and Paterson and the Soan Flake industry: a new perspective from the Soan Valley, Northern Pakistan. *Man and Environment, 16*(2), 90–99.

Dennell, R. W., & Hurcombe, L. (1993). Paterson, the British Clactonian and the Soan Flake industry: a re-evaluation of the Early Palaeolithic of Northern Pakistan. In C. Jarrige (Ed.), *South Asian archaeology* (pp. 69–72). Madison: Prehistory Press.

Dennell, R. W., & Roebroeks, W. (1996). The earliest colonization of Europe: the short chronology revisited. *Antiquity, 70*(269), 535–542.

Dennell, R. W., Rendell, H., Halim, M., & Moth, E. (1992). Site 55, Riwat: a 42, 000 yr.-bp. open-air Palaeolithic site from northern Pakistan. *Journal of Field Archaeology, 19*, 17–33.

Foley, R., & Lahr, M. (2003). On stony ground: lithic technology, human evolution, and the emergence of culture. *Evolutionary Anthropology, 12*, 109–122.

Gaillard, C. (2006). Les premiers peuplements d' Asie du Sud: vestiges culturels. *C.R. Palevol, 5*, 359–369.

Gaillard, C., & Mishra, S. (2001). The Lower Palaeolithic in South Asia. In under the direction of Sémah, F., Falgueres, C., Grimaud-Herve, D., Sémah, A., Origin of Settlements and Chronology of the Palaeolithic Cultures in Southeast Asia. Colloque International de la FONDATION SINGER POLIGNAC. Paris, June 3–5, 1998 (pp. 73–91). Paris: Semenanjung.

Gao, X., & Norton, C. J. (2002). A critique of the Chinese 'Middle Palaeolithic'. *Antiquity, 76*(292), 397–412.

Ghosh, A. K. (1970). The Paleolithic cultures of Singhbhum. *Transactions of the American Philosophical Society, 60*(1), 1–68.

Ghosh, A. K. (1975). Man evolving – in India. Sectional Presidential Address at the 62nd Science Congress. Indian Science Congress Association Calcutta. Part II, pp. 1–28.

Ghosh, A. K. (1985). An attempt to understand the Acheulian succession in India from spatial distributions. In V. N. Misra & P. Bellwood (Eds.), *Recent advances in Indo-Pacific prehistory* (pp. 29–34). New Delhi: Oxford/IBH.

Gill, G. S. (1983). Sedimentology of the Siwalik Group eposed between the rivers Ghaggar and Markanda-Northwestern Himalaya. *Publication of the Centre of Advanced Study in Geology, Panjab University, 13*, 274–312.

Guzder, S. (1980). *Quaternary environment and Stone Age cultures of the Konkan (Coastal Maharashtra, India)*. Pune: Deccan College Postgraduate and Research Institute.

Hurcombe, L. (2004). The stone artefacts from the Pabbi Hills. In R.W. Dennell (Ed.), *Early hominin landscapes in Northern Pakistan: investigations in the Pabbi Hills* (pp. 222–292). Oxford: British Archaeological Reports International Series 1265.

James, H., & Petraglia, M. D. (2005). Modern human origins and the evolution of behavior in the Later Pleistocene record of South Asia. *Current Anthropology, 46*, 3–27.

Jayaswal, V. (1978). *Palaeohistory of India - a study of the prepared core technique*. New Delhi: Agam Kala Prakashan.

Jayaswal, V. (1982). *Chopper-chopping component of Palaeolithic India*. Delhi: Agam Kala Prakashan.

Johnson, N. M., Opdyke, N. D., Johnson, N. D., Lindsay, E. H., & Tahirkheli, R. A. K. (1982). Magnetic polarity stratigraphy and ages of Siwalik Group rocks of the Potwar Plateau, Pakistan. *Palaeogeography, Palaeoclimatology, Palaeoecology, 37*, 17–42.

Karir, B. S. (1985). *Geomorphology and Stone Age culture of North-Western India*. New Delhi: Sundeep Prakashan.

Keates, S. G. (2002). The Movius Line: fact or fiction? *Bulletin of the Indo-Pacific Prehistory Association, 22*, 17–24.

Kennedy, K. A. R. (2003). *God–apes and fossil men: Palaeoanthropology of South Asia*. Ann Arbor: University of Michigan Press.

Khatri, A. P. (1963). Mahadevian: an Oldowan pebble culture of India. *Asian Perspectives, 6*, 186–197.

Khatri, A. P. (1966). Origin and evolution of hand-axe culture in the Narmada Valley (Central India). In D. Sen & A. K. Ghosh (Eds.), *Studies in prehistory (Robert Bruce Foote Memorial Volume)* (pp. 96–121). Calcutta: Firma K.L. Mukhopadhyay.

Khatri, A.P. (1975). Handaxe and chopper-chopping tool radiation: the ideas of two culture theory of Palaeolithic in Asia. Presented in the Pre-Congress symposium on Early Palaeolithic in South and East Asia. August 28–September 8, 1973, 3–14.

Kleindienst, M. R. (1961). Variability within the Late Acheulian assemblage in eastern Africa. *South African Archaeological Bulletin, 16*, 35–52.

Kumar, R., Ghosh, S. K., & Phadtare, N. R. (Eds.). (1994). *Siwalik Foreland Basin of Himalaya*. New Delhi: Oxford/IBH.

Kusumgar, S., & Yadava, M. G. (2002). Physical dating methods in South Asian archaeology: a brief review. In K. Paddayya (Ed.), *Recent Studies in India Archaeology* (pp. 403–453). New Delhi: Munshiram Manoharlal.

Lal, B. B. (1956). Palaeoliths from the Beas and Banganga Valleys, Panjab. *Ancient India, 12*, 58–92.

Larick, R., & Ciochon, R.L. (1996). *The African emergence and early Asian dispersals of the genus Homo. American Scientist (The Magazine of Sigma Xi, The Scientific Research Society)*. November–December.

Leakey, M. (1971). *Olduvai Gorge, Volume 3. Excavations in Beds I and II 1960–1963*. Cambridge: Cambridge University Press.

Lycett, S. (2007). Is the Soanian techno-complex a Mode 1 or Mode 3 phenomenon? A morphometric assessment. *Journal of Archaeological Science, 34*, 1434–1440.

Lycett, S.J., & von Cramon_Tabadel, N. (2008). Acheulean variability and hominin dispersals: a model-bound approach. *Journal of Archaeological Science, 35*(3), 553–562.

Marathe, A. (2006). Acheulian Cave at Susrondi, Konkan, Maharashtra. *Current Science, 90*(11), 1538–1544.

Mishra, S. (1992). The age of the Acheulian in India: new evidence. *Current Anthropology, 33*(3), 325–328.

Mishra, S. (1994). The South Asian Lower Palaeolithic. *Man and Environment, 19*(1–2), 57–72.

Mishra, S. (1995). Chronology of the Indian Stone Age: the impact of recent absolute and relative dating attempts. *Man and Environment, 20*(2), 11–16.

Misra, V. N. (1962). Problems of terminology in Indian prehistory. *The Eastern Anthropologist, 15*(2), 113–124.

Misra, V. N. (1985). The Acheulian succession at Bhimbetka, Central India. In V. N. Misra & P. Bellwood (Eds.), *Recent advances in Indo-Pacific prehistory* (pp. 35–47). New Delhi: Oxford/IBH.

Misra, V. N. (1987). Middle Pleistocene adaptations in India. In O. Soffer (Ed.), *The Pleistocene old world – regional perspectives* (pp. 99–119). New York: Plenum.

Mohapatra, G. C. (1981). Acheulian discoveries in the Siwalik frontal range. *Current Anthropology, 22*(4), 433–435.

Mohapatra, G. C. (1985). Cultural ecology of early man in India. In S. B. Deo & K. Paddayya (Eds.), *Recent advances in India Archaeology (Proceedings of the Seminar held in Poona in 1983* (pp. 23–38). Poona: Deccan College Post-Graduate and Research Institute.

Moran, E. (1982). *Human adaptability – an introduction to ecological anthropology*. Boulder: Westview Press.

Movius, H. L. (1948). The Lower Palaeolithic cultures of Southern and Eastern Asia. *Transactions of the American Philosophical Society, 38*(4), 329–420.

Movius, H. L. (1957). Pebble tool terminology in India and Pakistan. *Man in India, 37*(2), 149–156.

Movius, H. L. (1969). Lower Paleolithic archaeology in southern Asia and the Far East. Reprinted in W.W. Howells (Ed.), *Early man in the Far East, studies in physical anthropology No. 1* (pp. 17–82). New York: Humanities Press.

Nautyal, K. P., Khanduri, B. M., Nautiyal, V., & Rajput, D. L. (1982). A note on the early palaeolithics from the Alakananda valley, Garhwal Himalaya. *Man and Environment, 6*, 106–109.

Norton, C. J., Bae, K., Harris, J. W. K., & Lee, H. (2006). Middle Pleistocene handaxes from the Korean Peninsula. *Journal of Human Evolution, 51*(5), 527–536.

Oms, O., Paré, J. M., Martínez-Navarro, B., Agustíi, J., Toro, I., Martínez-Fernández, G., Turq, A. (2000). Early human occupation of Western Europe: Paleomagnetic dates for two Paleolithic sites in Spain. *Proceedings of the National Academy of Sciences of the United States of America, 97*(19), 10666–10667.

Ota, S. B. (1982–1983). New research designs in the prehistoric research of Orissa: with special reference to pebble industry of North Boudh-Khondmal District. *Manav*, (1), 166–180.

Paddayya, K. (1982). *The Acheulian culture of the Hunsgi Valley (Peninsular India) – a settlement system Perspective*. Pune: Deccan College Postgraduate and Research Institute.

Paddayya, K. (1987). The place of the study of site formation processes in prehistoric research in India. In D. T. Nash, M. D. Petraglia (Eds.), *Natural formation processes and the archaeological record. BAR International Series 352* (pp. 74–85). Oxford: Archaeopress.

Paddayya, K., Blackwell, B. A. B., Jhaldiyal, R., Petraglia, M. D., Fevrier, S., Chaderton, D. A., II, et al. (2002). Recent findings on the Acheulian of the Hunsgi and Baichbal valleys, Karnataka, with special reference to the Isampur excavation and its dating. *Current Science, 83*(5), 641–647.

Pant, P. C. (1982). *Prehistoric Uttar Pradesh*. New Delhi: Agam Kala Prakashan.

Pappu, R. S. (1974). *Pleistocene Studies in the Upper Krishna Basin*. Poona: Deccan College Postgraduate and Research Institute.

Pappu, R.S. (2001). *Acheulian culture in Peninsular India*. New Delhi: D.K. Printworld.

Paterson T. T., & Drummond, H. J. H. (1962). *Soan the Palaeolithic of Pakistan*. Karachi: Department of Archaeology (Government of Pakistan).

Patnaik, R., Chauhan, P. R., Rao, M. R., Blackwell, Skinner, A. R., Sahni, A., Chauhan M. S., Khan, H. S. (2009). New geochronological, palaeoclimatological and Palaeolithic data from the Narmada Valley hominin locality, central India. *Journal of Human Evolution, 56*, 114–133.

Petraglia, M. D. (1998). The Lower Palaeolithic of India and its bearing on the Asian record. In M. D. Petraglia & R. Korisettar (Eds.), *Early human behaviour in global context: the rise and diversity of the Lower Palaeolithic record* (pp. 343–390). Routledge: London.

Petraglia, M. D. (2002). Pursuing site formation research in India – a postscript. In S. Settar & R. Korisettar (Eds.), *Indian archaeology in retrospect: archaeology and historiography* (Vol. IV, pp. 217–241). New Delhi: Manohar.

Petraglia, M. D. (2006). The Indian Acheulean in global perspective. In N. Goren-Inbar & G. Sharon (Eds.), *Axe age: Acheulean tool-making from quarry to discard* (pp. 389–414). London: Equinox.

Pope, G. G., & Keates, S. G. (1994). The evolution of human cognition and cultural capacity – A view from the Far East. In R. S. Corruccini, R. L. Ciochon (Eds.), *Integrative paths to the past – Paleoanthropological advances in honor of F. Clark Howell* (pp. 531–567). Englewood Cliffs, NJ: Prentice Hall.

Rajendran, P. (1977). The prehistory of Kerala. *Journal of Kerala Studies, 4*(4), 497–502.

Rajendran, P. (1998–1999). Holocene cultures and their technological innovations in Kerala, South India. *Puratattva, 29*, 8–13.

Rao, K. V., Chakraborti, S., Rao, K. J., Ramani, M. S. V., Marathe, S. D., Borkar, B. T. (1997). Magnetostratigraphy of the Quaternary fluvial sediments and tephra of Narmada Valley, Central India. In S. G. Udhoji, M. P. Tiwari (Eds.), *Quaternary geology of the Narmada Valley – a multidisciplinary approach. Special publication No. 46 of the geological survey of India* (pp. 65–78). Calcutta: Geological Survey of India.

Reddy, K. T. (1969). Pebble tools from Cuddapah. *The Anthropologist, 16*(1–2), 47–56.

Reddy, K. T., Prakash, P. V., Rath, A., & Rao, Ch U B. (1995). A pebble tool assemblage on the Visakhapatnam Coast. *Man and Environment, 20*(1), 113–118.

Rendell, H. M., & Dennell, R. W. (1987). Thermoluminescence dating of an Upper Pleistocene site, Northern Pakistan. *Geoarchaeology, 2*(1), 63–67.

Rendell, H. M., Hailwood, E. A., & Dennell, R. W. (1987). Magnetic polarity stratigraphy of Upper Siwalik Sub-Group, Soan Valley, Pakistan: implications for early human occupation of Asia. *Earth and Planetary Science Letters, 85*, 488–496.

Rendell, H. M., Dennell, R. W., Halim, M. A. (1989). Pleistocene and Palaeolithic Investigations in the Soan Valley, Northern Pakistan. *British archaeological mission to Pakistan Series 2*. Oxford: British Archaeological Reports (International Series 544).

Roebroeks, W. (2001). Hominid behaviour and the earliest occupation of Europe: An exploration. *Journal of Human Evolution, 41*, 437–461.

Sankalia, H. D. (1974). *The Prehistory and Protohistory of India and Pakistan*. Pune: Deccan College and Postgraduate Research Institute.

Schick, K. D. (1994). The Movius Line reconsidered - Perspectives on the earlier Paleolithic of Eastern Asia. In R.S.Corruccini, R.L. Ciochon (Eds.), *Integrative paths to the past – Paleoanthropological advances in honor of F. Clark Howell* (pp. 569–596). Englewood Cliffs, NJ: Prentice Hall.

Schick, K., & Toth, N. (2006). An overview of the Oldowan Industrial Complex: the sites and the nature of their evidence. In N. Toth & K. Schick (Eds.), *The Oldowan: case studies into the Earliest Stone Age* (pp. 3–42). Gosport: Stone Age Institute Press.

Semaw, S., Renne, P., Harris, J. W. K., Feibel, C. S., Bernor, R. L., Fesseha, N., et al. (1997). 2.5-million-year-old stone tools from Gona, Ethiopia. *Nature, 385*, 333–336.

Sen, D. (1957). The Sohanian and the pebble-tool complex in India. *Man in India, 37*(2), 157–159.

Sharma, J. C. (1977). Palaeolithic tools from Pleistocene deposits in Panjab, India. *Current Anthropology, 18*(1), 94–95.

Sharma, T. C. (1991). Prehistoric situation in north-east India. In J. P. Singh & G. Sengupta (Eds.), *Archaeology of North-eastern India* (pp. 41–58). New Delhi: Har-Anand Publications in association with Vikas Publishing House.

Sharma, S. (2002). Geomorphological contexts of the Stone Age record of the Ganol and Rongram Valleys in the Garo Hills, Meghalaya. *Man and Environment, 27*(2), 15–29.

Sharma, H. C., & Roy, S. K. (1985). On the discovery of a pebble-tool industry in the Garo Hills, Meghalaya. In V. N. Misra & P. Bellwood (Eds.), *Recent Advances in Indo-Pacific Prehistory* (pp. 89–91). New Delhi: Oxford/IBH.

Singh, M. J. (1991). Pre-history of Manipur. In J. P. Singh & G. Sengupta (Eds.), *Archaeology of North-eastern India* (pp. 126–130). New Delhi: Har-Anand Publications in association with Vikas Publishing House.

Singh, O. K. (1997). *Stone Age Archaeology of Manipur*. Manipur: Amusana Institute of Antiquarian Studies.

Singh, M. P. (2003). First record of a Middle Pliocene hominid from the Siwalik Hills of South Asia. *Human Evolution, 18*, 213–228.

Singh, O. K., & Singh, R. K. R. (1990). A preliminary report of an archaeological exploration at Singtom, Chandel District, Manipur. *Man and Environment, 15*(1), 103–106.

Singhvi, A. K., Wagner, G. A., & Korisettar, R. (1998). Techniques for the chronometry of the Palaeolithic: evidence for global colonization. In M. D. Petraglia & R. Korisettar (Eds.), *Early human behavior in global context: the rise and diversity of the Lower Palaeolithic record* (pp. 23–83). New York: Routledge.

Soni, A. S., & Soni, V. S. (2005). Palaeolithic tools from the surface of optically stimulated luminescence dated alluvial fan deposits of Pinjaur Dun in NW sub-Himalayas. *Current Science, 88*(6), 867–871.

Stiles, D. (1978). Palaeolithic artifacts in Siwalik and post-Siwalik deposits. *Kroeber Anthropology Society Papers, 53–54*, 129–148.

Stiles, D. (1998). Raw material as evidence for human behavior in the Lower Pleistocene: the Olduvai case. In M. D. Petraglia & R. Korisettar (Eds.), *Early human behavior in global context – the rise and diversity of the Lower Palaeolithic record* (pp. 133–150). London: Routledge.

Supekar, S. N. (1968). Pleistocene Stratigraphy and Prehistoric Archaeology of the Central Narmada Basin. Ph.D. Dissertation, University of Poona.

Supekar, S. N. (1985). Some observations on the Quaternary stratigraphy of the Central Narmada Basin. In V. N. Misra & P. Bellwood (Eds.), *Recent Advances in Indo-Pacific Prehistory* (pp. 19–27). New Delhi: Oxford/IBH.

Tewari, R., Pant, P. C., Singh, I. B., Sharma, S., Sharma, M., Srivastava, P., et al. (2002). Middle Palaeolithic human activity and palaeoclimate at Kalpi in Yamuna Valley, Ganga Plain. *Man and Environment, 27*(2), 1–13.

Tieu, L. T. (1991). *Palaeolithic Pebble Industries in Europe*. Budapest: Akademiai Kiado.

Van Tan, H. (1997). The Hoabinhian and before. *Bulletin of the Indo-Pacific Prehistory Association (Chiang Mai Papers, Volume 3), 16*, 35–41.

Varma, R. K. (1975). A review of the Lower Palaeolithic Industries of India with special Reference to the Sohan Culture. In K.C. Chattopadhyaya Memorial Volume. Department of Ancient History, Culture and Archaeology (Allahabad University) (pp. 29–38). Allahabad: K.K. Dubey.

Verma, B. C. (1991). Siwalik Stone Age Culture. *Current Science, 61*(8), p. 496.

Verma, B. C., & Srivastava, J. P. (1984). Early Palaeolithic sites of the outer Siwalik belt in Sirmur District, Himachal Pradesh - their age and stratigraphic position. *Man and Environment, 8*, 13–19.

Wakanker, V. S. (1973). Bhim-betka excavations. *Journal of Indian History LI, 1*, 23–33.

Wakanker, V. S. (1985). Bhimbetka: the stone tool industries and rock paintings. In V. N. Misra & P. Bellwood (Eds.), *Recent advances in Indo-Pacific prehistory* (pp. 175–176). New Delhi: Oxford/IBH.

Zhang, S. S. (1985). The Early Palaeolithic of North China. In R. K. Wu & J. W. Olsen (Eds.), *Palaeoanthropology and Palaeolithic Archaeology in the People's Republic of China* (pp. 147–186). Orlando: Academic.

Chapter 10
South Asia as a Geographic Crossroad: Patterns and Predictions of Hominin Morphology in Pleistocene India

Sheela Athreya

Abstract Current models of Asian Pleistocene hominin evolution are based on finds that represent a small portion of the geographic expanse of that continent. This is due to the sparse nature of the fossil record as well as varying traditions of paleoanthropological inquiry in these countries. As a result, models of hominin evolution that have emerged for Asia as a whole are not necessarily appropriate characterizations for individual regional populations. The Indian subcontinent in particular is expected to be distinct from China and Indonesia, based in part on genetic studies that suggest this was one of the first regions to be occupied by dispersing African populations. This study evaluates patterns of hominin morphology through a morphometric comparative analysis of the Hathnora specimen found along the Narmada River, and the only fairly complete Pleistocene fossil from South Asia. This fossil is compared to a sample of Old World Pleistocene specimens as well as Early Holocene South Asians and Andaman Islanders to elucidate patterns of variation both across space and through time. The results indicate that the Narmada hominin from Hathnora exhibits a mix of Southeast Asian and sub-Saharan African features. These results are consistent with India's location between these two regions. A unique evolutionary model for Pleistocene *Homo* in India is proposed, incorporating the fact that this region was at a geographic crossroads, and predicting that the morphology found there will not conform to patterns found in other parts of Asia. Rather, morphological evolutionary trajectories for Pleistocene South Asian hominins are expected to reflect a blend of African and Asian traits.

Keywords Narmada • *Homo heidelbergensis* • Middle Pleistocene • *Homo erectus* • Multidirectional dispersals

S. Athreya (✉)
Department of Anthropology, Texas A&M University, 4352 TAMU, College Station, TX 77843
e-mail: athreya@tamu.edu

Introduction

It is a well-known fact in paleoanthropology that very few hominin fossil remains are known from India dating to the Early and Middle Pleistocene. As such, paleoanthropologists have a difficult time making inferences about the earliest hominin migrations and lifeways into this region based on the fossil evidence. The oldest known specimen from South Asia is the Narmada fossil from the Hathnora locality in central India (Sonakia 1984). It has been variably described as *Homo erectus* (de Lumley and Sonakia 1985) and "archaic" *Homo sapiens* (Wolpoff 1999; Cameron et al. 2004) but with the increasingly frequent use of the taxon *Homo heidelbergensis* over the past decade, it is relevant to evaluate Narmada in comparison to other specimens allocated to this species.

This paper is based on a previous study (Athreya 2007) of the morphology of the Narmada cranium that examined its most appropriate taxonomic allocation, taking into consideration historical trends in studies of the phylogeny of Middle Pleistocene *Homo* over the past decade. As an extension of that analysis, a new framework is proposed here for predicting and interpreting Pleistocene hominin morphology in South Asia. Evolutionary models for this region have historically been determined by those developed for East and/or Southeast Asia. Rather than group it with the broader Asian continent, the scenario proposed here takes into consideration India's unique geographic location as a subcontinent between Africa and Southeast Asia.

Background

The Narmada specimen (Fig. 10.1) was discovered in the central Indian state of Madhya Pradesh in 1982 (Sonakia 1984) (Fig. 10.2). It was found at the site of Hathnora, eroding out of a boulder conglomerate now known as the Surajkund formation (Tiwari and Bhai 1997) along the Narmada River, a major fluvial system that traverses north-central India. The fossil was originally assigned to the taxon *Homo erectus* by de Lumley and Sonakia (1985) based on a

C.J. Norton and D.R. Braun (eds.), *Asian Paleoanthropology: From Africa to China and Beyond*, Vertebrate Paleobiology and Paleoanthropology, DOI 10.1007/978-90-481-9094-2_10,

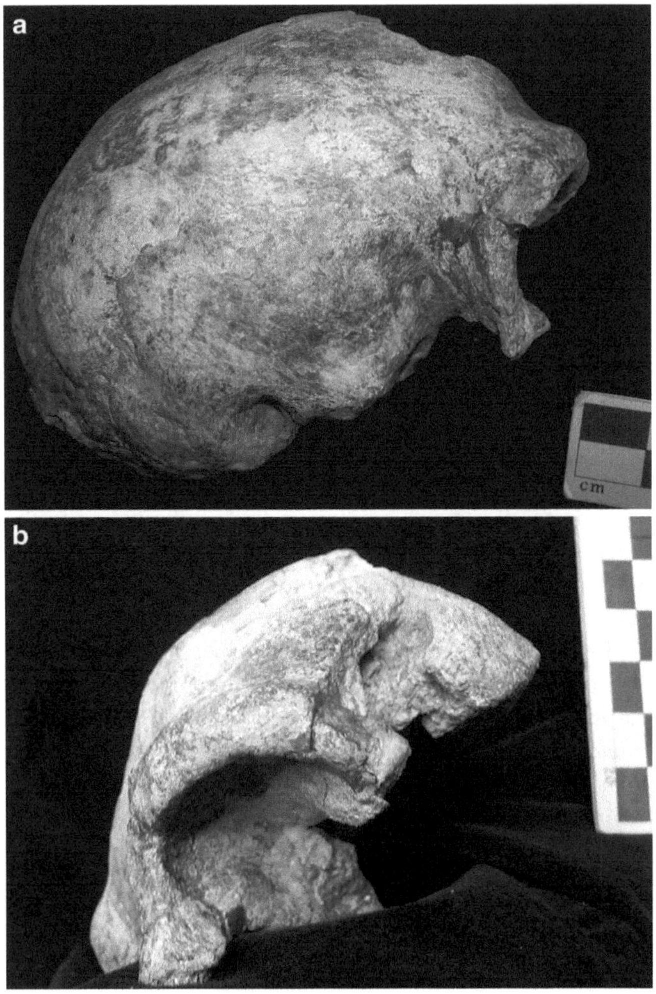

Fig. 10.1 Lateral and frontal views of Narmada fossil

morphometric analysis of continuous and categorical cranial traits. Kennedy and others (Kennedy and Chiment 1991; Kennedy et al. 1991) later performed a detailed multivariate morphological analysis and compared Narmada to a broad sample of hominins from the Early, Middle and Late Pleistocene. They demonstrated that it has certain morphological features that align it with *Homo erectus* such as a small mastoid process and well-developed supramastoid crest. It also has a prominent browridge and thick cranial bones, among the thickest when compared to a *H. erectus* sample with a value of 11.1 mm at bregma. But in other traits, they noted that Narmada has more modern morphology. For example, the height of the cranium at 138 mm is outside the range of *Homo erectus* and within that of *Homo sapiens*. In the occipital region, opisthocranion and inion are not coincidental as they are on African and Asian *Homo erectus*. Also, the estimated cranial capacity of the specimen is within the modern human range, with calculations ranging from 1,155 to 1,421 cc, with the average being 1,288 cc.

Based on this, Kennedy et al. (1991) declined to assign Narmada to a specific taxon but rather noted the inadequacy of the existing systematics of Middle Pleistocene *Homo* to accommodate its apparently transitional morphology. At the time of their study, the term *Homo heidelbergensis* was not widely used to describe such specimens. Rather it was more common to use terms such as "archaic" *Homo sapiens* or "advanced" *Homo erectus*. Objections to the utility and clarity of the available taxonomic categories at the time prompted these authors to settle on a morphological description of Narmada without a definitive allocation.

In conjunction with this taxonomic uncertainty has been an ongoing question as to the age of the specimen. The Surajkund formation in which the Narmada hominin was found is heavily reworked and contains temporally mixed faunal remains (Patnaik et al. 2009). The authors of this study propose a minimum age of the fossil at approximately 48,000 years ago (ka), with a more likely range placing it between 93 ± 5 to >236 ka. A uranium series date of 236 ka was obtained for a bovid scapular fragment found in association with the original cranial specimen (Cameron et al. 2004) but because of the mixed nature of the Surajkund formation, it is not possible to ascertain the contemporaneity of this fossil with the Narmada calvarium. So while the preponderance of evidence indicates that Narmada is most likely late-Middle Pleistocene or perhaps early-Late Pleistocene, it has been difficult to establish a definitive geological age.

Partly because of its insecure dating and partly because of its mixed morphology, there has been an ongoing debate as to what taxonomic group Narmada should be assigned. Previous studies have sought to determine whether it was *H. erectus* or an "archaic" *H. sapiens* (Sonakia 1985a, b; Kennedy and Chiment 1991; Kennedy et al. 1991). Now the resurgent use of the taxon *H. heidelbergensis*, which is commonly used to refer to fossils intermediary between these two groups, reconfigures the question of Narmada's taxonomic status because it too exhibits morphology that is intermediate between *H. erectus* and *H. sapiens*.

Middle Pleistocene Hominin Morphology in South Asia

In a previous study (Athreya 2007) the morphometric affinities of Narmada relative to a sample of other Pleistocene specimens were examined to determine if it was an Indian variant of *H. heidelbergensis*. The purpose of that study was simply to determine the most appropriate taxonomic position of this fossil. Building on that work, the analysis presented here reviews the implications of that previous study in order to provide a new model for predicting and interpreting Pleistocene hominin fossil morphology in South Asia.

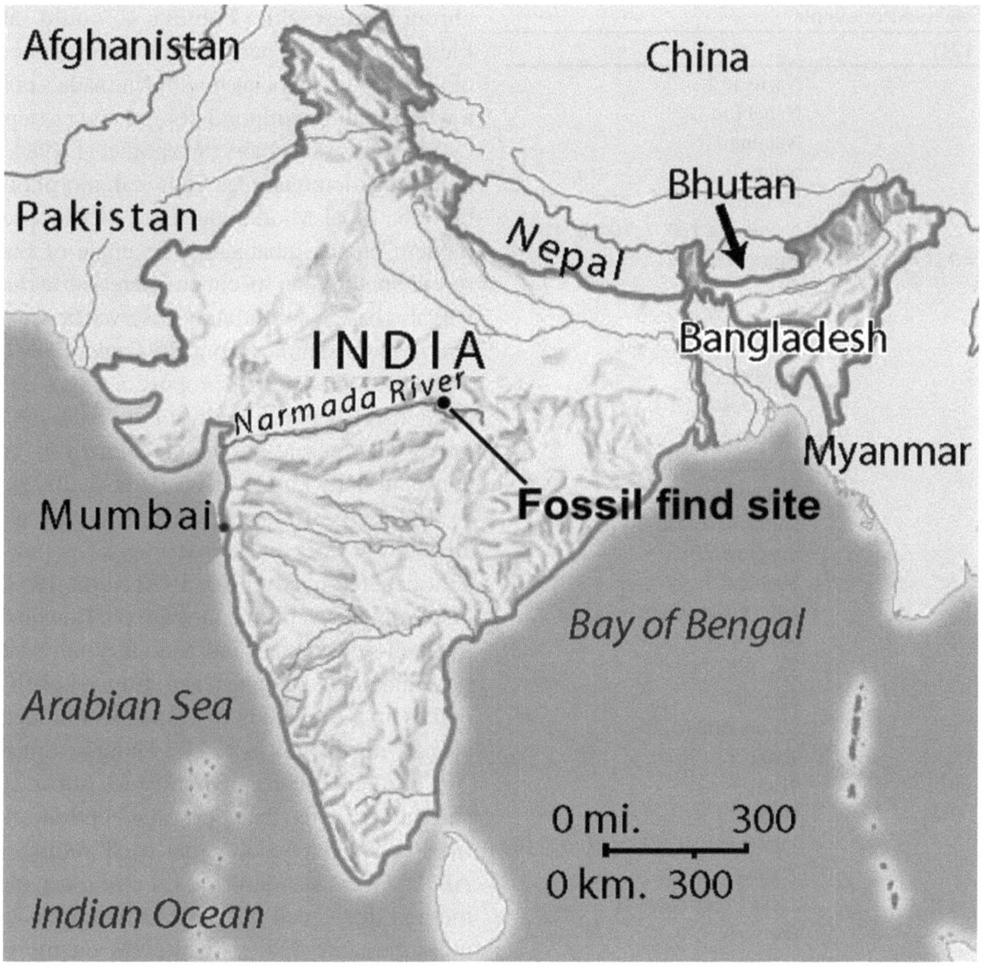

Fig. 10.2 Location of Narmada fossil discovery

In investigating the affinities of the Narmada hominin relative to other Middle Pleistocene specimens, particularly those considered *H. heidelbergensis*, it is necessary to define the taxonomic identification of each case in the comparative sample. However, there is as yet no consensus on what defines *H. heidelbergensis* or which specimens belong to it. Some see it as a separate evolutionary lineage (e.g., Rightmire 2001); others do not (e.g., Athreya 2006); some see it as exclusively European (Bermudez de Castro et al. 2003), while others consider it to have been in Africa and possibly Asia as well (Rightmire 1998). While the purpose of this study is not to resolve these issues, they do influence the configuration of the comparative sample, which will have an impact on the resulting group with which the Narmada specimen bears the strongest affinities. Therefore, four different configurations of this taxon were tested to see how Narmada is classified when *H. heidelbergensis* is defined according to different models of Middle Pleistocene hominin evolution and systematics. The purpose is to use this question as a vehicle to understand the nature of Narmada's morphology, as well as what its affinities can tell us about evolutionary trajectories in South Asia, and the relationship of this region to neighboring regions. Ultimately, this will contribute to a greater understanding of hominin migration and settlement patterns in the Indian subcontinent.

Materials and Methods

The comparative sample (Table 10.1) consists of fossil material dated to the Early, Middle and Late Pleistocene (n = 125). While the term "*H. heidelbergensis*" is used strictly to refer to certain fossils dated to the Middle Pleistocene, temporally neighboring specimens were included to evaluate the strength of taxonomic categories that are considered ancestral or descendant to the Middle Pleistocene populations. Since Narmada has previously been classified as *H. erectus* all specimens potentially belonging to this taxon, including the early African sample, are included in this study. Similarly, several Southeast Asian specimens considered to be *H. erectus* have been dated to the latter part of the Late Pleistocene

Table 10.1 Fossil and modern sample

Fossil Sample (n = 124)	
Narmada	Nacurrie 2
Amud	Nahal Ein Gev
Arago 21	Nanjing 1
Bodo	Ndutu
Cohuna	Neandertal
Combe Capelle	Ngandong 1, 6, 7, 10, 11, 1b2
Coobool Creek (N = 28)	Ngawi 1
Cro Magnon 1, 2, 3	OH9
Daka VP-2/66	Ohalo2
Dali	Omo 1, 2
Dmanisi 2280, 3444	Petralona
Elandsfontein	Predmost 1, 3, 4, 9, 10
ER3733, 3883	Qafzeh 6, 9
Florisbad	Saccopastore 1
Forbes Quarry	Sambungmacan 1, 3
Herto VP-16/1	Sambungmacan 4
Hexian	Sangiran 2, 17
Irhoud 1	Shanidar 1
Jinniushan	Sima de los Huesos 4, 5
Kabwe	Skhul 4, 5, 9
Kanalda	Spy 1, 2
Keilor 12	Steinheim
Kow Swamp 3, 5	Swanscombe
Krapina C, E	Tabun 1
La Chapelle aux Saints	Trinil
La Ferrassie	Wadjak 1
La Quina 5	Zhoukoudian 2, 3, 5, 10, 11, 12
Laetoli 18	Ziyang
Lake Nitchie 15	ZKD Upper Cave 101, 102, 103
Liujiang	Zuttiyeh
Minatogawa 1, 2, 4	
Mladec 1, 2, 5, 6	
Monte Circeo	
HISTORIC MODERN SAMPLE (n = 98)	RECENT MODERN SAMPLE (n = 56)
Bagor IV	Andaman Islanders (N = 54)
BellanPalassa (N = 4)	Onge L1, L5
Brahmagiri (N = 4)	
Burzahom BZH 1, 6	
Chanhu-Daro	
Harappa R37, H1	
Harappa (N = 26)	
Kodumanal (N = 4)	
KumharTekri (N = 16)	
Langhnaj (N = 4)	
Mahadaha (N = 5)	
Mohenjo-Daro 1	
Nagarjunakonda 1a	
Ruamgarh (N = 2)	
SaraiNaharRai1970-4	
Tekkalakota 15	
Timargarha (N = 18)	
Yeleswaram 1/61	

(Swisher et al. 1996), raising the possibility that the Narmada specimen – which was recovered from a poorly understood chronostratigraphic context – could also be of Late Pleistocene age. Therefore, in order to encompass the range of possibilities associated with Narmada's taxonomic affinities, a wide-ranging temporal comparative sample was used.

Since several authors (Wolpoff et al. 1984; Etler 2004) argue that Asia demonstrates regional morphological continuity between local Middle and Late Pleistocene populations and modern human inhabitants, a sample of Early Holocene and recent South Asian specimens were also included. This decision was also based on continuity observed between Late Pleistocene and modern samples from Sri Lanka (Hawkey 1998) as well as genetic studies that propose a direct ancestor-descendant relationship between south/southwest Asian Pleistocene populations and present-day Indian *adavasi* or "tribals" as they are conventionally known (Kivisild et al. 2003).

The Early Holocene data (n = 98) were derived from Kennedy's work on archaeological populations in the subcontinent (Kennedy 1965, 1972, 1982, 1993, 2000, 2002) and date from the Neolithic through the Roman times. In addition, a mixed-sex series of indigenous South Asians (n = 56) from the Andaman Islands taken from Howells' dataset (1973) makes up the recent human sample. Full details of the sample along with references can be found in Athreya (2007).

The data consist of published linear craniometric measurements taken from original reports and in some cases from secondary studies that used published measurements. All data were standardized for size using the geometric mean method developed by Darroch and Mossiman (1985). Only measurements that were present on the original Narmada fossil were included in the analysis. Thirteen variables were used, as shown in Table 10.2. Since the multivariate methods used to evaluate Narmada's morphological affinity required a complete dataset, and not all the comparative fossils were sufficiently complete to capture these thirteen variables, the missing data were imputed using multiple imputation (MI) and data augmentation (DA). Percentages of missing values for each variable are also shown in Table 10.2.

Table 10.2 Craniometric variables and percentage of missing values

Variable	Howells' Abbreviation	% Missing
Basion-bregma height	BBH	23.19
Biasterion breadth	ASB	31.16
Frontal sagittal arc	FAA	17.03
Frontal sagittal chord	FAC	5.43
Maximum cranial breadth	XCB	13.41
Maximum cranial length	GOL	1.45
Minimum frontal length	WFB	35.87
Occipital sagittal arc	OAA	27.17
Occipital sagittal chord	OCC	28.99
Orbit breadth	OBB	22.10
Orbit height	OBH	20.65
Parietal sagittal arc	PAA	16.30
Parietal sagittal chord	PAC	6.52

The missing-data imputation was conducted using NORM 2.3 (Schafer 1999b). This program has some precedence in physical anthropology for estimating missing values in craniodental studies (Stefan 2004; Rhode and Arriaza 2006; Scherer 2007). More conventional methods such as mean substitution or regression are known to bias the results by underestimating the variance in the population and inflating correlations between variables (Santos 1981; Schafer and Graham 2002). Multiple imputation is considered to be the least biased way of estimating missing data (Schafer and Olsen 1998; Schafer 1999a; Allison 2000). The goal is not to predict the missing values, but rather to configure the dataset in such a way as to preserve relationships among cases and variables in order to continue with more powerful methods of analysis that require complete datasets (Schafer 1997; Schafer and Graham 2002; Wayman 2003).

Multiple imputation calls for the creation of three to five datasets which are then combined for complete-case analysis. In this study four datasets were created using the Data Augmentation (DA) routine in NORM. Each individually imputed dataset was subjected to the same data analysis routines used in the final study, in order to determine if the outcomes were similar with respect to the allocation of Narmada and other fossil specimens. While the resulting values varied somewhat, each specimen was consistently assigned to the same group for each imputed dataset, demonstrating the stability of this method in producing repeatable results. The four imputed datasets were then averaged to form the one used in the final analyses. The data were reduced using principal components analysis, which were scored to create the variables that were statistically analyzed. The eigenvectors of the first six principal components, which represent 86% of the variance, are shown in Table 10.3.

In order to evaluate Narmada's taxonomic affinity, a probability-based classification/allocation procedure known as multinomial logistic regression was used. This method evaluates relationships among members of *a priori* defined groups using a categorical grouping variable, and then offers posterior probability values that each specimen belongs to one of the *a priori* groups. It is similar to discriminant function analysis (DFA), which was originally developed as a biostatistical procedure to determine the allocation of an unknown specimen (Rao 1948). However, while DFA uses a linear function of the variables that maximizes the separation among groups multinomial logistic regression (MLR) measures the contribution of an independent variable to variations in the dependent variable (Long 1997). The posterior probabilities computed from the linear discriminant functions tend to be overestimated when covariance matrices are unequal, and/or group sizes are small (Press and Wilson 1978). In paleoanthropological analyses, multinomial logit is superior to DFA because it is robust against violations of the assumptions of large and equal sample group sizes (Bull and Donner 1987). Like DFA, the maximum number of variables must not exceed the number of *a priori* groups. Since the data were reduced using principal components analysis, these variables were always the first several principal components which maximize variance among the original variables in a linear orthogonal manner. They therefore account for the majority of discriminatory information among measurements. In this study, the independent variables were the principal components derived from the craniometric data, and the dependent variable was "Group" for which there were between six to eight unordered outcomes, depending on the scenario tested (see below): *H. ergaster*, *H. erectus*, *H. heidelbergensis*, *H. rhodesiensis*, Neandertal, *H. sapiens*, Historic Modern or Recent Modern.

Typicality probabilities obtained from a discriminant function analysis were also analyzed. Unlike posterior probabilities, typicality probabilities are not affected by sample size bias. They are based on the *absolute* value of the Mahalanobis's distance from a group centroid, and are not

Table 10.3 Principal component eigenvectors

Eigenvectors	1	2	3	4	5	6
BBH	−0.3406	0.21406	0.01212	0.22518	−0.11977	0.19067
ASB	0.38175	−0.02344	−0.07309	0.06525	−0.14499	−0.31449
FAA	−0.18333	−0.26912	0.54139	0.02536	0.37056	−0.18127
FAC	−0.01225	0.19076	0.54647	0.5648	−0.00248	0.09346
XCB	0.32132	−0.18296	−0.06849	0.03495	0.37974	−0.06567
GOL	0.32164	0.23278	−0.00316	0.29242	−0.29768	0.05058
WFB	0.17964	0.30962	−0.21261	−0.0402	0.59657	0.50296
OAA	−0.22958	−0.5098	−0.24075	0.10841	0.08676	0.14501
OCC	−0.26335	−0.1081	−0.47741	0.40663	−0.15115	−0.10115
OBB	0.34815	0.04381	0.00181	−0.11475	−0.00531	−0.39541
OBH	0.17571	−0.29239	0.24188	−0.35717	−0.45337	0.54754
PAA	−0.38296	0.08159	0.08487	−0.35121	0.00064	−0.26915
PAC	−0.21148	0.54372	−0.03576	−0.30962	−0.05148	−0.05877

The first six eigenvectors were used. They represent 86% of the variance.

computed from the linear discriminant function. While posterior probabilities are reflective of a 'forced' allocation procedure, typicality probabilities are not (Campbell 1984). They allow for the fact that while Narmada may be closer to one group centroid than the others, it is not necessarily typical of that, or of any group in the analysis.

Evolutionary Scenarios for *H. heidelbergensis*

Four different configurations of *Homo heidelbergensis* were tested in order to account for the various models that exist for the phylogeny of Middle Pleistocene *Homo* (Table 10.4). Each was given an identifying name based on the regions that they propose supported *H. heidelbergensis* populations. The first configuration ("Eurasia/Africa") gave an *a priori* assignment to *H. heidelbergensis* of the so-called "transitional" (mixed archaic/modern) fossils from every region. This was the only scenario in which this species was considered to have been in Asia. In addition, the *erectus*-like fossils from Africa and Asia were combined into one group as opposed to being placed on two separate evolutionary lineages.

The second scenario ("Europe Only") is based on work by Bermudez de Castro et al. (2003), where the term *H. heidelbergensis* is restricted to refer to the specimens from Western Europe. In their view, the African transitional fossils are considered "*H. rhodesiensis.*" While these authors do not specify how to classify the Asian specimens, the implication is that those dated to the Middle Pleistocene are an evolutionary dead end – therefore, they are all placed in the category *H. erectus*. Given this assumption, the African fossils that show *erectus*-like features are classified as *H. ergaster* following a more speciose model that sees regionally unique evolutionary trajectories for Early and Middle Pleistocene hominins, and a central role for the African lineage in modern human evolution.

The third scenario ("Europe/Africa vs. *H. erectus*") gives an *a priori* identification of *H. heidelbergensis* to the transitional fossils from both Europe and Africa, but not Asia. Similar fossils from Asia are instead identified as *H. erectus*. In this configuration, the African and Asian *H. erectus* specimens are grouped together as one population. The fourth scenario ("Europe/Africa vs. Asian-only *erectus*") is more or less identical, with the exception that *H. ergaster* is again used to identify the African *erectus*-like fossils. Like the second scenario, this tested the notion that *H. erectus* populations from Asia and Africa were separate evolutionary lineages, with the Asian lineage to which Narmada presumably belonged, being an evolutionary dead-end. The *a priori* assignments of each individual specimen by scenario are in Table 10.5.

Results

For the first "Eurasia/Africa" scenario, where *H. heidelbergensis* is considered to have been in Asia, the results give Narmada a 67.8% posterior probability but just a 1% typicality probability of belonging to *H. heidelbergensis*. In this scenario, Narmada has a higher (32.1%) typicality probability of classifying with *H. erectus* (Table 10.6). In the second configuration ("Europe Only"), *H. heidelbergensis* is defined as an exclusively European lineage. The Narmada hominin is given a 98.5% posterior probability of being *H. erectus,* defined here as an exclusively Asian lineage. However, Narmada's typicality probability relative to this and all eight groups that were defined in this scenario is less than 1%. So while Narmada is closer to the *H. erectus* group centroid than to any other in the sample as reflected in the posterior probability, in absolute terms it is not typical of any of the eight groups defined here.

For the third scenario, "Europe/Africa vs. *H. erectus*", *H. heidelbergensis* is used to refer to the transitional Middle

Table 10.4 Scenarios used to determine a priori designations of comparative sample

Scenario Name[a]	Europe	Africa	Asia (including South Asia)	Summary
Eurasia/Africa	*H. heidelbergensis*	*H. heidelbergensis*	*H. heidelbergensis*	*H. heidelbergensis* was present in every Old World region including South Asia
Europe Only	*H. heidelbergensis*	*H. rhodesiensis*	*H. erectus*	*H. heidelbergensis* was present only in Europe
Europe/Africa vs. *H. erectus*	*H. heidelbergensis*	*H. heidelbergensis*	*H. erectus*	*H. heidelbergensis* was present in Europe and Africa. *H. erectus* was present in Africa and Asia, and is not an evolutionary dead-end
Europe/Africa vs. *H. ergaster/erectus*	*H. heidelbergensis*	*H. ergaster*→ *H. heidelbergensis*	*H. erectus*	*H. heidelbergensis* was present in Europe and Africa. *H. erectus* was present throughout the Middle Pleistocene in Asia only, and is an evolutionary dead-end.

[a]Based on location of *H. heidelbergensis* populations vs. *H. erectus*

Table 10.5 A priori assignments given to fossils in each scenario

Specimen/site name	Geological period	Eurasia/Africa	Europe Only	Europe/Africa vs. *H. erectus*	Europe/Africa vs. *H. ergaster/erectus*
Narmada					
Amud	Late Pleistocene	Neandertal	Neandertal	Neandertal	Neandertal
Arago 21	Middle Pleistocene	*H. heidelbergensis*	*H. heidelbergensis*	*H. heidelbergensis*	*H. heidelbergensis*
Bodo	Middle Pleistocene	*H. heidelbergensis*	*H. rhodesiensis*	*H. heidelbergensis*	*H. heidelbergensis*
Cohuna	Late Pleistocene	AMHS	AMHS	AMHS	AMHS
Combe Capelle	Late Pleistocene	AMHS	AMHS	AMHS	AMHS
Coobool Creek (N=28)	Late Pleistocene	AMHS	AMHS	AMHS	AMHS
Cro Magnon 1, 2, 3	Late Pleistocene	AMHS	AMHS	AMHS	AMHS
Daka VP-2/66	Early Pleistocene	*H. erectus*	*H. ergaster*	*H. erectus*	*H. ergaster*
Dali	Middle Pleistocene	*H. heidelbergensis*	*H. erectus*	*H. erectus*	*H. erectus*
Dmanisi 2280, 3444	Early Pleistocene	*H. erectus*	*H. ergaster*	*H. erectus*	*H. ergaster*
Elandsfontein	Middle Pleistocene	*H. heidelbergensis*	*H. rhodesiensis*	*H. heidelbergensis*	*H. heidelbergensis*
ER3733, 3883	Early Pleistocene	*H. erectus*	*H. ergaster*	*H. erectus*	*H. ergaster*
Florisbad	Middle Pleistocene	*H. heidelbergensis*	*H. rhodesiensis*	*H. heidelbergensis*	*H. heidelbergensis*
Forbes' Quarry	Late Pleistocene	Neandertal	Neandertal	Neandertal	Neandertal
Herto BOU VP-16/1	Middle Pleistocene	AMHS	AMHS	AMHS	AMHS
Hexian	Middle Pleistocene	*H. erectus*	*H. erectus*	*H. erectus*	*H. erectus*
Irhoud 1	Middle Pleistocene	*H. heidelbergensis*	*H. rhodesiensis*	*H. heidelbergensis*	*H. heidelbergensis*
Jinniushan	Middle Pleistocene	*H. heidelbergensis*	*H. erectus*	*H. erectus*	*H. erectus*
Kabwe	Middle Pleistocene	*H. heidelbergensis*	*H. rhodesiensis*	*H. heidelbergensis*	*H. heidelbergensis*
Kanalda	Late Pleistocene	AMHS	AMHS	AMHS	AMHS
Keilor 12	Late Pleistocene	AMHS	AMHS	AMHS	AMHS
Kow Swamp 3, 5	Late Pleistocene	AMHS	AMHS	AMHS	AMHS
Krapina C, E	Middle Pleistocene	Neandertal	Neandertal	Neandertal	Neandertal
La Chapelle aux Saints	Late Pleistocene	Neandertal	Neandertal	Neandertal	Neandertal
La Ferrassie	Late Pleistocene	Neandertal	Neandertal	Neandertal	Neandertal
La Quina 5	Late Pleistocene	Neandertal	Neandertal	Neandertal	Neandertal
Laetoli 18	Middle Pleistocene	*H. heidelbergensis*	*H. rhodesiensis*	*H. heidelbergensis*	*H. heidelbergensis*
Lake Nitchie 15	Late Pleistocene	AMHS	AMHS	AMHS	AMHS
Liujiang	Late Pleistocene	AMHS	AMHS	AMHS	AMHS
Minatogawa 1, 2, 4	Late Pleistocene	AMHS	AMHS	AMHS	AMHS
Mladec 1, 2, 5, 6	Late Pleistocene	AMHS	AMHS	AMHS	AMHS
Monte Circeo	Late Pleistocene	Neandertal	Neandertal	Neandertal	Neandertal
Nacurrie 2	Late Pleistocene	AMHS	AMHS	AMHS	AMHS
Nahal Ein Gev	Late Pleistocene	AMHS	AMHS	AMHS	AMHS
Nanjing 1	Middle Pleistocene	*H. erectus*	*H. erectus*	*H. erectus*	*H. erectus*
Ndutu	Middle Pleistocene	*H. heidelbergensis*	*H. rhodesiensis*	*H. heidelbergensis*	*H. heidelbergensis*
Neandertal	Late Pleistocene	Neandertal	Neandertal	Neandertal	Neandertal
Ngandong 1, 6, 7, 10, 11, 12	Late Pleistocene	*H. erectus*	*H. erectus*	*H. erectus*	*H. erectus*
Ngawi 1	Late Pleistocene	*H. erectus*	*H. erectus*	*H. erectus*	*H. erectus*
OH9	Early Pleistocene	*H. erectus*	*H. ergaster*	*H. erectus*	*H. ergaster*
Ohalo2	Late Pleistocene	AMHS	AMHS	AMHS	AMHS
Omo 1, 2	Middle Pleistocene	AMHS	AMHS	AMHS	AMHS
Petralona	Middle Pleistocene	*H. heidelbergensis*	*H. heidelbergensis*	*H. heidelbergensis*	*H. heidelbergensis*
Predmost 1, 3, 4, 9, 10	Late Pleistocene	AMHS	AMHS	AMHS	AMHS
Qafzeh 6, 9	Late Pleistocene	AMHS	AMHS	AMHS	AMHS
Saccopastore 1	Late Pleistocene	Neandertal	Neandertal	Neandertal	Neandertal
Sambungmacan 1	Middle Pleistocene	*H. erectus*	*H. erectus*	*H. erectus*	*H. erectus*
Sambungmacan 3, 4	Late Pleistocene	*H. erectus*	*H. erectus*	*H. erectus*	*H. erectus*
Sangiran 2, 17	Early Pleistocene	*H. erectus*	*H. erectus*	*H. erectus*	*H. erectus*
Shanidar 1	Late Pleistocene	Neandertal	Neandertal	Neandertal	Neandertal
Skhul 4, 5, 9	Late Pleistocene	AMHS	AMHS	AMHS	AMHS

(continued)

Table 10.5 (continued)

Specimen/site name	Geological period	Eurasia/Africa	Europe Only	Europe/Africa vs. *H. erectus*	Europe/Africa vs. *H. ergaster/erectus*
Sima de los Huesos 4, 5	Middle Pleistocene	*H. heidelbergensis*	*H. heidelbergensis*	*H. heidelbergensis*	*H. heidelbergensis*
Spy 1, 2	Late Pleistocene	Neandertal	Neandertal	Neandertal	Neandertal
Steinheim	Middle Pleistocene	*H. heidelbergensis*	*H. heidelbergensis*	*H. heidelbergensis*	*H. heidelbergensis*
Swanscombe	Middle Pleistocene	*H. heidelbergensis*	*H. heidelbergensis*	*H. heidelbergensis*	*H. heidelbergensis*
Tabun 1	Late Pleistocene	Neandertal	Neandertal	Neandertal	Neandertal
Trinil	Early Pleistocene	*H. erectus*	*H. erectus*	*H. erectus*	*H. erectus*
Wadjak 1	Late Pleistocene	AMHS	AMHS	AMHS	AMHS
Ziyang	Late Pleistocene	AMHS	AMHS	AMHS	AMHS
Zhoukoudian 2 3, 5, 10, 11, 12	Middle Pleistocene	*H. erectus*	*H. erectus*	*H. erectus*	*H. erectus*
ZKD Upper Cave 101, 102, 103	Late Pleistocene	AMHS	AMHS	AMHS	AMHS
Zuttiyeh	Middle Pleistocene	*H. heidelbergensis*	*H. rhodesiensis*	*H. heidelbergensis*	*H. heidelbergensis*

Table 10.6 Results of multinomial logistic regression (MLR) and discriminant function analysis (DFA)

	Allocation of Narmada Fossil		
Scenario tested	Highest posterior probability (MLR)	Highest typicality probability (DFA)	Typicality probability of belonging to *H. heidelbergensis (DFA)*
Eurasia/Africa	*H. heidelbergensis* 67.8%	*H. erectus* 32.1%	3.8%
Europe only	*H. erectus* 98.5%	*H. erectus* < 1.0%	
Europe/Africa vs. *H. erectus*	*H erectus* 63.7%	*H. erectus* 48.5%	
Europe/Africa vs. *H. ergaster/erectus*	*H. erectus* 93.7%	*H. ergaster* 6.0%	

Pleistocene fossils from Europe and Africa, but not Asia. In this scenario, Narmada has a 63.7% posterior probability and a 48.5% typicality probability of aligning with the *H. erectus* group, which is defined here in the broad sense as being both African and Asian. In the "Europe/Africa vs. *H. ergaster/erectus*" scenario, the results give Narmada a 93.7% posterior probability of belonging to *H. erectus* but it is given a <1% typicality probability relative to this group, which was defined here as strictly Asian. Its highest typicality probability is with *H. ergaster*, but that value is only 6.0%.

In addition, in all scenarios the Historic South Asian population groups with the anatomically modern Late Pleistocene sample, while the Andaman Islanders are morphologically quite distinct (Fig. 10.3). So, perhaps not surprisingly, the Narmada specimen is not similar to Late Pleistocene or early Holocene groups, but is strongly aligned morphologically with the Middle Pleistocene specimens. Given the uncertain age of the specimen, it is difficult to date it based on morphology. Such a finding could corroborate an older age of the specimen, but it could also point to a retention of archaic features in South Asia well into the Late Pleistocene. Without direct dating of the specimen, we cannot distinguish between these two possibilities.

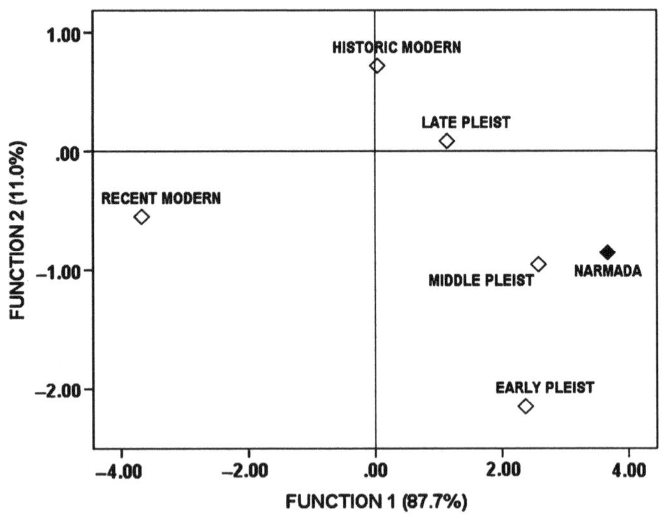

Fig. 10.3 Scatterplot of discriminant functions 1 and 2 showing Narmada fossil relative to temporal groups

Discussion

The "Eurasia/Africa" scenario is the only one in which some Asian specimens were given an *a priori* designation as *H. heidelbergensis*. In this analysis, while Narmada did group with the other specimens assigned to this taxon in its posterior probability value, it is an outlier as seen by its low typicality probability value. The second and fourth scenarios – "Europe Only" and "Europe/Africa vs. *H. ergaster/erectus*" are similar in that they divide the *erectus*-like sample into separate African (*H. ergaster*) and Asian lineages. In both of these scenarios, Narmada is given a high posterior probability of classifying with an exclusively Asian *H. erectus*, but has a very low typicality probability of belonging to this group, and in one case ("Europe/Africa vs. *H. ergaster/erectus*") Narmada actually has a higher typicality probability of classifying with *H. ergaster* though that value is likewise extremely low, at 6%. In theory, low typicality probability values relative to all groups in a scenario can suggest that an entirely new species is represented. But given Narmada's lack of unique characters – in all traits it resembles either *H. erectus* or *H. sapiens* – more parsimonious explanations can be found.

The most solid results statistically are seen in the third scenario, "Europe/Africa vs. *H. erectus*" where the term "*Homo heidelbergensis*" is used to classify the African and European transitional fossils, and "*Homo erectus*" is defined in the broad sense as being both African and Asian. In this scenario, Narmada has both a high typicality and posterior probability of grouping with *H. erectus*. This scenario is definitive in terms of a) the individually high probability values exhibited in both tests, and b) the consistency between the results of each test. Taking this alone, the results of this study suggest that the Narmada hominin is not an Indian *H. heidelbergensis* based on its continuous cranial traits. Instead, the patterning of classification for this specimen suggests that it is most strongly aligned with *H. erectus sensu lato*.

Models and Predictions for South Asia

Despite the statistical support for this interpretation, it would be premature to stop the discussion as to Narmada's affinity with the simple answer that Narmada is a *Homo erectus*. The broader implications of classifying Narmada as *Homo erectus* require that if we accept this result, we would have to do so with certain caveats. The definition of *H. erectus* would have to be expanded to include traits not present in the original definition based on the Asian fossils, and it would have to include a large brained hominin with certain traits not previously associated with this taxon. In other words, we would basically be calling Narmada an "advanced *H. erectus*" or an "archaic" *Homo sapiens*.

These caveats are so broad as to dilute the utility of the definition of *H. erectus*, and this was historically one of the criticisms of concepts such as "advanced *Homo erectus*" and "archaic *Homo sapiens*." These terms were once fairly widely used to refer to transitional fossils from the Middle Pleistocene (Bräuer 1984; Clarke 1990; Cronin et al. 1981; Santa Luca 1978), but their evolutionary meaning and taxonomic validity were so vague as to be considered objectionable except as a heuristic device (Tattersall 1986; Howell 1994). Their use has diminished in the past decade in favor of the term *Homo heidelbergensis*. In this context, Narmada actually fits in well with current tendencies to refer all "transitional" Middle Pleistocene fossils to *H. heidelbergensis*, based on its mosaic of *erectus*- and *sapiens*-like traits.

Previous studies have allocated Narmada to this group (regardless of its nomenclature) and not with *H. erectus* (Kennedy et al. 1991; Cameron et al. 2004). However, this presents a contradiction between the results of this study – which gave it a high probability of belonging to *Homo erectus* – and previous studies; it also contradicts existing models for classifying Middle Pleistocene *Homo*, which conventionally group all transitional fossils together in *H. heidelbergensis*. One problem in resolving this conflict is the fact that *H. heidelbergensis* is not yet well defined and broadly accepted as having a unique set of morphological traits shared by all members. While it is not the purpose of this paper to resolve issues of phylogeny and systematics of Middle Pleistocene *Homo*, this discussion highlights the absence of a solid evolutionary model by which to understand Narmada's morphology. In light of this fact, it is clear that the question of which lineage Narmada belongs to needs to be examined and understood in a new interpretive framework.

One consistent factor in these results is that Narmada's strength of affinity was influenced by the presence of a mixed African-Asian sample. When *H. heidelbergensis* or *H. erectus* included fossils from both of these regions, Narmada's affinity was with this mixed group. This is understandable given India's location between these two regions. In this context, the Narmada fossil's morphology can best be understood by a scenario whereby the hominin lineage in Asia was not an evolutionary dead end. Rather, South Asian populations were shaped by dispersals that entered the subcontinent from both East and Southeast Asia as well as Africa during the Middle Pleistocene. Such a proposition is consistent with the results of studies of both x- and y-linked haplotypes (Karafet et al. 2001; Comas et al. 2004) as well as archaeological evidence (Keates 2004).

The idea that India is a genetic crossroad is not new. Several decades ago, Dobzhansky (1963) proposed that India could serve as a crucial locality for clarifying the relationship between African and Asian Pleistocene populations.

> The problem is really how to discriminate between these two possibilities, a single species of *Homo*, or two species, an eastern and a western one in the mid-Pleistocene? One would like to know whether anywhere and at any time races of *H. erectus* and of *H. sapiens*, or of a western species ancestral to the latter, lived sympatrically but without production of intermediate or hybrid types.... One may also hope that populations bridging the gap between the supposed species may be discovered.

> No mid-Pleistocene hominid fossils are known from extensive territories in Asia, particularly from India and the surrounding countries, which may well provide the critical evidence. In the absence of such evidence, there is simply no foundation for a decision whether the mid-Pleistocene forms of eastern Asia and the western forms belonged to a single or to two or perhaps more species. (357)

The Narmada fossil is a piece of the critical evidence to which Dobzhansky refers. Kennedy et al.'s (1991), and Cameron et al.'s (2004) studies found Narmada to possess "transitional" morphology. The original morphometric analysis by de Lumley and Sonakia (1985) identified these traits, but explained them as being the result of an "evolved" *Homo erectus* with unique Indian features. In this study, Narmada grouped with *Homo erectus*, but only had a strong affinity with that group when it contained both African and Asian morphology. Recent studies support the unity of African and Asian *H. erectus*, strengthening this scenario (Baab 2008; Asfaw et al. 2002). When *H. heidelbergensis* was comprised of African and Asian fossils, Narmada grouped with that taxon. Regardless of the resulting taxonomic interpretations, in all studies Narmada was identified as having mixed morphology and did not fit within the typological definitions of the existing taxa.

The problems with determining Narmada's allocation stem not just from the unresolved systematics of the Middle Pleistocene, but also the broader evolutionary scenarios for this time period, particularly ones that view the Asian specimens as part of an evolutionary dead end. Most models consider the African and Asian fossils to represent separate lineages, with the latter conventionally believed to be an extinct one replaced by incoming anatomically modern *Homo sapiens* in the Middle or early Late Pleistocene (Andrews 1984; Bräuer and Mbua 1992; Swisher et al. 1996). However, the Regional Continuity model proposes an alternative scenario whereby East Asian and Australian populations were not replaced but experienced varying levels of gene flow with the incoming populations who dispersed from Africa (Thorne and Wolpoff 1981; Wolpoff et al. 1984; Etler 2004; Wu 2004). While this model emphasizes morphological continuity in a given locality through time, particularly geographical "edges" such as Australasia, it is less explicit about the issues of multidirectional dispersals and clinal variation across space. This is in part due to the lack of fossil evidence from important intermediate regions such as South and Central Asia. But as this study shows, the Narmada fossil from a transitional time period and locality exhibits not just a suite of *erectus* and *sapiens*-like features but also a combination of African and Asian features. It therefore supports scenarios that include east-west dispersals with admixture at geographical crossroad regions of Eurasia.

In light of the findings here, the existing phylogenetic models that treat the East Asian fossils as part of an evolutionary dead-end lineage may be elevating normal regional variation to species-level distinctions for the Early and Middle Pleistocene, and doing so as a result of convenient geographical gaps in our evidence. If a broad populational vs. typological approach to Middle Pleistocene variation is taken, morphology from regions such as India can be better understood. Based on these results, a new interpretive framework and associated set of predictions for India can be articulated as follows:

1. Hominin migration patterns into India during the Middle Pleistocene were the result of dispersals from both East/Southeast Asia as well as East Africa.
2. The morphology of these populations within India should therefore reflect a mosaic of African and Asian traits, which are regional distinctions that have developed due to drift, selection and migration but not speciation.
3. In this context, grouping India with East and/or Southeast Asia when formulating predictions, classifying specimens or interpreting morphology is inappropriate. Clearly the subcontinent was shaped by evolutionary forces that were quite different than the other two regions. Defining a broad "Asian" morphology is therefore an inaccurate homogenization of the patterning of morphology among these regions.
4. Future finds from the Middle Pleistocene should reflect these multidirectional dispersals and a mixed African/Asian morphology is expected for the populations of India during this time period.

Conclusion

It is important to reiterate that the purpose of this study was not to draw broad conclusions about the phylogeny of Middle Pleistocene *Homo*. It is possible that the scenario laid out in the Europe/Africa vs. *H. erectus* scenario is not reflective of the true evolutionary history of these hominins. Rather, the goal of this study was to investigate the affinities of the Narmada specimen and understand the potential evolutionary forces that shaped its morphology. In that sense, Scenario III was unequivocally the one that had the strongest statistical support and therefore merited deeper examination, particularly with reference to its implications for the phylogeny of South Asian hominins. However, if the predictions described above are not supported with the discovery of more fossils and another scenario were to bear out more evidence, it would still be important to explain the morphology of Narmada in that context instead rather than simply subsume it under a broader Asian model.

Currently, there are no models or predictions that uniquely address the expected morphological trajectories for Pleistocene *Homo* in India. Presumably this is due largely to the sparse fossil evidence from these early time periods of hominin evolution. Yet, the Narmada fossil does provide sufficient evidence to formulate preliminary predictions for this region of the Old World. Specifically, this study demonstrates that the mixed morphology of Narmada – specifically its suite of classic Asian *erectus*-like features combined with typical *sapiens*-like traits – does not comfortably fit into the traditional definitions of either species.

Similarly, the four scenarios tested here demonstrate that the existing systematic schemes for Middle Pleistocene *Homo* only allow for two possibilities for hominin occupation in South Asia: either these populations were *H. heidelbergensis* or *H. erectus*. Typically the term "*H. heidelbergensis*" is used for fossils such as Narmada which possess mixed *erectus-sapiens* morphology. Yet, several of the scenarios explored here were developed for specific regions, and these scholars do not support the presence of *H. heidelbergensis* in Asia. If these schemes are true, then Narmada would necessarily have to be allocated to *H. erectus*. This study demonstrated that morphologically it bears the strongest affinity to this group, when defined as an African-Asian species. However, as mentioned previously, allocating Narmada to *H. erectus* then requires us to expand the definition of that species.

So while the purpose of this study was not to seek support for any of the existing scenarios of Middle Pleistocene *Homo*, one contribution is to shed light on the contradictions inherent in some of the currently debated scenarios. Specifically, those studies that do not incorporate the Asian fossil evidence implicitly relegate the Asian transitional fossils to *H. erectus* without issue. This implication is hard to accept given the similarity of these fossils to other transitional Middle Pleistocene specimens. These models are likely to be invalidated as further evidence from various parts of Asia are uncovered. Similarly, other Asian fossils possesses this mixed morphology (e.g., Dali, Jinniushan). These specimens cannot be comfortably allocated to *H. erectus* which contradicts scenarios that propose a dead-end Middle Pleistocene hominin lineage in Asia, replaced by an advanced *H. sapiens* dispersal from elsewhere.

By studying the complex morphology of India's Narmada fossil in a broader temporal and geographic framework, the model offered here takes these important issues into consideration. It predicts that variation from East Africa to East Asia was continuous during the Middle Pleistocene. Like other scholars who work in Asia (Pope 1992; Etler 2004) have found, Asian *H. erectus* fossils were arguably not an evolutionary dead end. Rather these specimens suggest a scenario where Asian mid-Pleistocene hominins made contributions to the evolution of *Homo sapiens* not just in their local regions but also through multidirectional dispersals and admixture with migrating African populations, both of whom crossed through or occupied the subcontinent. Such a proposition that could be verified by future finds in India and other intermediate geographic regions such as Central Asia. The recovery of hominin specimens from these regions would clarify the true pattern of population distribution between Africa and East Asia. A resurgence of paleoanthropological research in these regions of the Old World will hopefully enable us to test these predictions soon. This will allow for a better understanding of the nature of hominin morphology in South Asia, as well as a clearer picture of the earliest migrations into the subcontinent.

Acknowledgments Thanks to Christopher Norton, David Braun and John Harris for the invitation to participate in the symposium at which this paper was first presented. I am grateful to Eric Delson for his suggestions, which improved the final version of this paper. Thanks also to Kenneth Kennedy for providing the comparative Holocene data, and to Mica Glantz, Hélène Rougier, and Laura Shackelford for the discussions and feedback which contributed to the ideas presented here. All errors or omissions are my own.

References

Allison, P. D. (2000). Multiple imputation for missing data: a cautionary tale. *Sociological Methods Research, 28*(3), 301–309.

Andrews, P. (1984). An alternative interpretation of the characters used to define *Homo erectus. Courier Forschungsintitut Senckenberg, 69*, 167–175.

Asfaw, B., Gilbert, W. H., Beyene, Y., Hart, W. K., Renne, P. R., WoldeGabriel, G., et al. (2002). Remains of *Homo erectus* from Bouri, Middle Awash, Ethiopia. *Nature, 416*(6878), 317–320.

Athreya, S. (2006). Patterning of geographic variation in Middle Pleistocene *Homo* frontal bone morphology. *Journal of Human Evolution, 50*(6), 627–643.

Athreya, S. (2007). Was *Homo heidelbergensis* in South Asia? A test using the Narmada fossil from Central India. In M. D. Petraglia & B. Allchin (Eds.), *The evolution and diversity of humans in South Asia* (pp. 137–170). New York: Springer.

Baab, K. L. (2008). The taxonomic implications of cranial shape variation in Homo erectus. *Journal of Human Evolution, 54*(6), 827–847.

Bermudez de Castro, J. M., Martinon-Torres, M., Sarmiento, S., & Lozano, M. (2003). Gran Dolina-TD6 versus Sima de los Huesos dental samples from Atapuerca: evidence of discontinuity in the European Pleistocene population? *Journal of Archaeological Science, 30*(11), 1421–1428.

Bräuer, G. (1984). The "Afro-European *sapiens* hypothesis," and hominid evolution in East Asia during the late Middle and Upper Pleistocene. *Courier Forshungsinstitut Senckenberg, 69*, 145–165.

Bräuer, G., & Mbua, E. (1992). *Homo erectus* features used in cladistics and their variability in Asian and African hominids. *Journal of Human Evolution, 22*, 79–108.

Bull, S. B., & Donner, A. (1987). The efficiency of multinomial logistic regression compared with multiple group discriminant analysis. *Journal of the American Statistical Association, 82*(400), 1118–1122.

Cameron, D., Patnaik, R., & Sahni, A. (2004). The phylogenetic significance of the Middle Pleistocene Narmada hominin cranium from Central India. *International Journal of Osteoarchaeology, 14*(6), 419–447.

Campbell, N. A. (1984). Some aspects of allocation and discrimination. In G. N. van Vark & W. W. Howells (Eds.), *Multivariate statistical methods in physical anthropology* (pp. 177–192). Dordrecht, The Netherlands: Reidel.

Clarke, R. J. (1990). The Ndutu cranium and the origin of *Homo sapiens*. *Journal of Human Evolution, 19*, 699–736.

Comas, D., Plaza, S., Wells, R. S., Yuldaseva, N., Lao, O., Calafell, F., et al. (2004). Admixture, migrations and dispersals in Central Asia: evidence from maternal DNA lineages. *European Journal of Human Genetics, 12*(6), 495–504.

Cronin, J. E., Boaz, N. T., Stringer, C. B., & Rak, Y. (1981). Tempo and mode in hominid evolution. *Nature, 292*, 113–122.

Darroch, J., & Mossiman, J. (1985). Canonical and principal components of shape. *Biometrika, 72*, 241–252.

de Lumley, M.-A., & Sonakia, A. (1985). First discovery of a *Homo erectus* on the Indian sub-continent, at Hathnora, in the Middle Valley of the Narmada River. *Anthropologie – Paris, 89*(1), 3–12.

Dobzhansky, T. (1963). Genetic entities in hominid evolution. In S. L. Washburn (Ed.), *Classification and human evolution* (pp. 347–362). Chicago: Aldine.

Etler, D. (2004). *Homo erectus* in East Asia: human ancestor or evolutionary dead-end? *Athena Review, 4*(1), 37–50.

Hawkey, D. E. (1998). *Out of Asia: dental evidence for affinities and microevolution of early populations from India/Sri Lanka. Anthropology* (p. 478). Tempe, AZ: Arizona State University.

Howell, F. C. (1994). A chronostratigraphic and taxonomic framework of the origins of modern humans. In M. H. Nitecki & D. V. Nitecki (Eds.), *Origins of anatomically modern humans* (pp. 253–319). New York: Plenum.

Howells, W. W. (1973). *Cranial variation in man: a study by multivariate analysis of patterns of difference among recent human populations*. Harvard University: Papers of the Peabody Museum. 67.

Karafet, T., Xu, L., Du, R., Wang, W., Feng, S., Wells, R. S., et al. (2001). Paternal population history of East Asia: sources, patterns and microevolutionary processes. *American Journal of Human Genetics, 69*(3), 615–628.

Keates, S. G. (2004). Home range size in Middle Pleistocene China and Human Dispersal Patterns in Eastern and Central Asia. *Asian Perspectives, 43*(2), 227–247.

Kennedy, K. A. R. (1965). Human skeletal material from Ceylon, with an Analysis of the Island's prehistoric and contemporary populations. *Bulletin of the British Museum of Natual History, Geology, 2*(4), 14–213.

Kennedy, K. A. R. (1972). Anatomical description of two Crania from Ruamgarh: an ancient site in Dhalbhum, Bihar. *Journal of the Indian Anthropological Society, 7*, 129–141.

Kennedy, K. A. R. (1982). Part II: Biological anthropology of human skeletal remains from Bagor: Osteology. In J. R. Lukacs, V. N. Misra, & K. A. R. Kennedy (Eds.), *Bagor and Tilwara: Late Mesolithic cultures of Northwest India* (pp. 27–60). Poona: Deccan College Postgraduate and Research Institute.

Kennedy, K. A. R. (1993). Recent discoveries and studies of the human skeletal remains of Ancient Sri Lankans: a paleoanthropological update. In P. K. Seth & S. Seth (Eds.), *New perspectives in anthropology* (pp. 299–341). New Delhi: MD Publications.

Kennedy, K. A. R. (2000). Yeleswaram revisited: the skeletal record. *Man and Environment, 35*(1), 35–57.

Kennedy, K. A. R. (2002). Megalithic Pochampad: the skeletal biology and archaeological context of an iron age site in Andhra Pradesh, India. *Asian Perspectives, 41*(1), 103–128.

Kennedy, K. A. R., & Chiment, J. (1991). The fossil hominid from the Narmada Valley, India: *Homo erectus* or *Homo sapiens*? *Indo-Pacific Prehistory Association Bulletin, 10*(1), 42–58.

Kennedy, K. A. R., Sonakia, A., Chiment, J., & Verma, K. K. (1991). Is the Narmada hominid an Indian *Homo erectus*? *American Journal of Physical Anthropology, 86*, 475–496.

Kivisild, T., Rootsi, S., Metspalu, M., Mastana, S., Kaldma, K., Parik, J., et al. (2003). The genetic heritage of the earliest settlers persists both in Indian tribal and caste populations. *American Journal of Human Genetics, 72*(2), 313–332.

Long, J. S. (1997). *Regression models for categorical and limited dependent variables*. Thousand Oaks, CA: Sage.

Patnaik, R., Chauhan, P. R., Rao, M. R., Blackwell, B. A. B., Skinner, A. R., Sahni, A., et al. (2009). New geochronological, paleoclimatological, and archaeological data from the Narmada Valley hominin locality, central India. *Journal of Human Evolution, 56*(2), 114–133.

Pope, G. (1992). Craniofacial evidence for the origin of modern humans in China. *Yearbook of Physical Anthropology, 35*, 243–298.

Press, S. J., & Wilson, S. (1978). Choosing between logistic regression and discriminant analysis. *Journal of the American Statistical Association, 73*(364), 699–705.

Rao, C. R. (1948). The utilization of multiple measurements in problems of biological classification. *Journal of the Royal Statistical Society Series B (Methodological), 10*(2), 159–203.

Rhode, M. P., & Arriaza, B. T. (2006). Influence of cranial deformation on facial morphology among prehistoric South Central Andean populations. *American Journal of Physical Anthropology, 130*(4), 462–470.

Rightmire, G. P. (1998). Human evolution in the Middle Pleistocene: the role of *Homo heidelbergensis*. *Evolutionary Anthropology, 6*(6), 218–227.

Rightmire, G. P. (2001). Patterns of hominid evolution and dispersal in the Middle Pleistocene. *Quaternary International, 75*, 77–84.

Santa Luca, A. P. (1978). A Re-examination of Presumed Neandertal-like Fossils. *Journal of Human Evolution, 7*, 619–636.

Santos, R. (1981). *Effects of imputation on regression coefficients* (pp. 140–145). American Statistical Association: Proceedings of the Survey Research Section.

Schafer, J. L. (1997). *Analysis of incomplete multivariate data*. Boca Raton, FL: Chapman & Hall/CRC Press.

Schafer, J. L. (1999a). Multiple imputation: a primer. *Statistical Methods in Medical Research, 8*, 3–15.

Schafer, J.L. (1999b). *NORM: multiple imputation of incomplete multivariate data under a normal model*. 2.3 edn. Pennsylvania, PA: Pennsylvania State University, State College.

Schafer, J. L., & Olsen, M. K. (1998). Multiple imputation for multivariate missing-data problems: a data analyst's perspective. *Multivariate Behavioral Research, 33*(4), 545–571.

Schafer, J. L., & Graham, J. W. (2002). Missing data: our view of the state of the art. *Psychological Methods, 7*(2), 147–177.

Scherer, A. K. (2007). Population structure of the classic period Maya. *American Journal of Physical Anthropology, 132*(3), 367–380.

Sonakia, A. (1984). The skull-cap of early man and associated mammalian cauna from Narmada Valley alluvium, Hoshangabad area, Madhya Pradesh (India). *Geological Survey of India Records, 113*(6), 159–172.

Sonakia, A. (1985a). Early *Homo* from Narmada Valley, India. In E. Delson (Ed.), *Ancestors: The Hard Evidence* (pp. 334–338). New York: Alan R. Liss.

Sonakia, A. (1985b). Skull cap of an early man from the Narmada Valley alluvium (Pleistocene) of Central India. *American Anthropologist, 87*, 612–615.

Stefan, V. H. (2004). Assessing intrasample variation: analysis of Rapa Nui (Easter Island) museum cranial collections example. *American Journal of Physical Anthropology, 124*(1), 45–58.

Swisher, C. C., Rink, W. J., Antón, S. C., Schwarcz, H. P., Curtis, G. H., Suprijo, A., Widiasmoro (1996). Latest *Homo erectus* of Java: potential contemporaneity with *Homo sapiens* in Southeast Asia. *Science, 274*, 1870–1874.

Tattersall, I. (1986). Species recognition in human paleontology. *Journal of Human Evolution, 15*, 165–175.

Thorne, A. G., & Wolpoff, M. H. (1981). Regional continuity in Australasian Pleistocene hominid evolution. *American Journal of Physical Anthropology, 55*, 337–349.

Tiwari, M. P., & Bhai, H. Y. (1997). Quaternary stratigraphy of the Narmada Valley. *Geological Survey of India Special Publication, 46*, 33–63.

Wayman, J. C. (2003). *Multiple imputation for missing data: what is it and how can I use it?* Chicago, IL: Annual Meeting of the American Educational Research Association.

Wolpoff, M.H. (1999). Paleoanthropology, 2nd edn. New York: McGraw-Hill.

Wolpoff, M. H., Wu, X., & Thorne, A. G. (1984). Modern *Homo sapiens* origins: a general theory of hominid evolution involving the fossil evidence from East Asia. In F. H. Smith & F. Spencer (Eds.), *The origins of modern humans: a world survey of the fossil evidence* (pp. 411–483). New York: Alan R. Liss.

Wu, X. (2004). On the origin of modern humans in China. *Quaternary International, 117*(1), 131–140.

Chapter 11
Cranial Morphology and Variation of the Earliest Indonesian Hominids

Yousuke Kaifu, Etty Indriati, Fachroel Aziz, Iwan Kurniawan, and Hisao Baba

Abstract The Early Pleistocene deposits of Sangiran, Central Java, have yielded the oldest hominid fossils in Indonesia. This Sangiran assemblage is the oldest sizable hominid fossil collection so far known from Asia, and along with the African and Georgian arrays is the best comparative collection of Early Pleistocene *Homo* in the world. For this reason, the Sangiran hominid materials are important for understanding the dispersal and paleobiology of earlier members of our genus. However, due to ambiguous contextual documentation and the fragmentary nature of many of the existing fossils, our understanding of the taxonomic affinities and morphological variation of the earliest Indonesian hominids remains unclear.

In this paper, we review recent chronostratigraphic data, and examine the Sangiran cranial remains. Contrary to previous arguments that the oldest Indonesian hominids are characterized by cranial robusticity, we propose that these hominids are actually highly variable, including both robust and gracile morphotypes. In overall cranial size and shape, and dentognathic morphology, the earliest Indonesian hominids appear to be comparable to c. 1.7 Ma early *Homo erectus* from East Africa. Evolutionary and taxonomic implications of these findings are discussed.

Keywords Out of Africa 1 • *Homo erectus* • Early Pleistocene • Java • Sangiran • Trinil • Human evolution

Introduction

Recent field research, dating, and morphological analyses of fossils suggest that the first colonization of East Asia by early hominids had been earlier than traditionally thought, and probably dates back to 1.6 Ma (Zhu et al. 2004, 2008; Antón and Swisher 2004; Kaifu et al. 2005b; but see Gao et al. 2005; Huffman et al. 2006; and Li et al. 2008). This raises many interesting questions: the exact dates of the initial colonization, the identity of the first settlers, their migration routes and evolutionary histories, and the number of early immigration events to name just a few (Gabunia et al. 2002; Mithen and Reed 2002; Antón et al. 2003; Dennell 2003; Langbroek 2004; Antón and Swisher 2004; Dennell and Roebroeks 2005; Shea 2006; Bettis et al. 2009). The oldest hominid fossil array from Sangiran, Java represents, along with the assemblage from Dmanisi, the only existing sizable collection of Early Pleistocene hominid remains from Asia and is well-suited to address many of these questions. Due to confusing contextual documentation (e.g., Arif et al. 2002; see below) and the fragmentary nature of many of the existing Sangiran fossils, our current understanding of the variation within this hominid group remains clouded. Opinions have varied regarding the Sangiran fossils' taxonomy, phylogenetic position, and the number of species recognized in the collection. Some researchers suggest that two or three distinct hominid groups existed, with some of the Sangiran fossils comparable to African early *Homo* or even to australopithecines. Other scientists have found little evidence to support these claims (see Rightmire 1990; Kramer 1994, for reviews). Recent studies on the Sangiran dentognathic remains provide a new perspective on this issue.

The Sangiran region has yielded the vast majority of Early Pleistocene hominid remains from Indonesia (Fig. 11.1). The hominid specimens from this area and Trinil have often been pooled to compare with other hominid groups. However, there is an old claim that significant temporal variation exists within the Sangiran hominid assemblage (e.g., Jacob 1976; Wolpoff 1999). This perspective has gained support from recent comprehensive analyses of the dentognathic remains (Kaifu et al. 2005b, 2007; Kaifu 2006). Compared to the cranial fossils, the dentognathic materials have a larger sample size, and were more appropriate in examining the temporal variation of the Early Pleistocene Indonesian hominids. According to Kaifu et al. (2005b), the approximate boundary of the two

Y. Kaifu (✉)
Department of Anthropology, National Museum of Nature and Science, 3-23-1 Hyakunincho, Shinjuku-ku, Tokyo 169-0073, Japan
e-mail: baba@kahaku.go.jp

C.J. Norton and D.R. Braun (eds.), *Asian Paleoanthropology: From Africa to China and Beyond*, Vertebrate Paleobiology and Paleoanthropology, DOI 10.1007/978-90-481-9094-2_11, © Springer Science+Business Media B.V. 2010

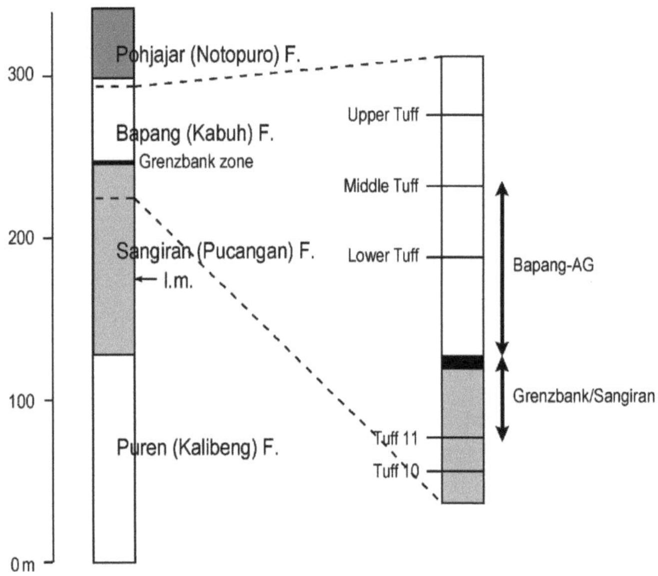

Fig. 11.1 Stratigraphy in the Sangiran area. The two hominid fossil bearing zones defined in this study are indicated on the right. "l.m." indicates the lowest mammalian fossil bearing horizon recorded in Watanabe and Kadar (1985). See Watanabe and Kadar (1985) for detailed descriptions of the sediments

chronological groups is recognized at the top of the Grenzbank zone (the lowermost part of the Bapang [Kabuh] Formation). The younger Sangiran dentognathic remains were labeled by these authors as Bapang-AG (Bapang above Grenzbank zone) and the older specimens were characterized as Grenzbank/Sangiran (Grenzbank zone and Sangiran Formation) assemblages, respectively.

The hominid fossils included in the Grenzbank/Sangiran assemblage are restricted to a relatively narrow stratigraphic zone, which consists of the Grenzbank zone and the underlying uppermost part of the Sangiran Formation (Matsu'ura 1982; Itihara et al. 1985; Larick et al. 2000; Kaifu et al. 2005a, b and references therein). At present, there is no compelling evidence for the presence of hominid fossils below Tuff 11 of the Sangiran Formation. Recent $^{40}Ar/^{39}Ar$ dating of several tuffs in the region indicates that the Grenzbank/Sangiran levels are older than 1.5 Ma (Larick et al. 2001), whereas magnetostratigraphy and other $^{40}Ar/^{39}Ar$ dates suggest younger chronologies (~0.9–1.1 Ma: Hyodo et al. 1993; Sémah 2001). Although the former chronology is more often-cited (Antón and Swisher 2004), some researchers are still cautious about accepting this long chronology (e.g., Matsu'ura et al. 2006).

The morphological differences between the Bapang-AG and the Grenzbank/Sangiran chronological samples of the dentognathic remains are substantial. While the younger Bapang-AG mandibular and dental specimens are morphologically advanced, showing a degree of dentognathic reduction comparable to the Middle Pleistocene northern Chinese *H. erectus*, the Grenzbank/Sangiran specimens display features that are equally or even slightly more primitive than African early *Homo erectus* (= *H. ergaster*) (e.g., large tooth size, comparatively narrow alveolar arcade shape, a robust mandibular corpus, and an anteriorly located, well-developed lateral prominence).[1]

In this study we explore morphological variation of the oldest hominid cranial assemblage from the Grenzbank/Sangiran stratigraphic levels with reference to the chronological scheme supported by the dentognathic evidence.

More than ten well-preserved adult cranial remains are known from the Early Pleistocene localities in Java (Indriati 2004; Kaifu et al. 2008). However, there have been few comprehensive studies on their chronological variation. In their descriptions of the Indonesian fossil records, Wolpoff (1999) and Antón (Antón 2003; Antón and Swisher 2004) compared older (Sangiran 4, 27, 31) and younger (Trinil 2, Sangiran 2, 3, 10, 12, 17, IX [Skull IX]) cranial specimens. Their definitions of chronological assemblages differed from each other. Wolpoff's definition was similar to the one proposed here by this study. However, Antón defined the boundary at the border between the Bapang and Sangiran Formations (bottom of the Grenzbank zone).[2] Despite this difference the chronologically delineated samples of Wolpoff and Antón share many commonalities. Both Wolpoff and Antón noted that the older crania seemed to be characterized by general robustness and possibly by small size. Antón emphasized the commonalities between the two cranial samples. The notion that a robust cranium is a primitive condition in Indonesian early hominids has its roots in the research of Weidenreich (1945) described Sangiran 4 in the 1940s.

In this study, we re-analyze the Grenzbank/Sangiran cranial assemblage. This assemblage includes, in addition to Sangiran 4, 27, and 31, some other cranial fragments as well as Perning 1 (Table 11.1). Furthermore, this definition also introduces the possibility of including Trinil 2 and Sangiran 2 in this oldest Indonesian hominid assemblage.

We examine the suggested antiquity of Trinil 2 and Sangiran 2 (see below) from three perspectives of morphology, with the null hypothesis being that the two crania had been derived from the Grenzbank/Sangiran stratigraphic levels. Although the question of stratigraphic position cannot be ultimately solved from fossil morphology, we attempt this approach to seek the most reasonable stratigraphic arrangement that fits the currently available evidence. In order to capture cranial morphology and variation of the oldest

[1]Wolpoff (1999) also defined two chronological groups of the Early Pleistocene Javanese hominin remains by setting a boundary at the top of the Grenzbank zone. However, he failed to find this temporal variation due to some errors in his stratigraphic allocation of the fossil specimens.

[2]To be exact, whether Sangiran 4 was unearthed from the uppermost part of the Sangiran Formation or overlying Grenzbank zone is currently indeterminate (Matsu'ura 1982; Itihara et al. 1985).

Table 11.1 Indonesian hominid cranial specimens unquestionably or potentially derived from the Grenzbank/Sangiran stratigraphic levels[a]

Specimen	Year of Discovery	Description	Stratigraphy[b]
Trinil 2	1891	Callote	Grenzbank zone?
Perning 1	1936	Calvaria	Upper Pucangan F. (Huffman et al. 2006)
Sangiran 2	1937	Callote	Lower Bapang F. or Grenzbank zone
Sangiran 4	1938/39	Posterior cranium and maxilla	Uppermost Sangiran F. or Grenzbank zone
Sangiran 26	1978	Temporal fragment	Sangiran F. (Jacob 1980)
Sangiran 27	1978	Crushed anterior cranium with face	Sangiran F.
Sangiran 31	1979	Crushed posterior cranium	Sangiran F.
Hanoman 1	1989	16 vault fragments	Uppermost Sangiran F. (Widianto et al. 1994)
Bp 9408	1994	Frontal fragment	Grenzbank zone
Bu 9604	1996	Occipital fragment	Uppermost Sangiran F. (or Grenzbank zone)

[a]This table lists only those specimens whose alleged find spots or areas have been recorded by scientists. See Jacob (1975), Rightmire (1990), Aziz (2001), Indriati (2004), for catalogue information of each specimen

[b] See text for those without references

Indonesian hominids, we include as wide a sample of Grenzbank/Sangiran fossils as possible. We also present a morphological description of a poorly known occipital fragment, Bu 9604 (Aziz 2001).

Materials and Methods

Materials

Ten cranial specimens are associated with the Grenzbank/Sangiran strata (or stratigraphic equivalents at Trinil and Perning) (Table 11.1). Of these, seven adult/young adult specimens are included in the present study. Most of the hominid remains from Sangiran have been collected by local villagers; thus, caution should be used when assigning these specimens to specific stratigraphic intervals. We describe these specimens below.

Trinil 2 Until the early 1980s, it had been widely believed (Weidenreich 1945; Jacob 1975; Le Gros Clark 1978; Day 1986; Wolpoff 1999; Antón 2003) that Trinil 2 was significantly younger than the Grenzbank/Sangiran hominid fossils (e.g., Sangiran 4). This belief, which persists today, was originally supported by Von Koenigswald's biostratigraphic reconstructions. Von Koenigswald (1934, 1935) believed that the fauna from Trinil (his "Trinil fauna") was younger than the Grenzbank/Sangiran fauna from the Sangiran region (his "Jetis fauna"). However, evidence presented by de Vos, Sondaar, and their colleagues (de Vos et al. 1982, 1994; Leinders et al. 1985) led them to conclude that Dubois's main bone bed at Trinil (Trinil H.K.), that had yielded Trinil 2, was correlated to the lower part of the Bapang Formation in the Sangiran region, including the Grenzbank zone. Field observations by Ithihara and his colleagues (IJRCP 1979) identified this same strata as a basal conglomeratic layer of the Kabuh Formation, which showed similar lithological characteristics with Grenzbank in Sangiran (Soeradi et al. 1985). Although similarities in lithofacies alone does not indicate contemporaneity, this stratigraphic interpretation may be supported by de Vos and colleague's faunal correlation. Bartstra (1982) suggested the possibility that Dubois had mixed fossils from the upper stratigraphic layers at the site, but there are counterarguments to this claim (de Vos and Sondaar 1982; Sondaar et al. 1983).

Sangiran 2 The alleged provenance of the Sangiran 2 cranium is in the Bapang-AG levels between the Middle and the Lower Tuffs (Itihara et al. 1985). However, the fluorine content of the bone suggests that it is more plausibly derived from the Grenzbank zone (Matsu'ura 1982; Matsu'ura et al. 2005).

Sangiran 4 Itihara et al. (1985) assigned this specimen to an area that stretches from the lowest Grenzbank zone to the uppermost part of the Sangiran Formation. This broad placement has been supported by fluorine analysis (Matsu'ura 1982).

Sangiran 27 and 31 These specimens were recovered during a canal excavation in the 1970s, and it is widely believed that they were unearthed from the Sangiran Formation that is exposed at the site (Indriati and Antón 2008). The crushed state of these specimens has been used to support this stratigraphic assignment (Antón and Swisher 2004), yet the exact provenance of these specimens remains unclear. The fossils have been ascribed to either the uppermost (Itihara et al. 1985; Larick et al. 2000; Indriati and Antón 2008) or lower (Jacob 1980; Sartono 1982; Swisher et al. 1994) parts of the Sangiran Formation.

Bp 9408 This is a surface find by a local inhabitant, Mr. Sutanto, from the riverbed of the Brangkal River. Field studies and the preservation of the specimen have led Baba et al. (1998, 2004) to suggest that it had been eroded out from the Grenzbank zone. These sediments have been observed in the outcrop about 100 m upstream from the collection site of this specimen. This inference has been supported by laboratory analyses of the fluorine and other trace elements in the bone (Sudijono et al. 2001).

Bu 9604 This specimen was discovered in April 1996 by a local person, from an outcrop near the Bukuran village.

The same series of outcrop yielded the Sangiran 1b mandible in 1936. Only the uppermost part of the Sangiran Formation (Tuff 10 and strata above it) is recorded in this area (V-4–6 of Fig. 7 in Itihara et al. 1985). This does not rule out the possibility that the overlying Grenzbank zone was once present in this area. According to the discoverer, the specimen was found in situ while he was digging into black clay sediments to collect shells. Stratigraphy of the area is complicated by landslides and mudfills, and it is possible that the specimen had been unearthed from a collapsed block of the sediments. Bu 9604 is well-mineralized and exhibits a dark color, a condition typical for the vertebrate bones from the Grenzbank zone. Larick et al. (2001) tentatively assigned the stratigraphic position of this specimen (Brn-1996.04 in their specimen number) to the uppermost part of the Sangiran Formation. The results of minor and trace element analysis of this fossil should determine whether it is from the Sangiran Formation or the Grenzbank zone (Matsu'ura, personal communication).

Comparative Samples and Data Collection

The Indonesian cranial specimens previously mentioned are compared to a chronologically younger sample from Indonesia, and chronologically older or penecontemporaneous Afro-Eurasian *Homo* samples. The first sample is comprised of specimens from the Bapang-AG stratigraphic levels in Sangiran (Sangiran 10, 12, 17, 38; Skull IX, Bukuran: Aziz 2001; Indriati 2004; Kaifu et al. 2005b, Table 11.1). The latter samples are crania of East African *H. habilis sensu lato* (KNM-ER 1470, 1805, 1813; OH 24), East African early *H. erectus* (*H. ergaster*: KNM-ER 3733, 3883), and Dmanisi *H. erectus* (D 2280, 2282, 3334). Only well-preserved, adult specimens are included in these comparative samples. Subadult specimens such as KNM-WT 15000 and D 2700 are complete, but do not exhibit fully mature features (Walker and Leakey 1993; Vekua et al. 2002; Rightmire et al. 2006).

The data for the Indonesian specimens were taken by us (Y.K. and H.B. with the assistance of other members) from the original specimens, except for Sangiran 31 where a cast was used to capture measurements. Due to the poor preservation, previous studies (e.g., Rightmire 1990; Antón 2002a) did not collect maximum cranial length data on Trinil 2 and Sangiran 2. However, we think the condition of these specimens is adequate enough to obtain reasonable estimates of their maximum lengths. The glabellar region always protrudes anteriorly in the known Trinil/Sangiran specimens (Sangiran 17, 18a; Skull IX, Bukuran, Bp 9408). Trinil 2 preserves the superior rim of the glabellar prominence, and Sangiran 2 retains a portion of the supraorbital torus above the left midorbital region. We first measured the lengths between the opisthocranion and these preserved parts (projected distance to the sagittal plane in the case of Sangiran 2). After reviewing other Sangiran frontals, we decided to add 3.0 mm and 3.5 mm to these direct lengths for Trinil 2 and Sangiran 2, respectively. The parietals of Sangiran 4 are broken along its coronal suture, and porion-bregma height cannot be measured directly. The projected distance between its porion and the forefront broken edge of the sagittal suture is 82 mm. Assuming that the bregma was 10–20 mm ahead from this breakage, we inferred that 82 + 1 mm was a reasonable estimate for the original porion-bregma height of this specimen. The data for the non-Indonesian specimens were taken from the literature (Wood 1991; Gabunia et al. 2000; Rightmire et al. 2006; Lordkipanidze et al. 2006) or casts.

Analysis

Hypotheses To Be Tested

Null hypothesis: Trinil 2 and/or Sangiran 2 originated from the Grenzbank/Sangiran strata.

Test 1: Morphology of Trinil 2 and Sangiran 2

If the morphology of Trinil 2 and Sangiran 2 come to the end of, or even outside, the variation range of the Bapang-AG cranial sample, the null hypothesis cannot be rejected. The hypothesis would receive further support if the observed differences indicate primitive aspects of Trinil 2 and Sangiran 2 relative to the crania from the Bapang-AG strata. Here, we focus on several cranial traits which show comparatively distinct differences between Afro-Eurasian early *Homo* (*H. habilis s.l.* and Dmanisi *H. erectus*) and Bapang-AG *H. erectus* (Table 11.2). This selection is based on previous

Table 11.2 Characters examined in the first test and their polarities between early *Homo* and Bapang-AG *H. erectus*

	Afro-Eurasian early *Homo* (Ancestral)	Bapang-AG *H. erectus* (Derived)
Overall cranial size	Small	Large
Frontal squama breadth	Narrow	Broad (variable)
Cranial bone thickness	Thin	Thick
Lateral projection of the mastoid region relative to the temporal wall	Strong	Weak
Course of the temporal lines on the parietals	Approach the midline	Parallel or divergent posteriorly
Supraorbital torus vertical thickness	Thin	Thick
Occipital torus development	Weak	Strong

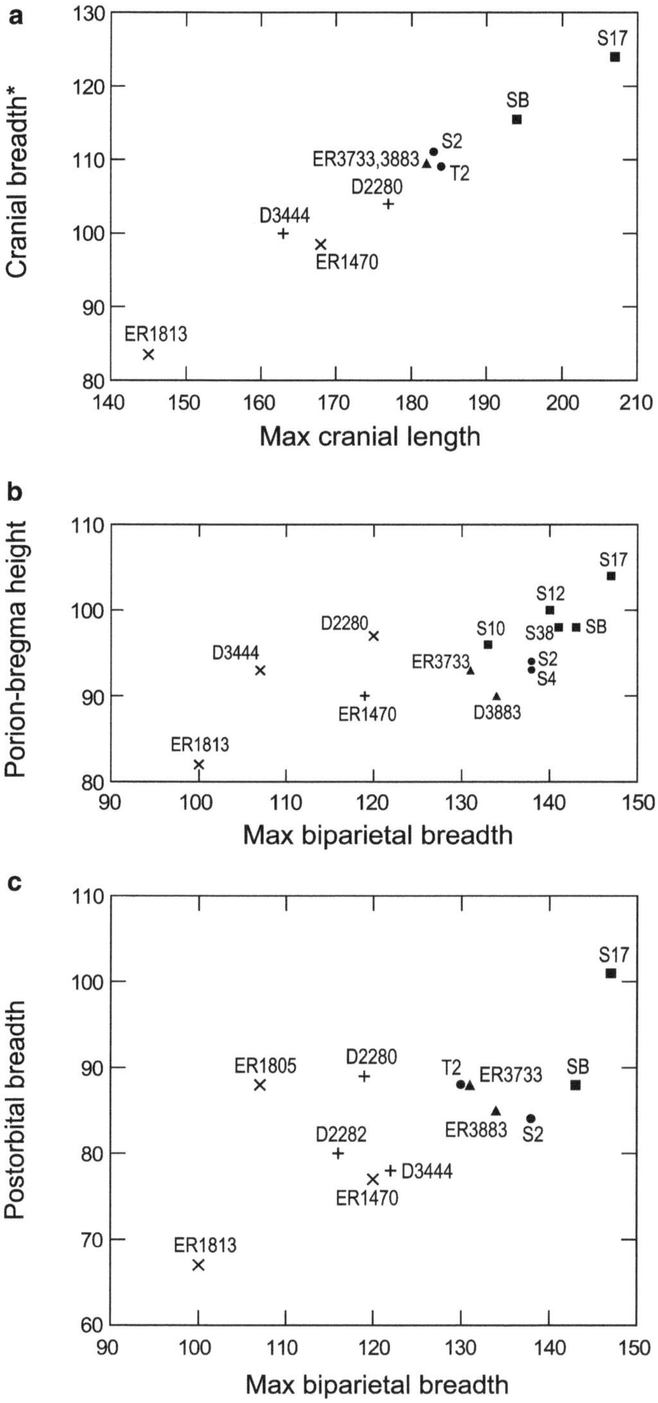

Fig. 11.2 Bivariate plots of cranial breadth, length, and height measurements (mm). *(maximum biparietal breadth + postorbital breadth)/2. ●, Grenzbank/Sangiran specimen (Sangiran 4), and Trinil 2 and Sangiran 2; ■, Bapang-AG *H. erectus*; ×, East African *H. habilis*; ▲, East African early *H. erectus*; +, Dmanisi *H. erectus*

studies (e.g., Wood 1991; Tobias 1991; Bräuer and Mbua 1992; Antón 2003; Rightmire et al. 2006; Lordkipanidze et al. 2006), but is in part modified by our own observations. Inferred character polarities between the two groups are presented in Table 11.2.

Table 11.3 Cranial bone thickness[a]

Landmark	*H. habilis*	Dmanisi	Turkana *H. erectus*	Bapang-AG
Bregma	3.5–5.8	7.0–8.0	8.0	9.0–10.0
Parietal eminence	6.5–8.0		7.5–10.0	9.0–12.0
Lambda	4.5–7.8	6.0–7.0	7.0–8.8	10.0–14.0
Asterion	5.7–7.7	6.7–10.0	10.2	14.0–19.5
Opisthocranion	11.0–19.0	10.0–14.0	14.0–24.0	15.0–24.0

[a] The range of variation is shown for each sample

Comparisons of cranial length, breadth, and height measurements indicate that Trinil 2 and Sangiran 2, as well as Sangiran 4, are some of the smallest specimens in Bapang-AG, in terms of gross cranial size (Fig. 11.2). Although the porion-bregma height of Trinil 2 cannot be measured because its poria are missing, it was probably as low as that of Sangiran 2, judging from the similarities in lateral profile between the two vaults.

Trinil 2 and Sangiran 2 do not display a relative frontal widening like Sangiran 17, one of the Bapang-AG crania (Fig. 11.2). A wide frontal squama characterizes chronologically later Indonesian *H. erectus* from Sambungmacan and Ngandong (Santa Luca 1980; Antón 2002a; Kaifu et al. 2008).

A thin cranial vault is an ancestral condition in early groups of *Homo* (Wood 1991; Bräuer and Mbua 1992; Rightmire et al. 2006). There are distinct differences in the cranial thickness between East African *H. habilis* and Dmanisi grouping together and the Bapang-AG crania grouping separately. The Turkana *H. erectus* occupies an intermediate position (Table 11.3). Trinil 2 and Sangiran 2 display contrasting trends in this aspect. While the latter falls within the variation range of Bapang-AG, the vault of the former is thinner and falls outside the observed range of variation for Bapang-AG (Fig. 11.3). Although Antón (1997a) suggested that Sangiran 2 may have suffered from hyperostosis calvariae internae, the resultant bone thickening was, according to her, restricted to small localized areas on the internal frontal and parietal surfaces.

As noted by Rightmire et al. (2006), when a skull is viewed from behind, the mastoid region projects laterally relative to the temporal wall in East African *H. habilis* and Dmanisi *H. erectus*; and this is probably an ancestral condition of *Homo*. This feature should be evident at least partially in the *MP projection over TW* (= *maximum bimastoid breadth – maximum biparietal breadth*) ratio (Table 11.4). *H. habilis* and Dmanisi *H. erectus* have large positive values for this variable because their cranial breadths are distinctly wider at the mastoid region than at the temporal squama region. The values of *MP projection over TW* are reduced in East African early *H. erectus* due to lateral expansion of their temporal walls, although they do not reach the condition of *H. sapiens* where the value tends to be negative (*maximum biparietal breadth* > *maximum bimastoid breadth*). Sangiran 2, 4, and

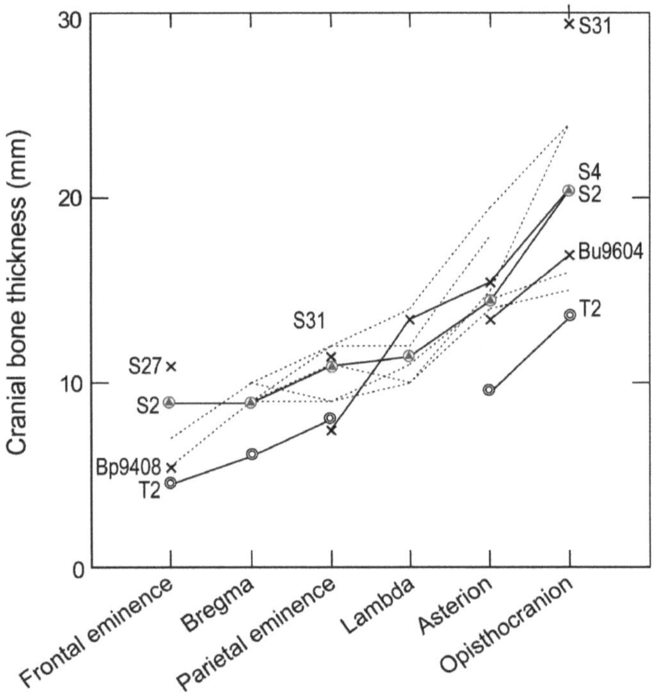

Fig. 11.3 Cranial bone thickness of the Indonesian specimens (mm). *Crosses* and *solid lines*: Grenzbank/Sangiran specimens, Trinil 2, and Sangiran 2. *Dotted lines*: Bapang-AG specimens. The measurements of Sangiran 31 are from a cast. Sangiran 27 and 31 are severely crushed (Antón 2003), but thickness measurements were taken at some undistorted regions of their frontal, parietal, and occipital bones. The frontal eminence of Sangiran 27 cannot be identified because of the damage. The measurement for Sangiran 27 presented here was taken at the location considered to be the right eminence. Our examination of the thickness distribution in Trinil 2, Sangiran 2, Sangiran 17, Bukuran, and Bp 9408 indicates that the thickness at and around the eminence does not vary significantly. Note the thin vaults of Trinil 2 and Bp 9408, and the great variation in the Grenzbank/Sangiran sample

the Bapang-AG crania display zero or small positive values, which may be a derived trait for these Sangiran hominids.

Earlier *Homo* (*H. habilis s.l.* and early *H. erectus* from Georgia and Africa) and Bapang-AG *H. erectus* show fairly distinct differences in the course of the temporal line on the frontal and parietal bones. In the former group, the right and left temporal lines approach toward the vault's midline on the parietals (Rightmire et al. 2006). However in Bapang-AG *H. erectus* the two lines keep a parallel relationship, or gradually diverge from each other as they extend posteriorly. This latter pattern also appears in chronologically later Indonesian *H. erectus* from Sambungmacan and Ngandong. The variation in temporal line direction is reflected in the relationships among three intertemporal line breadths in Table 11.4. The breadths decrease posteriorly in earlier *Homo*, whereas the identical *bistephanic breadth* and *minimum breadth on the parietal* in Bapang-AG *H. erectus* reflect a parallel or diverging relationship of the lines. In this respect, Sangiran 2 (and possibly Sangiran 10) shows a certain degree of affinity with earlier *Homo*. Trinil 2 is within the variation of Bapang-AG for this variable.

The supraorbital torus of *H. habilis* is thin and gracile particularly at its lateral ends. In contrast, the supraorbital torus of *H. erectus* is typically thick and displays strong anterior projection with the formation of a distinct supratoral plane or sulcus above the superior orbital rim. The distinct torus of Sangiran 2 is anteriorly protruding, but it is thinner and more delicate compared to the tori of four Bapang-AG crania (Fig. 11.4). Even the smallest cranium of the Bapang-AG, Sangiran 10, has a torus which is much thicker than that of Sangiran 2. The torus of Trinil 2 is damaged, and unfortunately its vertical development cannot be examined.

Poor expression or absence of the occipital torus is regarded as an ancestral condition of *Homo* (Lordkipanidze et al. 2006). The transversely straight, vertically thin occipital tori of Sangiran 2 and Trinil 2 are slightly less projecting posteriorly than the most delicate torus of Bapang-AG, Sangiran 10 (Rightmire 1990). The tori of Sangiran 12 and 17 exhibit marked posterior protrusion, and those of Sangiran 38, Skull IX, and Bukuran are vertically thicker compared to Sangiran 10 (Fig. 11.5). On the other hand, the torus morphology of Sangiran 2 and Trinil 2 is distinct compared

Table 11.4 Cranial measurements that reflect the projection of the mastoid process and the course of the temporal line[a]

				Sangiran Bapang-AG					*H. habilis*	Dmanisi	African early
	T 2	S 2	S 4	S 10	S 12	S 17	S 38	SB			*H. erectus*
Mastoid process lateral projection											
Maximum biparietal breadth	130	138	138	133	140	147	141	143	100–120	116–122	131–134
Maximum mastoid breadth	–	139	142	138	147	149	146	143	113–128	132–136[c]	135–142
MP projection over TW[b]	–	1	4	5	7	2	5	0	13–21	10–27	1–11
Intertemporal line breadths											
Minimum frontal breadth	85	69	–	–	–	88	–	74	65–88	66–75	80–83
Bistephanic breadth	85	69	–	73	(71)	(93)	86	80	–	–	–
Minimum breadth on the parietal	–	62	–	–	(71)	(93)	86	–	7–60?	51–70	66–70?

[a]Variation ranges are shown for the comparative samples

[b]Maximum mastoid breadth – maximum biparietal breadth

[c]The published supramastoid breadths are indicated here assuming that they approximate the maximum mastoid breadths in these specimens

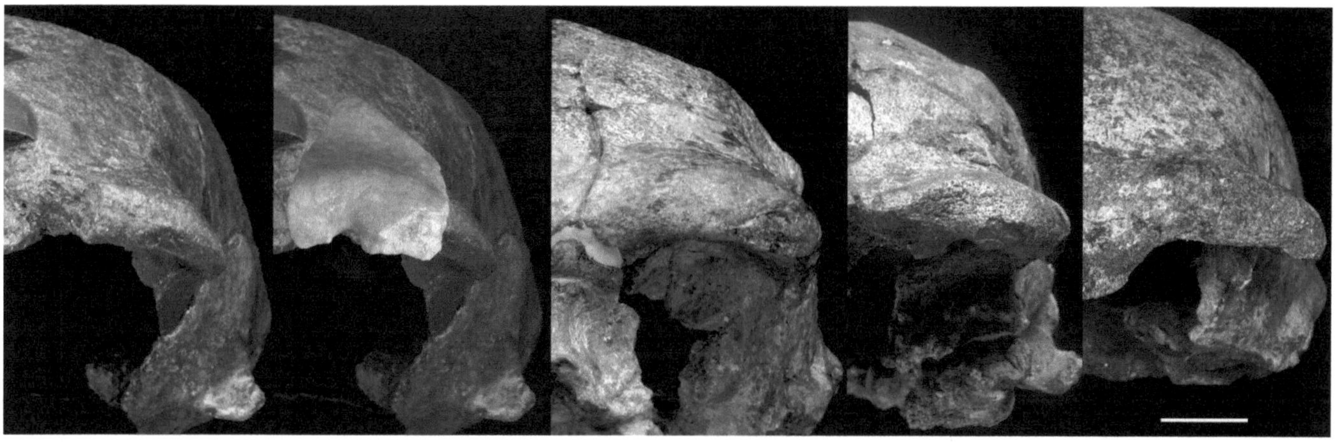

Fig. 11.4 Supraorbital tori of the Indonesian cranial specimens. From *left*, Sangiran 2, Sangiran 10 (flipped horizontally) superimposed on Sangiran 2 (cast), Sangiran 17, Skull IX, and Bukuran. Scale = 2 cm. Note the comparatively thin and delicate supraorbital torus of Sangiran 2

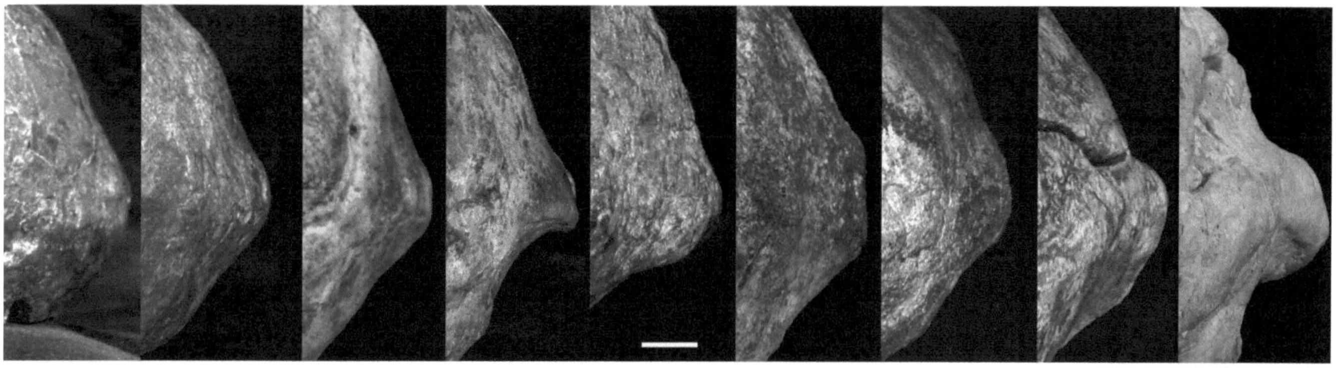

Fig. 11.5 Occipital tori of the Indonesian cranial specimens. From *left*, Trinil 2, Sangiran 2, 10, 12, 17, 38, Bukuran, Sangiran 4, and 31 (reconstructed cast by D. Tyler). Scale = 1 cm. The tori of Bapang-AG crania are variable, but tend to be thicker and more protruding compared to Trinil 2 and Sangiran 2. Two Grenzbank/Sangiran crania, Sangiran 4 and 31, show remarkable development in their tori

to the African and Dmanisi specimens. In these Indonesian specimens, a distinct external occipital protuberance is lacking, and the central part of the torus is even or slightly depressed relative to its right and left sides. The intact upper nuchal plane of Sangiran 2 is smooth, and there is minimal or no development of an external occipital crest. These characteristics are shared with some of the Bapang-AG crania (Sangiran 10, 38; Bukuran).

Trinil 2 and Sangiran 2 tend to show comparatively primitive morphology in some important aspects of the cranium. The null hypothesis cannot be rejected from the first test. It is possible that the small, thin, and gracile cranial morphology exhibited by these crania represent female characteristics within the Bapang-AG group. However, their male-like characteristic of a posteriorly sloping, flat frontal squama appears to contradict this supposition. The frontal squama of Trinil 2 and Sangiran 2 are flatter, and are inclined strongly posteriorly compared to those of Skull IX and Bukuran. Within the Bapang-AG and Ngandong assemblages of *H. erectus*, large possible male crania (Sangiran 17, Ngandong 6) tend to show posteriorly sloping, flat frontal squama, whereas the squamae of possibly small female specimens (e.g. Skull IX, Ngandong 7) stand more vertically and are more curved in lateral view.

Test 2: Bp 9408 and Bu 9604

The second test is whether Trinil 2 and Sangiran 2 falls within the variation range of the Grenzbank/Sangiran cranial sample. If there are some Grenzbank/Sangiran crania which are morphologically similar to Trinil 2 and Sangiran 2, the null hypothesis cannot be rejected.

As recognized from previous descriptive and comparative studies (Weidenreich 1945; Jacob 1976, 1980; Santa Luca 1980; Rightmire 1990; Sartono 1982; Sartono and Grimaud-Hervé 1983; Tyler et al. 1995; Wolpoff 1999; Antón 2003; Schwartz and Tattersall 2005; Indriati and Antón 2008), the three famous Grenzbank/Sangiran crania, Sangiran 4, 27 and 31, are very different from Trinil 2 and Sangiran 2, and do not serve to support the null hypothesis. However, Bp 9408 and Bu 9604 (Aziz 2001) are morphologically similar to Trinil 2 and Sangiran 2.

Bp 9408

Bp 9408 is a frontal fragment that preserves the lower central portion of the squama and the glabellar region, including the nasal root (Fig. 11.6). The overall size, anteroposterior development of the supraorbital torus, and nasal bone width (15 mm at the nasofrontal suture) of the specimen are largely comparable to other adult Sangiran frontals (Sangiran 2, Skull IX, Bukuran), suggesting its adult status. The basic morphological description has been provided in Baba et al. (2004), where the similarities between it and the Sangiran and Trinil *H. erectus* specimens (Trinil 2; Sangiran 2, 17) have been noted. Bp 9408 shares the following characteristics: an anteroposteriorly wide and flat supratoral plane, an anteriorly projecting glabellar region, and a strong posterior inclination of the frontal squama with a midline ridge.

Additional features are mentioned here. First, the squama of Bp 9408 is as thin as that of Trinil 2 (Fig. 11.3). The thickness of the frontal squama varies depending on the measured location, due to the gyral impressions of the frontal lobe. However, the range of the thickness in Bp 9408 is limited to 4.5–5.5 in the area more than 15 mm above the squama root (except the region of the frontal crest where the thickness increases toward the squama root up to ~8.0 mm). Second, Bp 9408 is below the range of the Bapang-AG crania (Sangiran 10, 17, Skull IX, Bukuran) in supraorbital torus thickness, as Sangiran 2. Third, as was observed in Trinil 2 and Sangiran 2, Bp 9408 does not show a steeply arising frontal squama like some of the Bapang-AG specimens (Skull IX, Bukuran).

Fig. 11.6 Anterior (*left*) and left lateral (*right*) views of Bp 9408. Scale = 1 cm

Bu 9604

The Bu 9604 occipital squama fragment preserves a significant portion of the nuchal plane, a part of the occipital plane, and the occipital torus (Fig. 11.7). A short segment of the posterior rim of the foramen magnum including the opisthion is also preserved. The surface preservation is excellent, and there is little sign of erosion.

A segment of the right lamboidal suture is preserved on the right side. This suture abruptly turns inferomedially at the point 17 mm above the superior nuchal line, and then again turns laterally just above the line before it disappears at the breakage. This strongly suggests the presence of an asterionic bone, which once held the asterion further down from the preserved end of the lamboidal suture. Presence of asterionic bone is not uncommon in Indoensian and Chinese *H. erectus*. (e.g., Bukuran; Sambungmacan 3; Ngandong 6, 7; Zhoukoudian Skull XII). The distance between the preserved lateral end of the right lamboidal suture and the corresponding point on the left side (missing) is estimated as 107 mm using a mirror image. The biasterionic breadth of Bu 9604 must have been larger than this estimate. Its opisthocranion-opisthion chord distance (54 mm) exceeds Sangiran 2 and approaches the lower limit of the four Bapang-AG crania. At the same time, Bu 9604 exceeds Trinil 2, and approaches Sangiran 2 in bone thickness (Fig. 11.3). Bu 9604 also shows a sharply flexed occipital squama as in other adult Trinil/Sangiran specimens. Analysis of the small existing sample of subadult *H. erectus* crania suggests that the strong occipital flexion in adult *H. erectus* developed during adolescence, probably its

Fig. 11.7 External (*left*), internal (*center*), and right lateral (*right*) views of Bp 9408. Scale = 1 cm

later phase (between the ages of KNM-WT 15000/D 2700 and Zhoukoudian Skull III: Antón 1997b, 2002b; Rightmire et al. 2006). These suggest that Bu 9604 should be considered as an adult or young adult individual, who had already or very nearly acquired its adult morphology.

The occipital torus of Bu 9604 is vertically restricted, and shows only a modest degree of posterior protrusion, as in Trinil 2 and Sangiran 2. Bu 9604 is also similar to the latter two specimens in the absence of a distinct external occipital protuberance, and a smooth upper nuchal plane with poor development of the external occipital crest. The blunt crest becomes more conspicuous on the lower nuchal plane, but this portion is missing in Trinil 2 and Sangiran 2.

Internally, the sinus impressions are clear and deep. Just below the upper breakage, the groove for superior sagittal sinus diverges into two grooves which run down side by side, until the lateral one turns right and continues to the groove for the right transverse sinus. The medial one of the former groove is fringed by a longitudinal, sharp crest that continues onto the internal occipital protuberance. This protuberance is located at some 25 mm below the opisthocranion. The distinct internal occipital crest runs for 17 mm from the protuberance, before it diverges into right and left sharp ridges which fringe the posterior rim of the foramen magnum.

Summary

Although Bp 9408 and Bu 9604, the two new Grenzbank/Sangiran cranial remains, are fragmentary they resemble Trinil 2 and Sangiran 2 as far as their preserved frontal (Bp 9408) and occipital (Bu 9604) portions are concerned. Thus, this comparison supports the view that Trinil 2 and Sangiran 2 fall within the morphological variation of hominids from the Grenzbank/Sangiran chronological group, indicating the null hypothesis cannot be rejected from the second test.

Test 3: Variation and Morphological Affinities of the Expanded Grenzbank/Sangiran Cranial Sample

If we include Trinil 2 and Sangiran 2 in the Grenzbank/Sangiran cranial assemblage, its within-group variation expands considerably. Then, the next question is whether this large variation is consistent with the dentognathic evidence.

The expanded Grenzbank/Sangiran cranial sample is highly variable particularly in cranial bone thickness, supraorbital torus and occipital torus developments, size of the mastoid process, and morphology of the sagittal keel (Figs. 11.3–11.5, and 11.8). In many of these traits, the variation range in the Bapang AG sample falls within the range of variation of the Grenzbank/Sangiran array.

Fig. 11.8 Posterior views of the right halves of Sangiran 2 (*left, flipped horizontally*) and Sangiran 4 (*right*). Note the differences in development of the sagittal keel, occipital torus, and mastoid process

In contrast to the weak expression of the occipital torus and the absence of an external occipital protuberance in Bu 9604, Trinil 2, and Sangiran 2, the torus of Sangiran 4 is vertically thick and posteriorly protruding particularly at its central area. This feature is associated with a strong external occipital crest below. A more exaggerated version of this torus morphology is seen in Sangiran 31 (Wolpoff 1999).

The mastoid process of Sangiran 4 is distinctly larger even compared to the process of the largest Bapang-AG cranium, Sangiran 17. The preserved right processes of Sangiran 27 is also large (Schwartz and Tattersall 2003; Indriati and Antón 2008), whereas Sangiran 2 has much slighter process which is comparable to one of the Bapang-AG crania, Bukuran.

A strong sagittal keel extends all the way to the vicinity of lambda along the sagittal suture of Sangiran 4, whereas the same keel appears to be restricted to the anterior half of the parietals in Trinil 2 and Sangiran 2 (Weidenreich 1945; Rightmire 1990). The damaged posterior parietals of Sangiran 31 seem to preserve two separate ridges along their medial borders. Opinions vary among researchers whether this structure should be regarded as part of the sagittal keel, the "double sagittal crests" comparable to the structure seen in KNM-ER 1805, or even an artifact of deformation (Sartono 1982; Bräuer and Mbua 1992; Tyler et al. 1995; Wolpoff 1999; Grimaud-Hervé 2001; Antón 2003). Interestingly, the presence of an unmistakable "double sagittal keel" has been reported for a possible male cranium from Dmanisi, D2280 (Rightmire et al. 2006). This does not conflict with the view that Sangiran 31 had a strong sagittal keel that bisects posteriorly.

The above observation suggests that two morphotypes are recognized in the Grenzbank/Sangiran cranial assemblage: one is the comparatively thin-vaulted, gracile type, and the other is the moderate to thick-vaulted, robust type. Even if Trinil 2 and Sangiran 2 are not included here the presence of the Bp 9408 frontal and Bu 9604 occipital fragments point to the existence of a gracile morphotype in the earliest Indonesian hominid assemblage. This is not to say that the above-listed differences between the morphological variation seen in the robust and gracile do not overlap with each other. For example, the cranial thicknesses are largely similar between Sangiran 2 and 4. However, the overall variation observed is more appropriately described as demonstrating a bimodal, rather than a unimodal pattern of distribution.

This bimodal pattern is also observed in the mandibular and dental assemblage (Kaifu et al. 2005a, b). The magnitude of variation in the Grenzbank/Sangiran mandibular corpus height and thickness are extensive, and comparable to East African *H. habilis s.l.* (Fig. 11.9). Some of the Grenzbank/Sangiran mandibles display a series of robust features (e.g., a thick and high mandibular corpus with marked lateral eversion, a strong lateral prominence, and a marked posterior marginal tubercle), together with other possibly ancestral (a single, large mental foramen, and double rooted premolars) or unique (a diagonal ridge along the posterior border of the lateral prominence, and lateral bulging of the lower lateral corpus associated with hollowing of its opposite lingual surface) features. On the other hand, the Grenzbank/Sangiran collection does include more gracile mandibles.

Previous studies indicate that when the modal condition of this highly variable sample is compared, Grenzbank/Sangiran hominids demonstrate close affinities with African early *H. erectus* in dental size and morphology, and mandibular dimensions and architecture (Kaifu et al. 2005b). When overall cranial dimensions are compared, Sangiran 2, Trinil 2, and Sangiran 4 are also similar to Turkana *H. erectus* (Fig. 11.2; see also Rightmire 1990). These crania are derived from *H. habilis s.l.* and Dmanisi *H. erectus*, in cranial size, and lateral expansion of the temporal wall relative to the mastoid region (Fig. 11.2, Table 11.4). Both the dentognathic and cranial collections show similarly great variation, and can be compared to African early *H. erectus* in gross evolutionary grade.

Discussion

Insufficient contextual documentation, small sample size, and the fragmentary nature of the existing fossil specimens still hinder the development of a convincing morphological profile of the earliest Indonesian hominids. In addition, a part of our morphological observations are based on casts, and portions of our comparative data are collected from several different publications. These are potential sources of error in the present analyses. Despite these limitations, the above

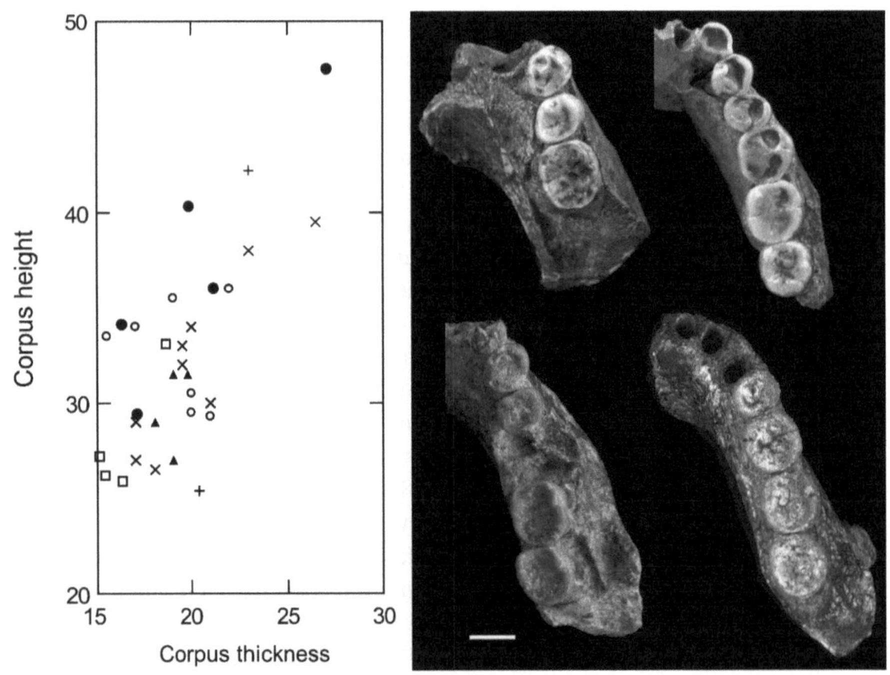

Fig. 11.9 Plots of mandibular corpus height and thickness at first molar (mm), and occlusal views of selected Grenzbank/Sangiran mandibles. *Symbols*: •, Grenzbank/Sangiran; ×, East African *H. habilis*; ▲, East African early *H. erectus*; ○, African late *H. erectus*; +, Dmanisi *H. erectus*; □, Zhoukoudian *H. erectus*. See Kaifu et al. (2005b) for the compositions and data sources of these samples. *Photographs*: clockwise from upper left, Sangiran 6a, 22, 1b, and 9. Scale = 1 cm

analyses indicate a need for the partial revision of our understanding of the earliest Indonesian hominids.

Since Weidenreich's initial study of Sangiran 4, and particularly after the discoveries of Sangiran 27 and 31, it has been widely believed that the oldest Indonesian hominid crania are best characterized by robustness and small size (Weidenreich 1945; Von Koenigswald 1950; Jacob 1976, 1979; Sartono 1982; Pope and Cronin 1984; Tyler et al. 1995; Wolpoff 1999; Antón 2003; Antón and Swisher 2004). However, a fundamental question of what reasonably defines the 'oldest' Indonesian cranial assemblage has so far not been addressed.

Based on the framework from the dentognathic evidence, we defined the earliest Indonesian hominid assemblage as consisting of the remains from the Grenzbank/Sangiran stratigraphic levels. If the two recently discovered cranial fragments, Bp 9408 and Bu 9604, are included in the Grenzbank/Sangiran assemblage, their morphology points to the existence of a gracile morphotype in the earliest Indonesian cranial assemblage. Additional gracile cranial specimens, Trinil 2 and Sangiran 2, may also belong to the Grenzbank/Sangiran assemblage.

The oldest Indonesian hominid crania are more variable than previously recognized. This finding is consistent with the tendency documented for the dentognathic remains (Kaifu et al. 2005a, b). In overall dimensions, the Grenzbank/Sangiran crania show the closest similarities to Turkana *H. erectus*. This is also consistent with the evidence from the mandibles and teeth, suggesting similar evolutionary grades between the oldest Indonesian hominids and East African early *H. erectus*.

Question of Great Variation

If we accept the relatively distinct bimodal pattern of morphological variation in the Grenzbank/Sangiran cranial, -mandibular, and -dental assemblage then it is possible that we must review the relationship between each cranial and mandibular specimen. The Sangiran 31 cranium has often been linked with the Sangiran 6a mandible (Sartono 1982; Tyler et al. 1995; Wolpoff 1999). This association between the most robust cranium and mandible from Indonesia seems to be reasonable. However, which cranial specimens should be associated with the gracile Grenzbank/Sangiran mandibles, Sangiran 1b and 22, is a question that needs to be considered. Weidenreich (1945) and Von Koenigswald (1950) originally assigned Sangiran 4 and 1b to the same hominid group. This scheme, which is followed even today by some researchers (e.g., Tyler et al. 1995; Wolpoff 1999), is based originally on Von Koenigswald's idea that the Trinil 2 cranium is chronologically younger than the Grenzbank/Sangiran hominid remains. However, given the above study, it is more reasonable to assume the association between the robust crania (Sangiran 4, 27, 31) with the robust mandibles (e.g., Sangiran 5, 6a, 8, 9), and the gracile mandibles (Sangiran 1b, 22) with the gracile crania (Bp 9408, Bu 9604, and possibly Trinil 2 and Sangiran 2).

The next question that needs to be considered is the reasons for the great variation within the Grenzbank/Sangiran collection. Two plausible explanations are a high degree of sexual dimorphism, and the presence of two groups of hominids with different evolutionary backgrounds. The observed variation presented here may be too great to conform to normal sexual dimorphism in one hominid species. For example, in his examination of metric variability of the mandibles and teeth, Kramer (1989) found that the great variation in the mandibular corpus breadth was an outstanding characteristic of the Grenzbank/Sangiran hominids (Sangiran 1, 5, 6a, 8, 9, 22) compared to *Gorilla*. Wolpoff (1999) assigned Sangiran 4 and 27 to the female sex, and Sangiran 31 to the male sex, but all these specimens would become male under the new grouping considered here. A general consensus from previous studies (e.g., O'Hoggins et al. 1990; Wood et al. 1991; Humphrey et al. 1999) is that the pattern of sexual dimorphism in craniomandibular morphology is not necessarily similar among the higher primate taxa, or even among geographic groups within a species. Therefore, the uniqueness of the dimorphic pattern would not be persuasive evidence against the sexual dimorphism hypothesis. In fact, Brown (1994) found that the Pleistocene Australians far exceeded modern Aboriginal Australians and other modern human groups in the degree of sexual dimorphism of cranial bone thickness.

Most of the late Early Pleistocene *Homo* collections from Africa and Asia display considerable degrees of morphological variation. Whether the c. 2.0 Ma *Homo* from Africa represents a highly-variable single (*H. habilis sensu lato*) or two different species (*H. habilis* and *H. rudolfensis*) has been an unresolved question for a long time (e.g., Wood 1991; Miller 2000). The *Homo* assemblage from Dmanisi is also known to include mandibles with a great size variation, which led some researchers to suggest the existence of two species (Gabunia et al. 2002; Skinner et al. 2006). However, Van Arsdale (2006) and Rightmire et al. (2008) concluded that both the large and small mandibles from Dmanisi should be regarded as belonging to a single taxon with a high degree of sexual dimorphism. Van Arsdale also suggested that this increased level of sexual dimorphism was an ancestral condition of *Homo*, which would explain the great variation seen in Pliocene African *H. habilis s.l.*. Furthermore, the recent discoveries of two small crania from Kenya, KNM-OL 45500 (0.97–0.90 Ma: Potts et al. 2004) and KNM-ER 42700 (~1.55 Ma: Spoor et al. 2007), raises the possibility that a large degree of sexual dimorphism characterizes the Early

Pleistocene populations of *H. erectus* in East Africa (Rightmire et al. 2008).

The findings presented here (see also Kaifu et al. 2005a, b) of the considerable degree of morphological variation in the oldest hominid cranial and dentognathic assemblages from Java further lends support to the emerging view that great variation is a persistent observation for the Early Pleistocene *Homo* in Africa and Eurasia. This may reflect a complex history of evolution and dispersal by multiple species of early *Homo*, but the occurrence of multiple species in all the three earlier *Homo* localities (East Africa, Dmanisi, and Java) appears to be unnatural. We feel the parsimonius explanation is that a high level of sexual dimorphism best characterizes earlier *Homo* in both Africa and Eurasia.

Questions of Taxonomy and Evolutionary Grade

The holotype of *H. erectus* is Trinil 2, a gracile cranium which may have been derived from the Grenzbank/Sangiran levels. Three other gracile crania examined in this study, Bp 9408, Bu 9604, and Sangiran 2, show close affinities with Trinil 2, and should be included in this species. Thus, irrespective of the question of Trinil 2's stratigraphic origin, *H. erectus* was present during the initial occupation of the Indonesian landmass.

Whether the other Grenzbank/Sangiran specimens belong to *H. erectus* depends on how we interpret their variation (see above) and phylogenetic relationships. If this assemblage represents a single population or deme, all of them, including the "*Meganthropus*" crania and mandibles (e.g., Sangiran 6a, 8, 27, 31), are *H. erectus* by definition. If the robust and gracile morphotypes of the Grenzbank/Sangiran assemblage represent two sympatric hominid groups with different ancestries, this indicates the presence of two different species unless they later intermixed and evolved together into Bapang-AG *H. erectus*.

Since the discovery of the '*Meganthropus*' mandible (Sangiran 6a) in the early 1940s, one of the major points of debate surrounding the earliest Indonesian hominids has been their identity: whether they were very primitive and comparable to australopithecine/early *Homo*, or just *H. erectus* of a broadly similar evolutionary grade as the Zhoukoudian hominids. Judging from the currently available morphological evidence, the correct answer seems to be intermediate between these two extremes. The Grenzbank/Sangiran hominids are derived compared to early *Homo* from Africa and Georgia, but more primitive than the Middle Pleistocene *H. erectus* from China. The Grenzbank/Sangiran hominid sample displays affinities with the Early Pleistocene *H. erectus* from the Turkana region. Assuming a modern human pattern of brain growth and a relative developmental age of 4–6 for the Perning 1 cranium, Antón (1997b) calculated its expected adult cranial capacity to be 740–860 cc. This figure falls within the range of the reported estimates for Sangiran 4, Trinil 2, and Sangiran 2 (813–940 cc), and KNM-ER 3733 and 3883 (804–848 cc) (Holloway et al. 2004), although Perning 1's developmental age is controversial (Coqueugniot et al. 2004).

Recent discoveries of *H. floresiensis* have led some researchers to suggest a possible link between these small-bodied hominids with very early forms of hominids such as early *Homo* or even australopithecines (Moorwood et al. 2005; Dennell and Roebroeks 2005). If this is correct, early *Homo* and/or australopithecines must have been present in the Early Pleistocene of the Sunda region, but such evidence is absent in the existing fossil record from Sangiran (Rightmire 1990; Antón 2003; Kaifu et al. 2005b; this study).

Dates of the Oldest Indonesian Hominids

With possible exception of Sangiran 27 and 31 (Swisher et al. 1994, but see Larick et al. 2000, 2001), the Grenzbank/Sangiran hominid remains are considered to have been derived from the levels approximately above the Tuff 11 of the Sangiran Formation. The $^{40}Ar/^{39}Ar$ dating suggest that Tuff 11 is >1.5 Ma (Larick et al. 2001), whereas magnetostratigraphy places it at ~1.1 Ma (Hyodo et al. 1993). The first appearance date for African *H. erectus/ergaster* represented by KNM-ER 3733 is estimated to be around 1.70 Ma (Feibel et al. 1989; Gathogo and Brown 2006; Suwa et al. 2007). If we accept the older date for Tuff 11 in Sangiran, the African and Indonesian hominids followed largely similar evolutionary grades and were penecontemporaneous. If the younger date for Tuff 11 is used, then this means the primitive morphotypes continued to persist in Java, long after the ancestral traits of early *H. erectus* were lost in Africa (Kaifu et al. 2005b).

Origins of the Robust Cranial Characteristics

Thick cranial bones and a strong development of ectocranial structures seen in Indonesian *H. erectus* are derived features relative to early *Homo*. We do not currently know when, where, and how these features evolved. It seems that a part of the earliest Indonesian hominids already had acquired these characteristics; thus its origins are to be looked for elsewhere in Asia. If we regard the Grenzbank/Sangiran hominids as a single taxon with a high level of sexual dimorphism, the remarkable cranial robusticity may have evolved first as a male character.

Conclusions

Based on the previously documented dentognathic evidence, we assign the oldest Indonesian hominid assemblage to the Grenzbank/Sangiran strata. Coupled with the trends observed in the dentognathic remains, the present examination of the Grenzbank/Sangiran crania, albeit preliminary, suggest the following points:

1. The earliest Indonesian hominids are comparable to c. 1.7 Ma *H. erectus* from East Africa in terms of cranio-mandibulo-dental morphology.
2. The earliest Indonesian hominids exhibit considerable morphological variation, as is indicated for East African *H. habilis s.l.* and Dmanisi *H. erectus*. This may reflect a complex history of evolution and dispersal by multiple species of early *Homo*, but it is also possible that an extremely high level of sexual dimorphism was a characteristic of early groups of *Homo*.
3. The stratigraphic assignment of Trinil 2 and Sangiran 2 are still indeterminate, but based on lithostratigraphic, biostratigraphic, chemical, and morphological evidence we suggest they originally derived from the Grenzbank/Sangiran strata; thus their affiliation to the earliest group of Indonesian hominids.

Acknowledgements We would like to thank Christopher Bae and David Braun for inviting us to participate in this volume. We are grateful to Friedmann Schrenk, Ottmar Kullmer, John de Vos, and Yahdi Zaim for access to the specimens in their care. Donald E. Tyler kindly provided us a cast of Sangiran 31, Shuji Matsu'ura and three anonymous reviewers made helpful comments, and Christopher Bae and David Braun assisted us in correcting the English. We also thank Sudijono, Iyep Saefudin, Eko Edi Susanto, Selamet, Sutanto, Shuichiro Narasaki, Koeshardjono, and Rusyad Adi Surijanto for field participation and laboratory assistance. This work was supported by the Japan Society for the Promotion of Science, and National Museum of Nature and Science, Tokyo.

References

Antón, S. C. (1997a). Endocranial hyperostosis in Sangiran 2, Gibraltar 1, and Shanidar 5. *American Journal of Physical Anthropolpology, 108*, 223–236.

Antón, S. C. (1997b). Developmental age and taxonomic affinity of the Mojoketro child, Java, Indonesia. *American Journal of Physical Anthropolpology, 102*, 497–514.

Antón, S. C. (2002a). Evolutionary significance of cranial variation in Asian *Homo erectus*. *American Journal of Physical Anthropology, 118*, 301–323.

Antón, S. C. (2002b). Cranial growth in *Homo erectus*. In N. Minugh-Purvis & K. J. McNamara (Eds.), *Human evolution through developmental change* (pp. 349–380). Baltimore: The Johns Hopkins University Press.

Antón, S. C. (2003). Natural history of *Homo erectus*. *Yearbook of Physical Anthropology, 46*, 126–170.

Antón, S. C., & Swisher, C. C., III. (2004). Early dispersal of *Homo* from Africa. *Annual Review of Anthropology, 33*, 271–296.

Antón, S. C., Leonard, W. R., & Robertson, M. L. (2003). An ecomorphological model of the initial hominin dispersal from Africa. *Journal of Human Evolution, 43*, 773–785.

Arif, J., Kaifu, Y., Baba, H., Suparka, M. E., Zaim, Y., & Setoguchi, T. (2002). Preliminary observation of a new cranium of *Homo erectus* (Tjg-1993.05) from Sangiran, Central Jawa. *Anthropological Science, 110*, 165–177.

Aziz, F. (2001). Hominin fossils housed at the Geological Research and Development Center, Bandung, Indonesia. In E. Indriati (Ed.), *A scientific life: papers in honor of Prof. Dr. T. Jacob* (pp. 53–66). Yogyakarta: Bigraf Publishing.

Baba, H., Kaifu, Y., Narasaki, S., Aziz, F., Sudijono, Saefudin, I., Susanto, E.E. (1998). Fossil frontal bone of *Homo erectus* found in 1994 from Sangiran, Java. *Anthropological Science, 107*, 46.

Baba, H., Aziz, F., Narasaki, S., Sudijono, Kaifu, Y., Saefudin, I., Susanto E.E. (2004). Frontal bone fragment of *Homo erectus* from Sangiran, Java. *Human Evolution, 19*, 197–202.

Bartstra, G.-J. (1982). The river-laid strata near Trinil, site of *Homo erectus erectus*, Java, Indonesia. *Modern Quaternary Research in Southeast Asia, 7*, 97–130.

Bettis III, E. A., Milius, A. K., Carpenter, S. J., Larick, R., Zaim, Y., Rizal, Y., Ciochon, R. L., Tassier-Surine, S. A., Murray, D., Suminto, Bronto, B. (2009). Way out of Africa: Early Pleistocene paleoenvironments inhabited by *Homo erectus* in Sangiran, Java. *Journal of Human Evolution, 56*, 11–24.

Bräuer, G., & Mbua, E. (1992). *Homo erectus* features used in cladistics and their variability in Asian and African hominins. *Journal of Human Evolution, 22*, 79–108.

Brown, P. (1994). Cranial vault thickness in Asian *Homo erectus* and *Homo sapiens*. *Courier Forschungsinstitut Senckenberg, 171*, 33–46.

Coqueugniot, H., Hublin, J.-J., Veillon, F., Houët, F., & Jacob, T. (2004). Early brain growth in *Homo erectus* and implications for cognitive ability. *Nature, 431*, 299–302.

Day, M. H. (1986). *Guide to Fossil Man* (4th ed.). Chicago: University of Chicago Press.

de Vos, J., & Sondaar, P. Y. (1982). The importance of the 'Dubois collection' reconsidered. *Modern Quaternary Research in Southeast Asia, 7*, 35–63.

de Vos, J., Sartono, S., Hardja-Sasmita, S., & Sondaar, P. Y. (1982). The fauna from Trinil, type locality of *Homo erectus*: a reinterpretation. *Geologie en Mijnbouw, 61*, 207–211.

de Vos, J., Sondaar, P. Y., Van den Bergh, G. D., & Aziz, F. (1994). The *Homo* bearing deposits of Java and its ecological context. *Courier Forschungsinstitut Senckenberg, 171*, 129–140.

Dennell, R. (2003). Dispersal and colonization, long and short chronologies: how continuous is the Early Pleistocene record for hominins outside East Africa? *Journal of Human Evolution, 45*, 421–440.

Dennell, R., & Roebroeks, W. (2005). An Asian perspective on early human dispersal from Africa. *Nature, 438*, 1099–1104.

Feibel, C. S., Brown, F. H., & McDougall, I. (1989). Stratigraphic context of fossil hominids from the Omo Group deposits; northern Turkana basin, Kenya and Ethiopia. *American Journal of Physical Anthropology, 78*, 595–622.

Gabunia, L., Vekua, A., Lordkipanidze, D., Swisher, C. C., III, Ferring, R., Justus, A., et al. (2000). Earliest Pleistocene hominin cranial remains from Dmanisi, Republic of Georgia: taxonomy, geological setting, and age. *Science, 288*, 1019–1025.

Gabunia, L., de Lumley, M. A., Vekua, A., Lordkipanidze, D., & de Lumley, H. (2002). Découverte d'un nouvel homininé à Dmanissi (Transcaucasie, Géorgie). *Comptes rendus Palevol, 1*, 243–253.

Gao, X., Wei, Q., Shen, C., & Keates, S. (2005). New light on the earliest hominin occupation in East Asia. *Current Anthropology, 46*, 115–120.

Gathogo, P. N., & Brown, F. H. (2006). Revised stratigraphy of Area 123, Koobi Fora, Kenya, and new age estimates of its fossil mammals, including hominins. *Journal of Human Evolution, 51*, 471–479.

Grimaud-Hervé, D. (2001). Taxonomic position of the Sangiran 31 hominin. In T. Simanjuntak, B. Prasetyo, & R. Handini (Eds.), *Sangiran: man, culture, and environment in Pleistocene times* (pp. 46–53). Jakarta: Yayasan Obor Indonesia.

Holloway, R. L., Broadfield, D. C., & Yuan, M. S. (2004). *The human fossil record, 3. Brain endocasts: the Paleoneurological evidence*. New Jersey: Wiley-Liss.

Huffman, O. F., Zaim, Y., Kappleman, J., Ruez, D. R., Jr., de Vos, J., Rizal, Y., et al. (2006). Relocation of the 1936 Mojokerto skull discovery site near Perning, East Java. *Journal of Human Evolution, 50*, 431–451.

Humphrey, L. T., Dean, M. C., & Stringer, C. B. (1999). Morphological variation in great ape and modern human mandibles. *Journal of Anatomy, 195*, 491–513.

Hyodo, M., Watanabe, N., Sunata, W., Susanto, E. E., & Wahyono, H. (1993). Magnetostratigraphy of hominin fossil bearing formations in Sangiran and Mojokerto, Java. *Anthropological Science, 101*, 157–186.

Indonesia-Japan Research Cooperation Programme – CTA-41 (IJRCP) (1979). Progress report of the Indonesia-Japan Joint Research Project on geology of human fossil bearing formations in Java. *Bulletin of Geological Research and Development Centre, 1*, 47–60.

Indriati, E. (2004). Indonesian fossil hominin discoveries from 1889–2003: problems and catalogue. *National Science Museum Monographs, 24*, 163–177.

Indriati, E., & Antón, S. C. (2008). Earliest Indonesian facial and dental remains from Sangiran, Java: a description of Sangiran 27. *Anthropological Science, 116*, 219–229.

Itihara, M., Sudijono, Kadar, D., Shibasaki, T., Kumai, H., Yoshikawa, S., Aziz, F., Soeradi, T., Wikarno, Kadar, A.P., Hashibuan, F., Kagemori, Y. (1985). Geology and stratigraphy of the Sangiran area. In N. Watanabe, D. Kadar (Eds.), *Quaternary geology of the hominin fossil bearing formations in Java*. Geological Research and Development Centre, Special Publication 4, pp. 11–43.

Jacob, T. (1975). Indonesia. In K. P. Oakley, B. G. Campbell, & T. I. Molleson (Eds.), *Catalogue of Fossil Hominins, Part III* (pp. 103–116). British Museum (Natural History): London.

Jacob, T. (1976). Early populations in the Indonesian region. In R. L. Kirk & A. G. Thorne (Eds.), *The origin of the Australians* (pp. 81–93). Canberra: Australian Institute of Aboriginal Studies.

Jacob, T. (1979). Hominine evolution in Southeast Asia. *Archaeology and Physical Anthropology in Oceania, 14*, 1–10.

Jacob, T. (1980). The Pithecanthropus of Indonesia: phenotype, genetics and ecology. In L. K. Königsson (Ed.), *Current Argument on Early Man* (pp. 170–179). Oxford: Pergamon.

Kaifu, Y. (2006). Advanced dental reduction in Javanese *Homo erectus*. *Anthropological Science, 114*, 35–43.

Kaifu, Y., Aziz, F., & Baba, H. (2005a). Hominin mandibular remains from Sangiran: 1952–1986 collection. *American Journal of Physical Anthropology, 128*, 497–519.

Kaifu, Y., Baba, H., Aziz, F., Indriati, E., Schrenk, F., & Jacob, T. (2005b). Taxonomic affinities and evolutionary history of the Early Pleistocene hominins of Java: Dento-gnathic evidence. *American Journal of Physical Anthropology, 128*, 709–726.

Kaifu, Y., Arif, J., Yokoyama, K., Baba, H., Suparka, E., & Gunawan, H. (2007). A new *Homo erectus* molar from Sangiran. *Journal of Human Evolution, 52*, 222–226.

Kaifu, Y., Aziz, F., Indriati, E., Jacob, T., Kurniawan, I., & Baba, H. (2008). Cranial morphology of Javanese *Homo erectus*: new evidence for continuous evolution, specialization, and terminal extinction. *Journal of Human Evolution, 55*, 551–580.

Kramer, A. (1989). The evolutionary and taxonomic affinities of the Sangiran mandibles of Central Java, Indonesia. Ph.D. dissertation, University of Michigan.

Kramer, A. (1994). A critical analysis of claims for the existence of Southeast Asian australopithecines. *Journal of Human Evolution, 26*, 3–21.

Langbroek, M. (2004). *'Out of Africa': an investigation into the earliest occupation of the old world. BAR international series 1244*. Oxford: Archaeopress.

Larick, R., Ciochon, R. L., Zaim, Y., Sudijono, S., Rizal, Y., & Aziz, F. (2000). Lithostratigraphic context for Kln-1993.05-SNJ, a fossil colobine maxilla from Jolotingkir, Sangiran Dome. *International Journal of Primatology, 21*, 731–759.

Larick, R., Ciochon, R. L., Zaim, Y., Sudijono, S., Rizal, Y., Aziz, F., et al. (2001). Early Pleistocene $^{40}Ar/^{39}Ar$ ages for Bapang Formation hominins, Central Jawa, Indonesia. *Proceedings of the National Academy of Sciences of the United States of America, 98*, 4866–4871.

Le Gros Clark, W. E. (1978). *The fossil evidence of human evolution: an introduction to the study of Paleoanthropology* (3rd ed.). Chicago: University of Chicago Press.

Leinders, J. J. M., Aziz, F., Sondaar, P. Y., & de Vos, J. (1985). The age of the hominin-bearing deposits of Java: state of the art. *Geologie en Mijnbouw, 64*, 167–173.

Li, H., Yang, X., Heller, F., & Li, H. (2008). High resolution magnetostratigraphy and deposition cycles in the Nihewan Basin (North China) and their significance for stone artifact dating. *Quaternary Research, 69*, 250–262.

Lordkipanidze, D., Vekua, A., Ferring, R., Rightmire, G. P., Zollikofer, C. P. E., Ponce de León, M. S., et al. (2006). A fourth hominin skull from Dmanisi, Georgia. *Anatomical Record, 288A*, 1146–1157.

Matsu'ura, S. (1982). A chronological framing for the Sangiran hominins: fundamental study by the fluorine dating method. *Bulletin of the National Science Museum, Tokyo Ser D, 8*, 1–53.

Matsu'ura, S., Kondo, M., Takeshita, Y., Kumai, H., Hyodo, M., Kamishima, Y., et al. (2005). Probable source horizon of the Pithechanthropus II (Sangiran 2) calotte and its possible palaeoanthropological implications. *Anthropological Science, 113*, 323.

Matsu'ura, S., Kondo, M., Hyodo, M., Danhara, T., Takeshita, Y., Kamishima, Y., et al. (2006). Toward the reliable age determination of the youngest homind-fossil bearing layers at Sangiran, Java. *Anthropological Science, 114*, 243.

Miller, J. M. A. (2000). Craniofacial variation in *Homo habilis*: an analysis of the evidence for multiple species. *American Journal of Physical Anthropology, 112*, 103–128.

Mithen, S., & Reed, M. (2002). Stepping out: a computer simulation of hominin dispersal from Africa. *Journal of Human Evolution, 43*, 433–462.

Moorwood, M. J., Brown, P., Jatmiko, Sutikna, T., Wahyu Saptomo, E., Westaway, K. E., Rokus Awe Due, Roberts, R. G., Maeda, T., Wasisto, S., Djubiantono, T. (2005). Further evidence for small-bodied hominins from the Late Pleistocene of Flores, Indonesia. *Nature, 437*, 1012–1017.

O'Hoggins, P., Moore, W. J., Johnson, D. R., & McAndrew, T. J. (1990). Patterns of cranial sexual dimorphism in certain groups of extant hominoids. *Journal of Zoology (London), 222*, 399–420.

Potts, R., Behrensmeyer, A. K., Deino, A., Ditchfield, P., & Clark, J. (2004). Small mid-Pleistocene hominin associated with East African Acheulean technology. *Science, 305*, 75–78.

Pope, G. G., & Cronin, J. E. (1984). The Asian Hominidae. *Journal of Human Evolution, 13*, 377–396.

Rightmire, G. P. (1990). *The evolution of Homo erectus: comparative anatomical studies of an extinct human species*. Cambridge: Cambridge University Press.

Rightmire, G. P., Lordkipanidze, D., & Vekua, A. (2006). Anatomical descriptions, comparative studies and evolutionary significance of the hominin skulls from Dmanisi, Republic of Georgia. *Journal of Human Evolution, 50*, 115–141.

Rightmire, G. P., Van Arsdale, A. P., & Lordkipanidze, D. (2008). Variation in the mandibles from Dmanisi, Georgia. *Journal of Human Evolution, 54*, 904–908.

Santa Luca, A. P. (1980). The Ngandong fossil hominins: a comparative study for a Far Eastern *Homo erectus* group. *Yale University Publications in Anthropology, 78*, 1–175.

Sartono, S. (1982). Sagittal cresting in *Meganthropus palaeojavanicus* von Koenigswald. *Modern Quaternary Research in Southeast Asia, 7*, 201–210.

Sartono, S., & Grimaud-Hervé, D. (1983). Les Pariétaux de l'homininé Sangiran 31. *L'Anthropologie (Paris), 87*, 465–468.

Schwartz, J. H., & Tattersall, I. (2003). *The human fossil record, 2. Craniodental morphology of genus Homo (Africa and Asia)*. New Jersey: Wiley-Liss.

Schwartz, J. H., & Tattersall, I. (2005). *The human fossil record, 4. Craniodental morphology of early hominins (Genera Australopithecus, Paranthropus, Orrorin), and overview*. New Jersey: Wiley-Liss.

Sémah, F. (2001). La position stratigraphique du site de Ngebung 2 (dôme de Sangiran, Java Central, Indonésia). In F. Sémah, C. Falguères, D. Grimaud-Hervé, & A.-M. Sémah (Eds.), *Origine des Peuplements et Chronologie des Cultures Paléolithiques dans le Sud-Est Asiatique* (pp. 299–329). Paris: Semenanjung.

Shea, J. (2006). Out of Africa 1: who, where, and when? *Evolutionary Anthropology, 15*, 1–2.

Skinner, M. M., Gordon, A. D., & Collard, N. J. (2006). Mandibular size and shape variation in the hominins at Dmanisi, Republic of Georgia. *Journal of Human Evolution, 51*, 36–49.

Soeradi, T., Shibasaki, T., Kadar, D., Sudijono, Itihara, M., Kumai, H., Hayashi, T., Furuyama, K., Aziz, F., Siagian, H., Furutani, M., Suminto, Yoshikawa, S. (1985). Geology and stratigraphy of the Trinil area. In N. Watanabe, D. Kadar (Eds.), *Quaternary geology of the hominin fossil bearing formations in Java*. Geological Research and Development Centre, Special Publication 4, pp. 45–48.

Sondaar, P. Y., De Vos, J., & Leinders, J. J. M. (1983). Facts and fiction around the fossil mammals of Java. *Geologie en Mijnbouw, 62*, 339–343.

Spoor, F., Leakey, M. G., Gathogo, P. N., Brown, F. H., Antón, S. C., McDougall, I., et al. (2007). Implications of new early Homo fossils from Ileret, east of Lake Turkana, Kenya. *Nature, 448*, 688–691.

Sudijono, Matsu'ura, S., Kondo, M., Aziz, F. (2001). *Stratigraphic allocation of the hominin frontal fragment (Bp 9408) from Sangiran by the multielement approach*. Geological Research and Development Centre, Special Publication 27, pp. 81–85.

Suwa, G., Asfaw, B., Haile-Selassie, Y., White, T., Katoh, S., WoldeGabriel, G., et al. (2007). Early Pleistocene *Homo erectus* fossils from Konso, southern Ethiopia. *Anthropological Science, 115*, 133–151.

Swisher III, C. C., Curtis, G. H., Jacob, T., Getty, A. G., Suprijo, A., Widiasmoro (1994). Age of the earliest known hominins in Java, Indonesia. *Science, 263*, 1118–1121.

Tobias, P. V. (1991). *Olduvai Gorge, 4: the skulls endocasts and teeth of Homo habilis*. Cambridge: Cambridge University Press.

Tyler, D. E., Krantz, G. S., & Sartono, S. (1995). The taxonomic status of the '*Meganthropus*' cranial (Sangiran 31) and the '*Meganthropus*' occipital fragment III. In J. R. F. Bower & S. Sartono (Eds.), *Evolution and ecology of Homo erectus* (pp. 189–201). Leiden University: Pithecanthropus Centennial Foundation.

Van Arsdale, A. P. (2006). Mandibular variation in early homo from Dmanisi, Georgia. Ph.D. Dissertation, University of Michigan.

Vekua, A., Lordkipanidze, D., Rightmire, G. P., Agusti, J., Ferring, R., Maisuradze, G., et al. (2002). A new skull of early *Homo* from Dmanisi, Georgia. *Science, 297*, 85–89.

Von Koenigswald, G. H. R. (1934). Zur Stratigraphie des Javanischen Pleistocän. *De Ingenieur in Nederlandsch-Indie, 1*, 185–201.

Von Koenigswald, G. H. R. (1935). Die fossilen Säugetierfaunen Javas. Proceedings of the Koninklijke Nederlandse Akademie van Wetenschappen, Amsterdam 38, 188–198.

Von Koenigswald, G. H. R. (1950). Fossil hominins from the Lower Pleistocene of Java. Report of 18th International Geological Congress 1948, London, Part 9, 59–61.

Walker, A., & Leakey, R. (Eds.). (1993). *The Nariokotome Homo erectus Skeleton*. Cambridge: Harvard University Press.

Watanabe, N., & Kadar, D. (Eds.) (1985). Quaternary geology of the hominin fossil bearing formations in Java. Geological Research and Development Centre, Special Publication 4.

Weidenreich, F. (1945). Giant early man from Java and South China. *Anthropological Papers of American Museum of Natural History, 40*, 1–134.

Widianto, H., Sémah, A. M., Djubiantono, T., & Sémah, F. (1994). A tentative reconstruction of the cranial human remains of Hanoman 1 from Bukuran, Sangiran (Central Java). *Courier Forschungsinstitut Senckenberg, 171*, 47–59.

Wolpoff, M. H. (1999). *Paleoanthropology* (2nd ed.). Boston: McGraw-Hill.

Wood, B. A. (1991). *Koobi Fora research project, 4. Hominin cranial remains from Koobi Fora*. Oxford: Clarendon.

Wood, B. A., Li, Y., & Willoughby, C. (1991). Intraspecific variation and sexual dimorphism in cranial and dental variables among higher primates and their bearing on the hominin fossil record. *Journal of Anatomy, 174*, 185–205.

Zhu, R. X., Potts, R., Xie, F., Hoffman, K. A., Deng, C. L., Shi, C. D., et al. (2004). New evidence on the earliest human presence at high northern latitudes in northeast Asia. *Nature, 431*, 559–562.

Zhu, R. X., Potts, R., Pan, Y. X., Yao, H. T., Lu, L. Q., Zhao, X., et al. (2008). Early evidence of the genus Homo in East Asia. *Journal of Human Evolution, 55*, 1075–1085.

Chapter 12
Central-East China – A Plio-Pleistocene Dispersal Corridor: The Current State of Evidence for Hominin Occupations

Christopher J. Norton, Xing Gao, Wu Liu, David R. Braun, and Xiujie Wu

Abstract Early hominins moved out of Africa and arrived in China sometime after 1.8 Ma. Using Majuangou, North China (1.71–1.66 Ma) and Yuanmou, South China (1.7 Ma) as the two tentative reference points for early hominin activity, we argue that similar paleoanthropological evidence should be present between these two localities in Central-East (CE) China. In particular, since the floral and faunal data strongly indicate that the region served as a continuous migration corridor throughout the Plio-Pleistocene, CE China should contain a diversity of evidence of early hominin activity. In this paper, we review the current state of Early and Middle Pleistocene paleoanthropological evidence in CE China. We note that there are questions/problems with some of the current evidence or interpretations of this evidence. However, the dataset forms a solid foundation from which to develop more detailed research programs and hypotheses to test. The key to investigating this region more thoroughly is systematic field and laboratory research conducted by a multi-disciplinary team of scientists.

Keywords Pleistocene • China • Paleoanthropology • *Homo erectus* • Archaic *Homo sapiens* • Dispersal corridor

Introduction

The success of early hominins to emigrate from Africa into Asia sometime after 2 Ma was strongly predicated on a variety of morphological and behavioral adaptations to new environments (Potts 1998; Anton and Swisher 2004; Dennell and Roebroeks 2005; Anton 2007; Klein 2009). Of the many novel landscapes that early *Homo erectus sensu lato* may have encountered, East Asia is one of the most important, particularly because it is considered the origin of subsequent dispersals to Japan, Australia, and eventually the New World. China forms the bulk of the eastern Eurasian landmass and was home to a diversity of environments and many unique floras and faunas throughout the Pleistocene. These factors would have prompted early hominins to develop new, or modify old, behaviors in order to survive.

The earliest currently accepted hominin occupation outside of Africa is Dmanisi in Georgia. This site is dated to ~1.7 Ma by Ar/Ar analysis of an underlying basalt (Gabunia et al. 2000). Two sites exist in China that are of comparable age: Yuanmou in South China (Zhu et al. 2001, 2008; but see Hyodo et al. 2002) and Majuangou in North China (Zhu et al. 2004). In order to reach these specific points, it is possible that Early Pleistocene hominins dispersed across Central Asia (Dennell 2003, 2004, 2010; Dennell and Roebroeks 2005; Holmes 2007), crossing or skirting the Himalayan Mountain Range that has been hypothesized to have been much lower during this time period. However, there is currently no evidence of pre-1 Ma hominin occupations in Central Asia (Glantz 2010). Furthermore, it seems likely that the major uplift of the Asian mountain ranges occurred during the Miocene, rather than during the Pleistocene (see Cao 1994; Coleman and Hodges 1995; Fort 1996; Rowley and Currie 2006 for discussion); thus, the formidable mountain barriers would have already been in place before hominins arrived in the region. In fact, current evidence indicates humans may not have reached the higher altitudes (e.g., Qinghai-Tibetan Plateau) until late in Marine Isotope Stage ("MIS") 3 (~30 ka) (Norton and Jin 2009). A more feasible region to investigate for presence/absence of paleoanthropological evidence is Central-East ("CE") China, where the lower-lying Qinling Mountain Range formed a less formidable barrier.

The focus of this paper is to review the CE Chinese paleoanthropological record (Fig. 12.1; Table 12.1). We then present an array of hypotheses that are currently being, or will be, tested to address existing questions in Chinese paleoanthropological and paleoenvironmental research. CE China comprises the following provinces as well other provinces to the east: Shanxi, Henan, Hubei, and Hunan (Fig. 12.1). Shaanxi Province is also included here because of the important hominin fossil localities (Gongwangling, Chenjiawo).

C.J. Norton (✉)
Department of Anthropology, University of Hawaii,
2424 Maile Way, Saunders Hall 346, Honolulu, HI 96822, USA
e-mail: cjbae@hawaii.edu

C.J. Norton and D.R. Braun (eds.), *Asian Paleoanthropology: From Africa to China and Beyond*, Vertebrate Paleobiology and Paleoanthropology, DOI 10.1007/978-90-481-9094-2_12,

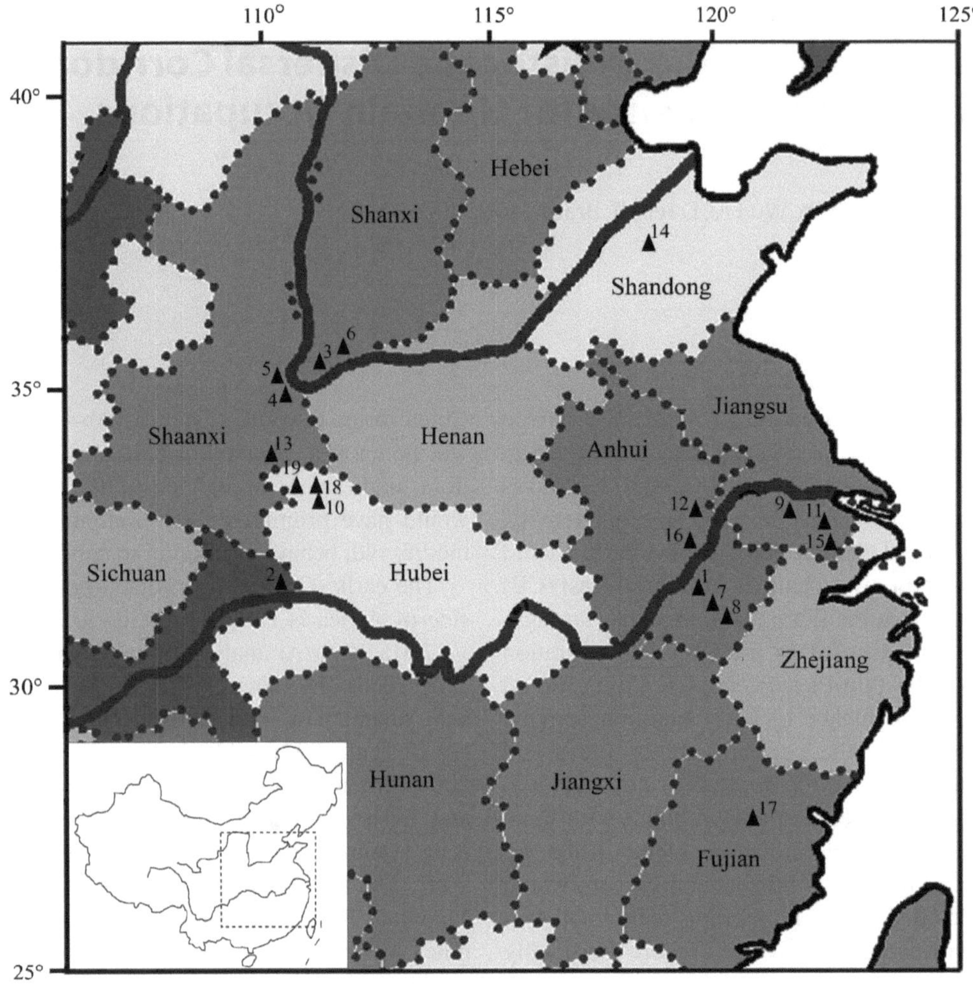

Fig. 12.1 Map of China with the provinces for the CE China region and the primary Early and Middle Pleistocene sites labeled. The various shades of *gray* are used to distinguish the different provinces. Site descriptions are presented in Table 12.1

A number of additional important Early and Middle Pleistocene paleoanthropological localities are present in South (e.g., Baise Basin, Panxian Dadong) and North China (e.g., Xiaochangliang, Donggutuo, Zhoukoudian, Dingcun, Xujiayao, Dali), but are covered in other chapters in this volume or in detail elsewhere. Throughout this paper, the Early Paleolithic represents all sites earlier than ~40 ka and lack blade/microblade technologies, since the western Old World three stage cultural model (Lower, Middle, Upper Paleolithic) has been found to be inapplicable to much of East Asia (e.g., Gao and Norton 2002; Norton et al. 2006, 2009).

Natural Barriers and Possible Migration Corridors

Encompassing an area 123,000 km^2, the Qinling Mountain Range forms a natural barrier separating the Chinese Loess Plateau in North China and the tropical and subtropical forests in South China (Ji 1980, 1982; Wang et al. 1997; Xie et al. 2004). The faunal and floral complexes from North and South China are distinct: Palearctic fauna (e.g., *Hipparion*, *Equus*, *Coelodonta*) and warm-temperate deciduous broadleaf forests and open grasslands in the north; Oriental fauna (e.g., *Ailuropoda*, *Stegodon*) and subtropical dense evergreen forests in the south (Wang et al. 1997; Xie et al. 2004; Norton et al. 2010). The Qinling Mountain Range averages 2,000–3,000 m above sea level ("MASL"), with the highest peak, Taibaishan, approximately 3,700 MASL. During the Pleistocene the mountain range is estimated to have been as much as 1,000 m lower than the present day and would have formed a less formidable barrier, as indicated by the presence of Oriental faunas north of the mountain range (Ji 1980, 1982; Wang et al. 1997). Oriental faunas reach the Korean Peninsula during the Middle Pleistocene (Norton 2000; Norton et al. 2010).

Table 12.1 Background of primary early hominin (only Early and Middle Pleistocene) sites from Central-East China. Huanglongdong falls outside the range of this review, but is a good representative example of an ongoing multi-disciplinary research project conducted primarily by Chinese researchers

# on Fig. 1	Site	Province	Site context	Relative age	Chronometric age	Dating method	Hominin fossils	Early paleolithic artifacts	Fauna	Existing questions	Reference
1	Renzidong	Anhui	Cave/Fissure	Early Pleistocene	2.4–2.0 Ma or 1.7–0.8 Ma	Biostratigraphy, ESR	None	?	Yes	Site context. Artifacts or ecofacts?	Jin et al. (2000)
2	Longgupo (Wushan)	Chongqing	Cave/Fissure	Early Pleistocene	2.0–1.8 Ma	Magnetostratigraphy, biostratigraphy, ESR	?	?	Yes	Hominin or hominoid fossils? Artifacts or ecofacts?	Huang et al. (1995); Etler et al. (2001)
3	Xihoudu	Shanxi	Open-air	Early Pleistocene	1.27 Ma	Magnetostratigraphy	None	?	Yes	Artifacts or ecofacts?	Zhu et al. (2003)
4	Gongwangling (Lantian)	Shaanxi	Open-air	Early Pleistocene	1.15	Magnetostratigraphy	*Homo erectus*	Yes	Yes	None reported	Wu and Poirier (1995)
5	Chenjiawo (Lantian)	Shaanxi	Open-air	Middle Pleistocene	650 ka	Magnetostratigraphy	*Homo erectus*	Yes	Yes	None reported	Wu and Poirier (1995)
6	Kehe	Shanxi	Open-air	Middle Pleistocene	N/A	Biostratigraphy	None	?	Yes	Artifacts or ecofacts?	Schick and Dong (1993)
7	Chenshan	Anhui	Open-air	Middle Pleistocene	817–126 ka	ESR	None	Yes	No	None reported	Fang (1997)
8	Maozhushan	Anhui	Open-air	Middle Pleistocene	665–420 ka	ESR	None	Yes	No	None reported	Fang et al. (2004)
9	Huludong, Tangshan	Jiangsu	Cave/Fissure	Middle Pleistocene	620–400 ka	U-series	*Homo erectus*	No	Yes	None reported	Liu et al. (2005)
10	Quyuanhekou (Yunxian)	Hubei	Open-air	Middle Pleistocene	581 ka	ESR	*Homo erectus*	Yes	Yes	None reported	Li and Etler (1992)
11	Fangniushan	Jiangsu	Open-air	Middle Pleistocene	432–209 ka	ESR	None	Yes	No	None reported	Zhang et al. (2005)
12	Hexian	Anhui	Cave/Fissure	Middle Pleistocene	412 ka	ESR, U-series	*Homo erectus*	No	Yes	None reported	Grun et al. (1998)
13	Longyadong	Shaanxi	Cave/Fissure	Middle Pleistocene	356–273 ka	TL	None	Yes	Yes	Site context	Wang (2005)
14	Yiyuan	Shandong	Cave/Fissure	Middle Pleistocene	N/A	Biostratigraphy	*Homo erectus*	No	Yes	Age	Lu et al. (1989); Etler and Li (1994)
15	Heshangdun	Jiangsu	Open-air	Middle Pleistocene	300 ka	Magnetostratigraphy	None	Yes	No	None reported	Zhu et al. (2007)
16	Chaoxian	Anhui	Cave/Fissure	Middle Pleistocene	200–160 ka	U-series	archaic *Homo sapiens*	No	Yes	Age	Wu and Poirier (1995)
17	Lingfengdong (Wanshouyan)	Fujian	Cave/Fissure	Middle Pleistocene	200–185 ka	U-series	None	Yes	Yes	None reported	Li et al. (2001); Cultural Relic Bureau of Fujian Province et al. (2006)
18	Bailongdong	Hubei	Cave/Fissure	Early Pleistocene	N/A	Biostratigraphy	*Homo erectus*	Yes	Yes	Age	Wu and Poirier (1995)
19	Huanglongdong	Hubei	Cave/Fissure	Late Pleistocene	100 ka	U-series	*Homo sapiens*	Yes	Yes	None reported	Wu et al. (2006)

ESR = Electron Spin Resonance; U-series = Uranium series; TL = Thermoluminescence

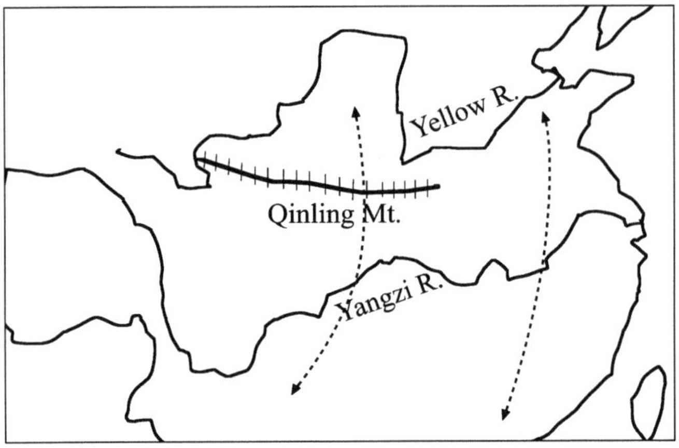

Fig. 12.2 Possible migration corridors between North and South China

We propose that at least two possible migration corridors existed between North and South China during the Pleistocene (Fig. 12.2). Due to the presence of deposits bearing hominin remains north and south of the Qinling Mountain Range, a migration corridor would have allowed hominins to move directly through the mountains following the low-lying river valleys. The other migration corridor skirts the Qinling Mountain Range, allowing hominins to follow the more open, low-lying areas to the east, much of which has been less than 1,000 MASL throughout the Quaternary.

Through the Qinling Mountain Range

Perhaps the best known Early Pleistocene site south of the Qinling Mountain Range is Longgupo (also referred to as Wushan; Huang and Fang 1991; Huang et al. 1995). In 1985, a purported hominin (*Homo erectus* or *H. habilis*) mandibular fragment with intact P2 and M1, along with an isolated upper incisor, was recovered in the same sediments as Early Pleistocene mammalian fossils (e.g., *Homotherium*, *Hyaena licenti peii*, *Nestoritherium*). Paleomagnetic and electron spin resonance ("ESR") analyses of the deposits and associated materials indicated an age bracket between 2.0 and 1.8 Ma (Huang et al. 1995; Chen et al. 2000), which is in accordance with the known age ranges of the associated fauna.

Many scientists (e.g., Schwartz and Tattersall 1996; Wu 2000; Etler et al. 2001) have argued that the Longgupo mandible is actually representative of *Lufengpithecus*, a hominoid that lived in the region during the time interval, rather than an early form of *Homo*. In addition, the incisor is not distinguishable from modern *Homo sapiens*, suggesting to some (e.g., Etler et al. 2001), that at least that area of the site contained reworked deposits. Early Paleolithic artifacts were said to be found in association with the hominin fossils (Huang 1998; Hou et al. 1999; Li 1999), though they remain unconvincing to some (e.g., Wei 2000; Chen 2003). A joint Sino-French team has recently renewed fieldwork, where a diversity of lithics have been reportedly excavated in situ from multiple stratigraphic layers.

Hubei Province, located just south of the Qinling Mountain Range, is best known for two crushed fully adult male *Homo erectus* crania that were excavated in 1989 from Quyuanhekou (Yunxian County). The crania exhibit a combination of features displayed by western Old World and Zhoukoudian Locality 1 hominins (Li and Etler 1992; Etler and Li 1994; Etler 1996). Electron spin resonance (ESR) analysis on associated teeth indicates the age of the deposits to be 581 ± 93 ka (Chen et al. 1996), though a more recent paleomagnetic study suggests a late Early Pleistocene age (Li et al. 1998). Associated fauna indicates an Early-Middle Pleistocene transitional period. Two hundred lithic artifacts were excavated, including cores and a diversity of flake tools. The primary raw material is locally available quartz and sandstone cobbles. Ten percent of the lithics were refit, suggesting that stoneknapping occurred on site, with minimal fluvial disturbance (Wang 2003).

Two cave sites in Yunxi County, Hubei Province are Bailongdong and Huanglongdong (Wu and Poirier 1995; Wu et al. 2006). Bailongdong was excavated in 1977 by researchers from the Institute of Vertebrate Paleontology and Paleoanthropology ("IVPP"). Investigations led to the recovery of six *Homo erectus* teeth: molars (3) and premolars (3), representing at least two individuals, in association with late Early Pleistocene fauna and a few stone artifacts. Research at Bailongdong was renewed in 2007. Huanglongdong is located roughly 20 km from Bailongdong and is best known for the presence of a few modern *H. sapiens* teeth (Wu et al. 2006b). The fossils were found in association with Late Pleistocene fauna and a diversity of Early Paleolithic core and flake stone tools. Uranium-series (U-series) and ESR dating of associated speleothems and teeth have produced ages of 103 ka and 44 ka (Wu et al. 2006b), and it is believed that the human teeth are more closely associated with the older date. If substantiated, Huanglongdong would represent one of the earliest occurrences of modern *H. sapiens* outside of Africa. The research at Bailongdong and Huanglongdong is particularly important because of the rich contextual information that can be captured by the multi-disciplinary nature of the research at this locality. Specialists from the IVPP and other scientific institutes are actively involved with different aspects of the research.

The remaining sites that may be evidence of early hominin movement through the Qinling Mountain Range are all located in the Chinese Loess Plateau. Xihoudu and Kehe are two representative localities. They are open-air sites located in Shanxi Province and excavated in the early 1960s (Zhang 1989; Schick and Dong 1993; Chen 2003). Xihoudu was initially dated to the bottom of the Olduvai Subchron (1.9 Ma)

(Chia and Wang 1978), but more recent studies indicate an age of 1.27 Ma (Zhu et al. 2003). Kehe lacks solid chronometric or geochronometric dates, though associated fauna suggests a Middle Pleistocene age (Schick and Dong 1993). Small lithic assemblages comprised of large cores and flakes were recovered (Xihoudu = 30; Kehe = 138). The lithics were rolled and abraded and the assemblages lack small flakes and/or debitage, suggesting fluvial transport influenced site formation at these localities (Schick 1986; Schick and Dong 1993; Zhang 1998; Chen 2003). Sieving of archaeological deposits was not common until the late 1980s in China; therefore it is possible that small flakes and debitage were not recovered because of the excavation techniques. A few of the faunal remains from Xihoudu are said to display butchery marks (Chia and Wang 1978; Huang and Hou 1996), though this evidence has been questioned (e.g., Pei and Zhang 1985; Zhang 1998).

In 1963 and 1964, the Lantian hominin fossils were excavated from two localities, Gongwangling and Chenjiawo in Shaanxi Province. The Gongwangling *Homo erectus* assemblage comprise a nearly complete calotte, maxilla with intact right M2 and M3, and an isolated left upper M2 (Etler and Li 1994; Wu and Poirier 1995). The cranium was subjected to substantial postdepositional deformation from sediment compaction making it difficult to use in comparative anatomical studies with other Asian *H. erectus* fossils (Anton 2002). The importance of the Gongwangling hominin fossils is the associated paleomagnetic date of 1.15 Ma, which is just below the Jaramillo Subchron (An and Ho 1989; Zhu et al. 2003). The paleomagnetic date is in concordance with the associated Early Pleistocene fauna. Twenty artifacts (cores, flakes, scrapers) are reported (Wu and Poirier 1995). Less than 30 km from Gongwangling, a *H. erectus* mandible was excavated at Chenjiawo. The sediments containing the mandible have a paleomagnetic date of 650 ka (An and Ho 1989). The Chenjiawo fossil was found in association with typical Middle Pleistocene fauna, in addition to four stone artifacts (Wu and Poirier 1995).

Systematic archaeological surveying began in the late 1990s in the Luonan Basin, Shaanxi Province, close to Lantian. Fifty open-air sites (or lithic scatters) and one cave site, Longyadong were identified (Wang 2005). A *Homo erectus* tooth collected from local village children has been attributed to Longyadong. Thermoluminescence (TL) dates from the cave site range between 356 and 210 ka, while two TL dates from the open-air deposits are 251 and 182 ka (Wang 2005). The faunal assemblage comprises a mixture of late Middle Pleistocene Oriental and Palearctic faunas, which corroborates the TL dates. Perhaps the most interesting aspect of the Luonan Basin finds is the presence of surface collected handaxes and cleavers from the open-air deposits, but absent in Longyadong (Wang 2005). Determinations of Middle Pleistocene hominin behavior in the Luonan Basin still require further investigation (see Hosfield 2006).

East of the Qinling Mountain Range

The primary sites in this region can be divided into two types: caves/fissures and open-air (Table 12.1). The cave and fissure deposits contain a diversity of Pleistocene faunas. Many of these localities are associated with hominin skeletal materials. However, many of the deposits contain no trace of hominin activity (manuports, lithics, hominin-modified bone). At other localities the association between the hominin body and trace fossils is questionable. Due to the acidic soil, almost all of the open-air sites in this region retain only lithic artifacts; a situation not unlike the nearby Korean Peninsula (Norton 2000, 2007; Norton et al. 2006).

Renzidong is currently the only purported Early Pleistocene paleoanthropological site east of the Qinling Mountain Range (Jin et al. 2000). It was first discovered in 1998 in Anhui Province, just south of the Yangtze River. Renzidong is a cave fissure with a stratigraphic profile roughly forty meters deep. Contained within the sediments were many intact skeletons of Early Pleistocene animals (e.g., *Homotherium*, *Sinomastodon*, *Nestoritherium*, *Tapirus*). The Renzidong deposits have an age bracket between 2.4 and 2.0 Ma based on biostratigraphy (Han et al. 1999; Jin et al. 2000), though an array of ESR dates suggests a younger age bracket: 1.24–0.8 Ma, with one date at 1.70 Ma (Chen et al. 2003). A paleomagnetic study is currently being conducted that should clarify the age of these sediments. Even though no hominin fossils were reported, several stone and bone artifacts were found in association with the faunal remains (Han et al. 1999; Jin et al. 2000; Zhang et al. 2000). Since the discovery however, many scientists (e.g., Wei 2000; Chen 2003) have questioned the authenticity of the artifacts and of the association of the fauna and artifacts. Several artifacts lack diagnostic features of conchoidal fracture associated with hominin modified stone tools. No debitage was recovered and the primary raw material, ferrolite, is intractable and not known to have been used by hominins to make stone artifacts at other sites in China (Chen 2003). Further study of this site is necessary in order to determine whether these stones are truly artifacts. In addition, more detailed analyses are necessary to determine whether the deposits were reworked, with the artifacts originating from a more recent time period.

Even if Renzidong is a natural accumulation, the vertebrate paleontological remains from the site are important because they form a foundation to reconstruct Early Pleistocene faunal movements east of the Qinling Mountain Range. Most importantly, taxonomic diversity indicates that Palearctic and Oriental faunas are almost equally represented in the deposit (Jin et al. 2000). This patterning likely indicates a combination of two factors: (1) The Renzidong fauna is representative of a continuous migration corridor between the Palearctic and Oriental biogeographic zones; and/or (2) there was some post-depositional reworking of different stratigraphic horizons, resulting in specimens from different biogeographic zones found in the same horizon.

In 1993 two partial *Homo erectus* crania and one tooth were discovered during excavations at Huludong (Hulu Cave), Tangshan, near Nanjing (Wu et al. 2002). The first cranium discovered (Skull No. 1) has been identified as a female, while the second cranium (Skull No. 2) has been characterized as a male. Skull No. 1 comprises pieces of the left and right parietals and the left upper part of the facial region. Skull No. 1 has a cranial capacity of 860 cc, which falls at the lower end of the Zhoukoudian *H. erectus* range, though multivariate comparisons indicate they are more closely aligned to each other than with the Indonesian hominins (Liu et al. 2005). It has been suggested that Skull No. 1 displays evidence of periostitis on the frontal bone (Wu et al. 2002; but see Shang et al. 2008). Skull No. 2 is a crushed, but more or less intact, calotte. It is considered to be more derived than Skull No. 1, being associated with late *H. erectus* or even transitional between *H. erectus* and *H. sapiens* (Wu et al. 2002). Initial ESR dates indicated an age of 350 ka (Chen et al. 1996). However, more recent mass spectrometric U-series dating of uranium isotopes in associated speleothems suggests an age between 620 and 500 ka (Zhao et al. 2001; Liu et al. 2005). A diversity of Middle Pleistocene faunas were excavated in association with the hominin crania. No hominin postcrania or trace fossils have been reported from the site (Wu et al. 2002).

Longtandong (Longtan Cave) was discovered in Hexian County, Anhui Province in 1973 by local farmers during construction of an irrigation canal (Wu and Poirier 1995). Vertebrate paleontologists from the IVPP identified the associated faunal remains as indicative of the Middle Pleistocene faunal community in China. A field team comprised of local Anhui Province archaeologists and IVPP researchers excavated Longtandong from 1980 to 1981. Among the many Middle Pleistocene fossils excavated were a fragmented calotte (primarily frontal and parietals), a partial mandible with M2 and M3, and five teeth, all representing *Homo erectus* (Wu and Poirier 1995). A combination of ESR and U-series dates indicated an age of 412 ka for mammalian teeth found in association with the hominin crania (Grun et al. 1998). Recent comparative studies of the Hexian endocast with Zhoukoudian, Indonesian, and East African *H. erectus* endocasts indicate that Hexian aligns most closely with Zhoukoudian (X.J. Wu et al. 2006a), which correlates with previous studies (e.g., Wu and Dong 1982; but see Bae 2010). No archaeological materials are reported from the site (Wu and Poirier 1995).

Four fossiliferous cave localities were identified on Qizianshan (Qizian Hill) in Yiyuan County in Shandong Province in 1981 and 1982 (Lu et al. 1989). Researchers from Yiyuan County and Peking University excavated a partial fragmented *Homo erectus* cranium and two teeth from Locality 1 and five teeth from Locality 3. An eighth tooth was recovered during sieving that is associated with the material from Locality 3. Variability in the robusticity of the supraorbital fragments suggests the hominin cranium represents two individuals (Etler and Li 1994). Variation in wear and coloration of the teeth supports the hypothesis that this collection represents two individuals. As the specimens were discovered in two separate caves there is good reason to believe the presence of two individuals at this locality. A non-metric analysis of the Yiyuan hominin fossil material indicates that it is very similar to Zhoukoudian *H. erectus* (Etler and Li 1994). The associated vertebrate paleontological remains are Middle Pleistocene. No chronometric studies have been conducted yet on the Yiyuan deposits. No hominin trace fossils were reported (Lu et al. 1989).

In 1981 the Geological Survey of the Anhui Bureau of Geology found late Middle Pleistocene fossils on Yinshan Hill, Chaoxian County, Anhui Province. In 1982 and 1983 researchers from the IVPP and Anhui Province excavated the Chaoxian cave site (Wu and Poirier 1995). The primary excavated fossils were archaic *Homo sapiens* occipital and maxilla fragments. Comparative studies with Zhoukoudian and Hexian *H. erectus*, indicated that the Chaoxian materials represent a more advanced hominin; thus the archaic *H. sapiens* designation (Etler 1990; Wu and Poirier 1995). The initial U-series date was 200–160 ka (Chen et al. 1987), although more recent U-series analysis suggests an older age bracket between 360 and 310 ka (Shen et al. 2010). No archaeological materials were reported from these deposits (Wu and Poirier 1995). A few meters from the Chaoxian site, a second collapsed cave exists. This second cave has yet to be investigated for hominin body and trace fossils, though based on the findings from the first cave the potential of finding additional paleoanthropological materials is quite high.

Since the late 1980s a series of Middle Pleistocene open-air sites [Maozhushan, Chenshan (Anhui Province); Fangniushan, Heshangdun (Jiangsu Province)] were surveyed and excavated in CE China (Fang 1997; Fang et al. 2001; Fang et al. 2004; Zhu et al. 2007). Despite the distance (~500 km apart), the two areas in Anhui and Jiangsu have many similarities in artifact composition and depositional histories; thus are discussed together here. All of the artifacts were found in aeolian clays laid down during the Middle Pleistocene. The Maozhushan and Chenshan materials were excavated from "fish net patterned" red clay deposits that formed under particularly humid conditions in South China (Fang et al. 2004; Yin and Guo 2006). The Fangniushan and Heshangdun materials were found in the slightly younger Xiashu Loess, which overlies the "fish net patterned" red clay deposits in the region (Zhang et al. 2005). Electron spin resonance and paleomagnetic reconstructions indicate an age bracket for the "fish net patterned" red clay and Xiashu Loess that spans the entire Middle Pleistocene (Table 12.1; Yin and Guo 2006).

The Early Paleolithic stone tool assemblages from the Anhui and Jiangsu open-air sites are comprised of chunks, cores, and flakes produced principally on locally available quartzitic and feldspathic sandstone (Fang et al. 1992, 2001; Fang 1997;). The number of artifacts from these sites vary (e.g., Fangniushan: 54; Heshangdun: ~200), as do the artifact densities (Fangniushan: 0.02/m^3; Chenshan: 0.06/m^3) and the overall site sizes (Y.S. Fang, pers. comm. 2007). Artifact typologies and densities are not unlike those from penecontemporaneous archaeological deposits from the Korean Peninsula, which lies directly across the Yellow Sea/West Sea (Norton 2000; Norton et al. 2006), except that the Chinese sites lack bifacially worked heavy duty tools. In Korea, Middle Pleistocene hominins were also producing lithics on locally available quartz and quartzite river cobbles (Norton 2000; Norton et al. 2006).

At the southern part of the designated CE China region is Lingfengdong (Lingfeng Cave), a cave site located on Wanshouyan Hill, in Sanming, Fujian Province (Li et al. 2001; Cultural Relic Bureau of Fujian Province et al. 2006). During a 4 month excavation season between late 1999 and spring 2000, ninety-nine artifacts and an array of late Middle Pleistocene vertebrate fossils were collected. Uranium series dating of associated travertine indicated an age bracket between 200 and 185 ka. Lithics (hammerstones, cores, flakes, burins) were produced on locally available quartzite and sandstone. Lingfengdong is currently considered the earliest paleoanthropological occurrence in Fujian Province. This situation may mean that Middle Pleistocene hominins did not reach that region until the late Middle Pleistocene. Only through more systematic field and laboratory investigations can this hypothesis be tested. A second cave, Chuanfandong, is located on the western slope of Wanshouyan. This second cave also contains evidence of hominin activity (300 artifacts), though the AMS dates on associated fauna indicate a Late Pleistocene age (37–36 ka) (Cultural Relic Bureau of Fujian Province et al. 2006).

Discussion

Two general observations can be drawn from this review. The first is that there are currently relatively few Early and Middle Pleistocene sites in CE China. The second is that even of the sites currently known from this region much work is needed to extrapolate the movements, migratory, behavioral, and morphological patterning of hominin populations in this region. In the following section we will identify the questions that may be answered through further analysis of these localities.

Relatively Few Early and Middle Pleistocene Sites Have Been Identified in CE China

Central-East China is a region representing roughly one third of the Chinese mainland. Nevertheless, the number of primary Early and Middle Pleistocene paleoanthropological sites in this vast region is relatively few. This paucity of sites may be evidence for absence of early hominin activity; thus, supporting suggestions that early East Asian hominin population sizes were much smaller than African populations during the Early and Middle Pleistocene (Schick 1994; Lycett 2007; Lycett and Norton 2010). However, the small number of paleoanthropological sites over such a large region of eastern Eurasia is also likely due to a current absence of evidence.

We use the nearby Korean paleoanthropological record as an analogy for suggesting the presence of yet many unknown sites in CE China. The Korean peninsula was part of the eastern Eurasian landmass throughout the Plio-Pleistocene. Systematic surveying and excavation of many different regions of the Korean Peninsula, particularly South Korea, has led to the identification of more than one hundred and fifty Pleistocene and Early Holocene cave and open-air sites (Norton 2000, 2007; Norton et al. 2006; Bae 2010). South Korea (99,000 km^2) is approximately 1/97th the size of China (9,600,000 km^2), but due to its booming economy, a tremendous amount of resources has been allocated to scientific investigation, including human evolutionary studies. This increased interest has facilitated many detailed paleoanthropological investigations throughout the southern half of the peninsula. Very little is known of the North Korean paleoanthropological record due to poor dissemination of research results by North Korean scholars and lack of access given to foreign scientists to visit sites and examine materials directly (see Norton 2000 for review). As such, the primary reason why so few sites or regions have been identified in CE China is probably not the result of limited hominin activity, but rather the lack of research funding and accessibility to investigate regions that surely served as faunal and floral migration corridors throughout the Plio-Pleistocene.

Many of the Existing Sites Require More Detailed Study

Many of the CE China hominin-bearing caves (e.g., Yiyuan, Hexian, Chaoxian) lack associated cultural materials. For instance, the *Homo erectus* crania from Huludong were found in association with Middle Pleistocene faunas. However, there are no reported hominin trace fossils. A possible explanation for this situation is that the hominin fossils were found deep inside Huludong, which runs about 80 m in length.

Since the entrance of the cave is only 91 MASL, it is possible that during periods of high water level, the cave could have been inundated with water, sweeping in the crania during the flooding process. Since most of the currently known hominin-bearing sites in CE China are from low lying caves/fissures (all lower than 100 MASL), it may be the reason why no trace fossils have been found in association with the hominin body fossils (i.e., the faunal materials were washed into the caves and fissures). Alternatively, the crania could have been carried in by carnivores. It may also be possible that early hominins were living in open areas (e.g., Maozhushan) rather than caves, thus early hominin trace fossils would not be expected to be present in the latter sites (Brooks 2007). Further research will clarify these questions.

Some of the hominin "cave" sites in CE China are in fact fissure localities. A good example of this is Renzidong, which is a fissure deposit that contains complete skeletons of many different taxa (Jin et al. 2000). Presence of many complete skeletons usually implies some type of natural trap (see Norton and Gao 2008), and even though there are purported stone tools at Renzidong, most are questionable. These questions about site context and formational history not only apply to sites like Renzidong, but other caves and fissures in CE China as well. More thorough investigations of the geological and site formation processes are critical.

Conclusions

Until the paleoanthropological and paleoenvironmental records of China are better understood, caution is probably warranted until various lines of evidence detail the complexity of these records. Presently, the investigation into the Pleistocene occupation of CE China requires more systematic multi-disciplinary studies of previously excavated and newly discovered Pleistocene sites and materials. Furthermore, in the face of isolated sites and small assemblages we should begin to try and form hypotheses as to why hominin dispersals would have included eastern Asia during the Pleistocene (e.g., see Potts 1998; Dennell 2004, 2010; Dennell and Roebroeks 2005; Anton 2007). This is the only way to attain an accurate understanding of early hominin morphological and behavioral variability in Pleistocene eastern Eurasia. The multi-disciplinary research projects currently being conducted in China are necessary initiatives to truly appreciate the important role the Chinese human evolutionary record plays in Old World paleoanthropology.

Acknowledgements The writing of this manuscript was supported in part by the Major Basic Research Projects of MST of China Grant (2006CB806400). We thank Julie Cormick, Dennis Etler, and Sari Miller-Antonio for commenting on an earlier draft of this manuscript. We take full responsibility for any errors that may be present here.

References

An, Z. S., & Ho, C. K. (1989). New magnetostratigraphic dates of Lantian Homo erectus. *Quaternary Research, 32*, 213–221.

Anton, S. C. (2002). Evolutionary significance of cranial variation in Asian Homo erectus. *American Journal of Physical Anthropology, 118*, 301–324.

Anton, S. C. (2007). Climatic influences on the evolution of early Homo? *Folia Primatol, 78*, 365–388.

Anton, S. C., & Swisher, C. C., III. (2004). Early dispersals of Homo from Africa. *Annual Review of Anthropology, 33*, 271–296.

Baab, K. L. (2010). Cranial Shape in Asian *Homo erectus*: Geographic, Anagenetic, and Size-Related Variation. In C. J. Norton, D. Braun (Eds.), *Asian Paleoanthropology: From Africa to China and beyond. Vertebrate paleobiology and Paleoanthropology series*, Springer: Dordrecht, The Netherlands.

Bae, K. D. (2010). Peopling in the Korean Peninsula. In: C. J. Norton, D. Braun (Eds.), *Asian paleoanthropology: From Africa to China and beyond. vertebrate paleobiology and paleoanthropology series*, Springer Press, Dordrecht, The Netherlands.

Brooks, A. (2007). Discussion. Asian Paleoanthropology symposium, Association of American Physical Anthropologists Conference, Philadelphia, PA.

Cao, B. X. (1994). Quaternary. In H. F. Yin (Ed.), *The Palaeobiogeography of China* (pp. 293–304). Oxford: Clarendon.

Chen, C. (2003). Retrospect of fifty years of Palaeolithic archaeology in China. In C. Shen, S. Keates (Eds.), *Current research in Chinese Pleistocene archaeology. BAR series 1179* (pp. 21–36). Oxford: Archaeopress.

Chen, T. M., Yuan, S. X., Gao, S. J., & Hu, Q. Y. (1987). Uranium-series data-lists on Hexian and Chaoxian Anhui Palaeolithic- and Palaeo-anthropological sites. *Acta Anthropologica Sinica, 6*, 249–254 (in Chinese).

Chen, T. M., Yang, Q., Hu, Y., & Li, T. Y. (1996). ESR dating of hominid fossil strata at Yunxian Hubei. *Acta Anthropologica Sinica, 15*, 114–118 (in Chinese).

Chen, T. M., Yang, Q., & Chen, Q. (2000). ESR dating of Longgupo profile Wushan. *Acta Anthropologica Sinica, 19*, 17–20 (in Chinese).

Chen, Q., Chen, T. M., Yang, Q. (2003). ESR dating of Early Pleistocene archaeological sites in China. In C. Shen, S. Keates (Eds.), *Current research in Chinese Pleistocene archaeology. BAR series 1179* (pp. 119–126). Oxford: Archaeopress.

Chia, L. P., Wang, J. (1978). Hsihoutu – a culture site of Early Pleistocene in Shansi Province. Beijing: Wenwu Press (in Chinese).

Coleman, M., & Hodges, K. (1995). Evidence for Tibetan plateau uplift before 14 Myr ago from a new minimum age for east-west extension. *Nature, 374*, 49–52.

Cultural Relic Bureau of Fujian Province, The museum of fujian province, The cultural relic administrative office of sanming city, fujian province (Eds.) (2006). *Wanshouyan Paleolithic Cave Site in Sanming, Fujian Province*. Beijing: Cultural Relics Publishing House (in Chinese).

Dennell, R. (2003). Dispersal and colonization, long and short chronologies: how continuous is the Early Pleistocene record for hominids outside East Africa? *Journal of Human Evolution, 45*, 421–440.

Dennell, R. (2004). Hominid dispersals and Asian biogeography during the Lower and early Middle Pleistocene, c. 2.0–0.5 Mya. *Asian Perspectives, 43*, 205–226.

Dennell, R. (2010). The colonization of "Savannahstan": issues of timing(s) and patterns of dispersal across Asia in the Late Pliocene and Early Pleistocene. In C.J. Norton, D. Braun (Eds.), *Asian Paleoanthropology: from Africa to China and beyond. Vertebrate paleobiology and paleoanthropology series.* Dordrecht, The Netherlands: Springer.

Dennell, R., & Roebroeks, W. (2005). An Asian perspective on early human dispersal from Africa. *Nature, 438*, 1099–1104.

Etler, D. A. (1990). A case study of the "erectus"-"sapiens" transition in Asia: hominid remains from Hexian and Chaoxian counties, Anhui Province, China. *Kroeber Anthropological Papers, 71–72*, 1–19.

Etler, D. A. (1996). The fossil evidence for human evolution in Asia. *Annual Review of Anthropology, 25*, 275–301.

Etler, D. A., & Li, T. Y. (1994). New archaic human fossil discoveries in China and their bearing on hominid species definition during the Middle Pleistocene. In R.S. Corruccini, R.L. Ciochon (Eds.), *Integrative paths to the past: paleoanthropological advances in honor of F. Clark Howell* (pp. 639–675). Englewood Cliffs, NJ: Prentice Hall.

Etler, D. A., Crummett, T. L., & Wolpoff, M. H. (2001). Longgupo: early Homo colonizer or Late Pliocene Lufengpithecus survivor in South China. *Human Evolution, 16*, 1–12.

Fang, Y. S. (1997). A report on excavation of Chenshan locality in 1988 Anhui Province. *Acta Anthropologica Sinica, 16*, 96–106 (in Chinese).

Fang, Y. S., Yang, D. Y., Han, H. Y., & Zhou, L. F. (1992). Taphonomic study of Paleolithic site group of Shuiyang River. *Acta Anthropologica Sinica, 11*, 134–142 (in Chinese).

Fang, Y. S., Huang, Y. P., Liang, R. Y., Chen, Y., & Peng, D. Q. (2001). Discovery of Early Paleolithic relic at Maozhushan site in Ningguo Anhui Province. *Acta Anthropologica Sinica, 20*, 115–124 (in Chinese).

Fang, Y. S., Huang, Y. P., & Shen, C. (2004). Pebble semicircle structure from a Lower Paleolithic site in southern China. *Eurasian Prehistory, 2*, 3–12.

Fort, M. (1996). Late Cenozoic environmental changes and uplift on the northern side of the central Himalaya: a reappraisal from field data. *Palaeogeography, Palaeoclimatology, Palaeoecology, 120*, 123–145.

Gabunia, L., Vekua, A., Lordkipanidze, D., Swisher, C. C., Ferring, R., Justus, A., et al. (2000). Earliest Pleistocene hominid cranial remains from Dmanisi, Republic of Georgia: taxonomy, geological setting, and age. *Science, 288*, 1019–1025.

Gao, X., & Norton, C. J. (2002). A critique of the Chinese 'Middle Palaeolithic'. *Antiquity, 76*, 397–412.

Glantz, M. M. (2010). The history of hominin occupation of Central Asia in review. In C.J. Norton, D. Braun (Eds.), *Asian paleoanthropology: from Africa to China and beyond. Vertebrate paleobiology and paleoanthropology series*. Dordrecht, The Netherlands: Springer.

Grun, R., Huang, P. H., Huang, W. P., McDermott, F., Thorne, A., Stringer, C. B., et al. (1998). ESR and U-series analyses of teeth from the palaeoanthropological site of Hexian, Anhui Province, China. *Journal of Human Evolution, 34*, 555–564.

Han, L.G., Jin, C.Z., Wei, G.B. (1999). A preliminary study on the stone and bone artifacts from Renzidong site, Fanchang, Anhui Province. In IVPP (Eds.), *Abstracts of international symposium on Palaeoanthropology in commemoration of the 70th anniversary of the discovery of the first skull of Peking Man at Zhoukoudian* (p. 8). Beijing: Chinese Academy of Sciences.

Holmes, K. M. (2007). Using Pliocene palaeoclimatic data to postulate dispersal pathways of early hominins. *Palaeogeography, Palaeoclimatology, Palaeoecology, 248*, 96–108.

Hosfield, R. (2006). Book review: "Perspectives on Hominid Behaviour and Settlement Patterns: a study of the Lower Palaeolithic sites in the Luonan Basin, China" by Shejiang Wang. Proceedings of the Prehistoric Society.

Hou, Y. M., Xu, Z. Q., & Huang, W. P. (1999). Newly discovered stone artifacts in 1997 at the Longgupo site. *Longgupo Prehistoric Culture, 1*, 69–78 (in Chinese).

Huang, W. W. (1998). Dealing seriously with scientific contributions made by previous researchers and yourselves – response to the articles in Science. *Acta Anthropologica Sinica, 17*, 322–324 (in Chinese).

Huang, W. P., & Fang, Q. (Eds.) (1991). *Wushan Hominid Site*. Beijing: Ocean Press (in Chinese).

Huang, W. W., & Hou, Y. M. (1996). On the human activities around circum-western Pacific area. *Journal of Chinese Antiquity, 1*, 68–74.

Huang, W. P., Ciochon, R., Gu, Y. M., Larick, R., Fang, Q. R., Schwarcz, H., et al. (1995). Early Homo and associated artifacts from Asia. *Nature, 378*, 275–278.

Hyodo, M., Nakaya, H., Urabe, A., Saegusa, H., Xue, S. R., Yin, J. Y., et al. (2002). Paleomagnetic dates of hominid remains from Yuanmou, China, and other Asian sites. *Journal of Human Evolution, 43*, 27–41.

Ji, H. X. (1980). The classification of Quaternary mammalian fauna in Lantian area. *Vertebrata PalAsiatica, 18*, 220–228 (in Chinese).

Ji, H. X. (1982). The living environment of the Quaternary mammalian faunas of South China. *Vertebrata PalAsiatica, 20*, 148–154 (in Chinese).

Jin, C. Z., Dong, W., Liu, J. Y., Wei, G. B., Xu, Q. Q., Zheng, J. J., Zheng, L.T., Han, L. G., Wang, F. Z. (2000). A preliminary study on the Early Pleistocene deposits and the mammalian fauna from the Renzi Cave, Fanchang, Anhui, China. *Acta Anthropologica Sinica, 19*(Suppl.), 235–245.

Klein, R. G. (2009). *The human career* (3rd ed.). Chicago: University of Chicago Press.

Li, Y. X. (1999). Stone artifacts discovered in 1985–1988 at Longgupo. *Longgupo Prehistoric Culture, 1*, 21–24 (in Chinese).

Li, T. Y., & Etler, D. A. (1992). New Middle Pleistocene hominid crania from Yunxian in China. *Nature, 357*, 404–407.

Li, Y., Ji, H., Li, T., Feng, X., & Li, W. (1998). The stone artifacts from the Yunxian Man site. *Acta Anthropologica Sinica, 17*, 94–120.

Li, J., Chen, Z., & Yu, S. (2001). Lingfengdong – the first Lower Paleolithic site found in Fujian Province. *Acta Anthropologica Sinica, 20*, 248–255 (in Chinese).

Liu, W., Zhang, Y. Y., & Wu, X. Z. (2005). A Middle Pleistocene human cranium from Tangshan, Nanjing, of southeast China: a comparison with Homo erectus from Eurasia and Africa based on new reconstruction. *American Journal of Physical Anthropology, 127*, 253–262.

Lu, Z., Huang, Y., Li, P., & Meng, Z. (1989). Yiyuan fossil man. *Acta Anthropologica Sinica, 8*, 301–313 (in Chinese with English abstract).

Lycett, S. J. (2007). Why is there a lack of Mode 3 Levallois technologies in East Asia? A phylogenetic test of the Movius-Schick hypothesis. *Journal of Anthropological Archaeology, 26*, 541–575.

Lycett, S. J., & Norton, C. J. (2010). A demographic model for Palaeolithic technological evolution: the case of East Asia and the Movius Line. *Quaternary International, 211*, 55–65.

Norton, C. J. (2000). The current state of Korean paleoanthropology. *Journal of Human Evolution, 38*, 803–825.

Norton, C. J. (2007). Sedentism, territorial circumscription, and the increased use of plant domesticates across Neolithic – Bronze Age Korea. *Asian Perspectives, 46*, 133–165.

Norton, C. J., & Gao, X. (2008). Zhoukoudian Upper Cave revisited: a taphonomic perspective. *Current Anthropology, 49*, 732–745.

Norton, C. J., & Jin, J. (2009). The evolution of modern humans in East Asia: behavioral perspectives. *Evolutionary Anthropology, 18*, 247–260.

Norton, C. J., Bae, K. D., Harris, J. W. K., & Lee, H. Y. (2006). Middle Pleistocene handaxes from the Korean Peninsula. *Journal of Human Evolution, 51*, 527–536.

Norton, C. J., Gao, X., & Feng, X. W. (2009). The criteria defining the East Asian Middle Paleolithic reexamined. In M. Camps & P. R. Chauhan (Eds.), *Transitions in the Paleolithic* (pp. 245–254). Dordrecht, The Netherlands: Springer.

Norton, C. J., Jin, C. Z., Wang, Y., & Zhang, Y. Q. (2010). Rethinking the Palearctic-Oriental biogeographic boundary in Quaternary China. In: C. J. Norton, D. R. Braun (Eds.). *Asian paleoanthropology: From Africa to China and beyond. Vertebrate paleobiology and paleoanthropology series*. Dordrecht, The Netherlands: Springer.

Pei, W. C., & Zhang, S. S. (1985). *A Study on the Lithic Artefacts of Sinanthropus*. Beijing: Science.

Potts, R. (1998). Variability selection in hominid evolution. *Evolutionary Anthropology, 7*, 81–96.

Rowley, D. B., & Currie, B. S. (2006). Palaeo-altimetry of the late Eocene to Miocene Lunpola basin, central Tibet. *Nature, 439*, 677–681.

Schick, K. (1986). *Stone Age sites in the making: experiments in the formation and transformation of archaeological occurrences. BAR international series 319*. Oxford: Archaeopress.

Schick, K. (1994). The Movius Line reconsidered. In R.S. Corruccini, R.L. Ciochon (Eds.), *Integrative paths to the past* (pp. 569–596). Englewood Cliffs, NJ: Prentice Hall.

Schick, K., & Dong, Z. (1993). Early Paleolithic of China and eastern Asia. *Evolutionary Anthropology, 2*, 22–35.

Schwartz, J. H., & Tattersall, I. (1996). Whose teeth? *Nature, 381*, 201–202.

Shang, H., Trinkaus, E., Liu, W., Wu, X., & Zhu, Q. (2008). Neurocranial abnormalities of the Gongwangling Homo erectus from Lantian, China. *Journal of Archaeological Science, 35*, 2589–2593.

Shen, G. J., Fang, Y. S., Bischoff, J. L., Feng, Y. X., & Zhao, J. X. (2010). Mass spectrometric U-series dating of Chaoxian hominin site at Yinshan, eastern China. *Quaternary International, 211*, 24–28.

Wang, Y. P. (2003). New Palaeolithic discoveries in the middle Yangzi River Region, Southern China. In: C. Shen, S.G. Keates (Eds.), *Current research in Chinese Pleistocene archaeology. BAR international series 1179* (pp. 57–66). Oxford: Archaeopress.

Wang, S. J. (2005). *Perspectives on hominid behaviour and settlement patterns: a study of the Lower Palaeolithic sites in the Luonan Basin, China. BAR international series 1406*. Oxford: Archaeopress.

Wang, H., Ambrose, S. H., Liu, C. L., & Follmer, L. R. (1997). Paleosol stable isotope evidence for early hominid occupation of East Asian temperate environments. *Quaternary Research, 48*, 228–238.

Wei, Q. (2000). Early Pleistocene paleoliths in China. *Wenwu Chunqiu, 2*, 1–14 (in Chinese).

Wu, X. Z. (2000). Longgupo hominoid mandible belongs to ape. *Acta Anthropologica Sinica, 19*, 1–10 (in Chinese).

Wu, R. K., & Dong, X. R. (1982). Preliminary study of Homo erectus remains from Hexian Anhui. *Acta Anthropologica Sinica, 1*, 2–13 (in Chinese).

Wu, X. Z., & Poirier, F. E. (1995). *Human Evolution in China*. Oxford: Oxford University Press.

Wu, R. K., Li, X. X., Wu, X. Z., Mu, X. N. (Eds.) (2002). *Homo erectus from Nanjing*. Nanjing: Jiangsu Science and Technology Publishing House (in Chinese and English).

Wu, X. J., Schepartz, L. A., Falk, D., & Liu, W. (2006a). Endocranial cast of Hexian Homo erectus from South China. *American Journal of Physical Anthropology, 130*, 445–454.

Wu, X. Z., Liu, W., Gao, X., & Yin, G. M. (2006b). Huanglong Cave, a new Late Pleistocene hominid site in Hubei Province, China. *Chinese Science Bulletin, 51*, 2493–2499.

Xie, Y., Mackinnon, J., & Li, D. M. (2004). Study on Biogeographical divisions of China. *Biodiversity and Conservation, 13*, 1391–1417.

Yin, Q. Z., & Guo, Z. T. (2006). Mid-Pleistocene vermiculated red soils in southern China as an indication of unusually strengthened East Asian monsoon. *Chinese Science Bulletin, 51*, 213–220.

Zhang, S. S. (1989). The Lower Paleolithic cultures in North China. In R. Wu, X. Wu, & S. Zhang (Eds.), *Early Humankind in China* (pp. 97–158). Beijing: Science (in Chinese).

Zhang, S. S. (1998). A discussion on the so-called cultural relics found at Xihoudu site in Ruicheng County Shanxi Province. *Acta Anthropologica Sinica, 17*, 81–93 (in Chinese).

Zhang, S. S., Han, L. G., & Jin, C. J. (2000). Artefacts discovered in 1998 at the Renzidong Paleolithic site Fangchang. *Acta Anthropologica Sinica, 19*, 169–181 (in Chinese).

Zhang, Q., Zhu, C., Jiang, T., & Becker, S. (2005). Mid-Pleistocene environmental reconstruction based on Xiashu loess deposits in the Yangtze Delta, China. *Quaternary International, 135*, 131–137.

Zhao, J. X., Hu, K., & Collerson, K. D. (2001). Thermal ionization mass spectrometry U-series dating of a hominid site near Nanjing, China. *Geology, 29*, 27–30.

Zhu, R. X., Hoffman, K. A., Potts, R., Deng, C. L., Pan, Y. X., Guo, B., et al. (2001). Earliest presence of humans in northeast Asia. *Nature, 413*, 413–417.

Zhu, R. X., An, Z. S., Potts, R., & Hoffman, K. A. (2003). Magnetostratigraphic dating of early humans in China. *Earth-Science Reviews, 61*, 341–359.

Zhu, R. X., Potts, R., Xie, F., Hoffman, K. A., Deng, C. L., Shi, C. D., et al. (2004). New evidence on the earliest human presence at high northern latitudes in northeast Asia. *Nature, 431*, 559–562.

Zhu, C., Zhang, Y. Y., Ma, C. M., Fang, Y. S., & Wu, C. L. (2007). Paleomagnetism and magnetic susceptibility of the Quaternary sediments at the Heshangdun Paleolithic site in Jintan, Jiangsu Province. *Journal of Stratigraphy, 31*, 35–44 (in Chinese).

Zhu, R. X., Potts, R., Pan, Y. X., Yao, H. T., Lu, L. Q., Zhao, X., et al. (2008). Early evidence of the genus Homo in East Asia. *Journal of Human Evolution, 55*, 1075–1085.

Chapter 13
The Earliest Hominin Occupations in the Nihewan Basin of Northern China: Recent Progress in Field Investigations

Chen Shen, Xing Gao, and Qi Wei

Abstract The Nihewan Basin of northern China has been a focus of Pleistocene geological, paleontological, and archaeological studies since the early part of the last century. In the past 10 years, the ROM-IVPP research team has investigated Early Pleistocene sites in the region with an aim of investigating the earliest occupations of East Asian hominins. Several Early Pleistocene sites have been identified with abundant lithic artifacts and associated faunal remains. This new archaeological evidence reveals that the Nihewan Basin hosts the earliest settlements of hominin migrations as early as 1.7 million year ago. In this study, we report the recent progress of the field investigations at four sites: Goudi, Dachangliang, Xiaochangliang, and Donggutuo, with reviews of previous studies at these sites. Our study concentrates on the lithic technology of the Early Pleistocene in northern China, which illustrates the use of expedient flake tools through multi-platform core reduction by early hominins. This study sheds new light on our understanding of early hominin behavior in East Asia.

Keywords Early Pleistocene • Lithic Technology • Nihewan • Northern China

Introduction

Since the first discovery of *Homo erectus* at Zhoukoudan in the early part of the last century, data from China have been important to human evolutionary studies in East Asia. With new field discoveries and the employment of systematic field and laboratory research, our knowledge about the nature of early hominin dispersals into China continues to be refined. Here we discuss recent developments from Early Pleistocene research in the Nihewan Basin, northern China.

The uplift of the Himalayan Mountains and the Qinghai-Tibet Plateau, and the development of Eurasian steppe deserts may have inhibited hominin interaction on both sides of the Eurasian landmass (Dennell and Roebroeks 2005). However, these geological barriers do not appear to keep early African hominins from entering East Asia sometime during the Early Pleistocene. Many Early Pleistocene hominins have been identified in central China (Wu and Dong 1985; Jia 1985; Wu et al. 1999; Zhu et al. 2003). However, due to questions about the materials and context of many of these central Chinese Early Pleistocene sites it is difficult to reconstruct hominin behavioral activities. For instance, both the Yuanmou and Lantian materials were the product of surface collections and there is currently no detailed information about their provenience. The sites from the Nihewan Basin currently represent the earliest securely dated in situ hominin occupation in northern China (Olsen 2000; Shen and Chen 2003; Zhu et al. 2004; Gao et al. 2005). Since 1998, the Royal Ontario Museum of Canada (ROM) has established a long term research program with the Institute of Vertebrate Paleontology and Palaeoanthropology (IVPP), Chinese Academy of Sciences in the Nihewan Basin. The aim of this research is to identify evidence of early hominin behaviors in the region, with a focus on stone tool production and use, as well as site formation processes and vertebrate taphonomy. In this paper, we will concentrate on the field investigations of four primary sites that were newly discovered (Goudi and Dachangliang) and were re-excavated and re-studied (Xiaochangliang and Donggutuo) (Fig. 13.1).

Background

The Nihewan Basin, covering an area of some 150–200 km^2 of Cenozoic fluvial-lacustrine deposits, is located in northeastern Hebei province, western Shanxi province, and in the southwestern portion of the Inner Mongolia Autonomous Region. The Nihewan Basin can be divided into three subbasins from west to east: Datong, Yuxian, and Yangyuan.

C. Shen (✉)
Department of World Cultures, Royal Ontario Museum, Toronto, ON M5S 2C6, Canada
e-mail: chens@rom.on.ca

C.J. Norton and D.R. Braun (eds.), *Asian Paleoanthropology: From Africa to China and Beyond*, Vertebrate Paleobiology and Paleoanthropology, DOI 10.1007/978-90-481-9094-2_13,

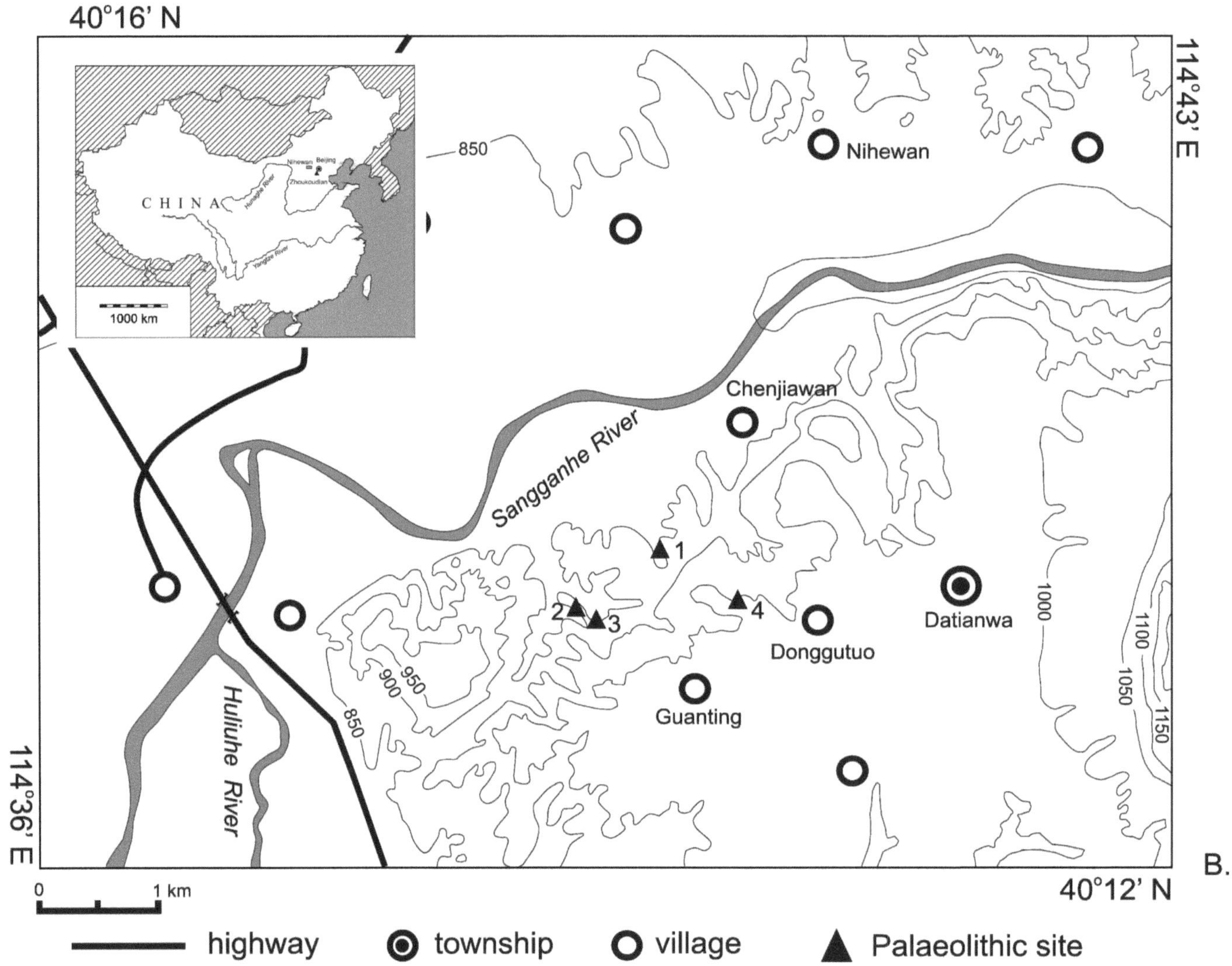

Fig. 13.1 Site location in the Nihewan. 1. Goudi; 2. Dachangliang; 3. Xiaochangliang; 4. Donggutuo

Our archaeological investigations concentrate in Yangyuan at the eastern end of the Nihewan Basin.

Since the early part of the last century, the Nihewan Basin has been a major area for paleontological and geological research (Barbour 1924, 1925; Barbour et al. 1927; Teilhard de Chardin and Piveteau 1930). These early field investigations established the geological significance of the Nihewan Beds because of the presence of a diversity of Pliocene and Early Pleistocene fossil remains. The fossil fauna, collected primarily by the French missionaries and paleontologists Licent and Teihard de Chardin, are now housed at the Tianjin Museum of Natural History. The Nihewan fauna include some late Teriary/early Quaternary forms such as *Postschizoherium, Hipparion, Zygolophodon,* and *Proboscidipparion* (Teilhard de Chardin and Piveteau 1930; Qiu 2000). Breuil (1935) claimed to have discovered a hominin-made stone chopper in these sediments. This was subsequently dismissed by Teilhard de Chardin (1935) and Pei (1935) who considered it a geofact (Teilhard de Chardin 1935; Pei 1935). From the 1930s until the 1970s little work was done in the region, although paleontological prospecting did confirm the presence of archaeological materials in the region.

Lower Palaeolithic archaeology studies in the Nihewan were initially based around the discovery of Xiaochangliang (You et al. 1978) in 1978 and Donggutuo in 1981 (Jia 1985; Wei 1985). For the first time, these two sites confirmed the presence of large numbers of artifacts associated with Early Pleistocene faunal remains. These discoveries initiated more detailed archaeological surveys in the eastern end of the Nihewan, leading to the discovery of additional sites such as Cenjiawan and Maliang in the mid-1980s. These four sites were initially excavated by Chinese archaeologists from the IVPP and Hebei Provincial Institute of Archaeology and Cultural Relics.

In the 1990s, an international collaboration for field research in the Nihewan Basin was initiated by J. Desmond Clark (UC Berkeley) and Lanpo Jia (IVPP) (Schick et al.

1991). Excavations were conducted at Donggutuo from 1990 to 1992 and Feiliang in 1996 (Xie et al. 2006). The ROM-IVPP collaboration in the Nihewan Basin began in 1998 and included re-excavating the Xiaochangliang site in 1998, Donggutuo site in 2000–2001, Dachangliang 2000, and Maliang in 2006 (Xie et al. 1994; Xie and Li 1995; Shen and Chen 1999, 2000, 2003; Keates 2000; Shen and Wei 2004; Gao et al. 2005).

As a result of these field investigations, more than a dozen Early to Middle Pleistocene localities have been identified and subsequently excavated by archaeologists from the IVPP and Hebei Province. All of these sites are concentrated in the eastern end of the basin, within a 5 km radius of the Donggutuo village, Yangyuan County (Wei 2000, 2004; Xie et al. 2006). In addition to the well-known and artifact-rich sites such as Xiaochangliang and Donggutuo, other sites include Goudi, Majuangou, Banshan, Putaoyuan, Feiliang, Madigou, Cenjiawan, Shanshenmiaozui, Xujiapu, and Dachangliang.

Chronology

The chronology for these sites was first established based on biostratigraphic and lithostratigraphic evidence (Tang et al. 1995; Cai and Li 2003). Early paleomagnetic studies suggested that the age bracket for the artifact-bearing deposits was between 1.8 and 0.78 Ma (from the onset of the Olduvai subchron to the onset of the Brunhes chron) (Chen et al. 1978; Qiang et al. 1983; Wei 1991, 1997; Yuan et al. 1996). Paleomagnetic and ESR analyses were applied to date Xiaochangliang. These early attempts provided rough age estimates but precise chronostratigraphic control was not possible (Chen et al. 2003; also see Shen and Chen 2003:67).

Recently Zhu and his colleagues from the Institute of Geology, Chinese Academy of Sciences, have systematically tested a series of palaeomagnetic samples from the Nihewan using more precise techniques. These studies have led to magnetostratigraphic ages for Majuangou III at 1.66 Ma (Zhu et al. 2004; but see Gao et al. 2005), Majuangou I at 1.55 Ma (Zhu et al. 2004), Xiaochangliang at 1.36 Ma (Zhu et al. 2001), Dachangliang at 1.36 Ma (Deng et al. 2006), Banshan at 1.32 Ma (Zhu et al. 2004), Feiliang at 1.2 Ma (Deng et al. 2007), Donggutuo at 1.1 Ma (Wang 2005; Wang et al. 2005), Cenjiawan at 1.1 Ma (Wang 2005; Wang et al. 2006), and Maliang at 0.78 Ma (Wang et al. 2005). These data suggest that the early hominins arrived in northern China as early as 1.7–1.6 Ma. This remarkable discovery also suggest that early hominins were able to survive in high latitudes for much of the Early and Middle Pleistocene.

The Lower Palaeolithic Sites

Xiaochangliang

Discovered in 1978, Xiaochangliang attracted great attention from international scientists (including archaeologists, geologists, and paleontologists) because of its purported antiquity (You et al. 1978). The first excavation in 1978 produced abundant lithic artifacts and faunal remains. Unfortunately the three dimensional provenience of the finds was not recorded. A joint research team of archaeologists and geologists from China and North America (U.S.A., Canada and England) re-surveyed the site in 1990 and carried out an exploratory excavation at an adjacent promontory of Xiaochangliang (Pope et al. 1990; Pope and Keates 1994; Keates 2000:38–42). Their reported test-excavation area (four 2 × 2 m test its), which they referred to as 'Xiaochangliang Dong Yang' (Pope et al. 1990:69), is adjacent to the site of Xiaochangliang. Subsequent studies have determined that this area may represent an area of slope erosion from the Shanshenmiaozhui site which is stratigraphically above their excavation (Wei 2004:89). Their surveys confirmed that the site represents one of the rare Early Pleistocene hominin activity sites in East Asia (Pope and Keates 1994; Keates 2000). From 1990 to 1997, Xiaochangliang was continuously investigated by geologists and palaeontologists from the IVPP. During this time, five seasons of field surveys and uncontrolled excavations took place with a focus on collecting additional fossils and well-made stone tools. Thousands of lithic artifacts and faunal remains were collected, but, as with the original 1978 materials, three dimensionsal coordinates were not recorded during the excavation of materials (Tang Yingjun, 1998, personal communication).

In 1998, the ROM-IVPP team conducted a joint excavation of Xiaochangliang. For the first time the collaborative project employed new methods and techniques (Shen and Chen 2003). A 4 × 4 m excavated unit was exposed in Area A (see Fig. 13.2 in Shen and Chen 2003), recovering a total of 901 artifacts, of which 560 pieces were recovered with full three dimensional coordinates while the rest were recovered from sieving. More than half of the lithic artifacts (n = 611) are classified as debris, which has no potential for further use as tools. Forty nodules and 71 cores were recovered *in situ*. Flakes (n = 172) display no obvious edge modification by retouch; however, these pieces are suitable to be used directly as tools. Only seven artifacts show intentional edge modification (Table 13.1).

The artifact collection recovered during the 1998 Xiaochangliang field season is dominated by small, irregular shaped flakes. On the basis of 531 examined pieces, the majority of flakes are in the range of 10–30 mm length and 10–20 mm width. The raw materials at Xiaochangliang are

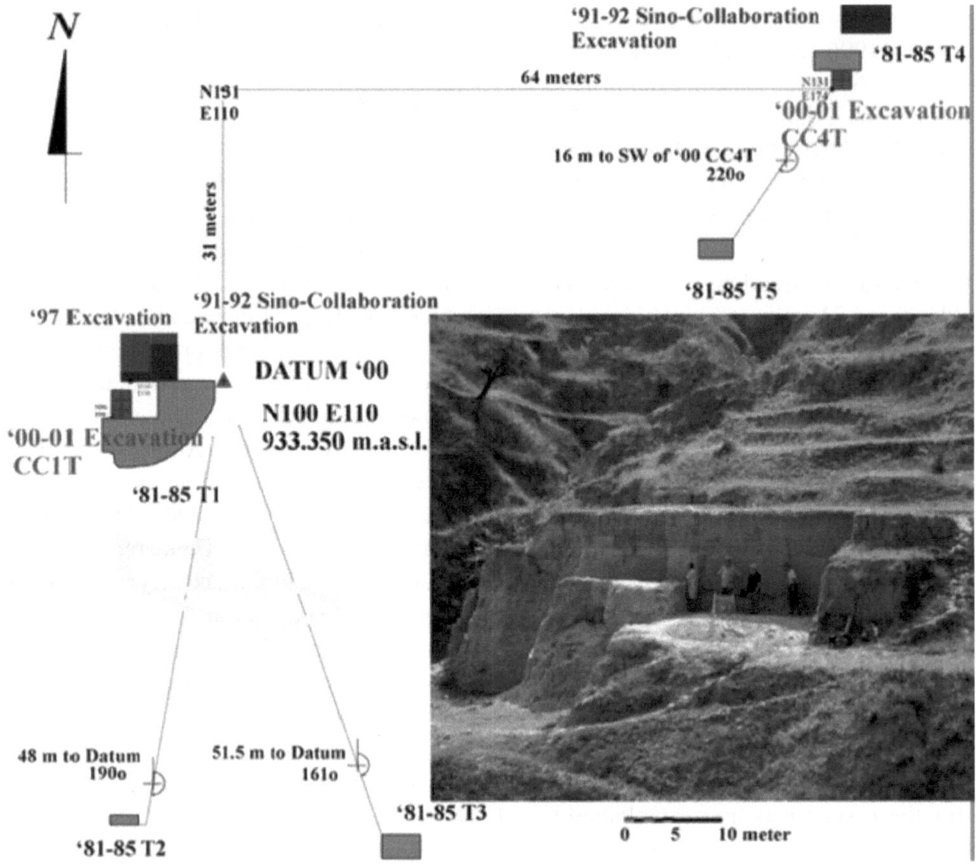

Fig. 13.2 Excavation units at the Donggutuo site

Table 13.1 Lithic artifacts from Xiaochangliang and Donggutuo

		Xiaochangliang		Donggutuo	
		98 Excavation		2000–2001 Excavation T1 layer 3	
Class	Type	N	%	N	%
Nodule		40	4.4		
Core		17	7.9	105	14.2
	Amorphous	39		35	
	Bipolar	27		12	
	Fragment	5		58	
Formed type		7	0.8	86	11.7
	Chopper	1			
	Modified flake	6		55	
	Retouch flake			23	
	Notch			3	
	Side-scraper			1	
	Endscraper			4	
Debitage		172	19	410	55.6
	Bipolar	29		28	
	Blade-flake	3			
	Flake	140		381	
	Primary flake			1	
Debris		611	67.9	137	18.6
	Chip	197			
	Chunk	414		137	
TOTAL		901	100	738	100

mainly quartzite with chert-like quality. Some artifacts are produced on volcanic rocks of various types. Since these materials have a tendency to fracture in an unpredictable manner, most of the flakes might have been produced from shattering the raw material as opposed to the more prevalent hard hammer percussion seen in many other Early Pleistocene sites. The debitage from this sort of artifact production does not leave the characteristic attributes of conchoidal fractures such as striking platforms or bulbs of percussion (see also Schick et al. 1991:21–22; Shen and Chen 2003).

During the 1998 season a total of 3,291 faunal remains were recovered, of which, more than 90% were fragmentary, and non-identifiable to taxon. Dental remains were identified to *Viverra* sp., *Felis* sp., *Coelodonta antiquitatis*, *Palaeoloxodon* sp., *Hipparion* sp., and *Bison* sp. Bone fragments were microscopically examined for modifications by natural and/or hominin agencies. We randomly selected 145 pieces for the taphonomic study; some of these fossils were studied using scanning electron microscopy (SEM) (Peterson et al. 2003). The conclusion drawn from this study was that the distribution of faunas and lithics together were the result of hydraulic processes. The evidence indicates that the site was located on the shoreline of a paleo-lake.

Donggutuo

Donggutuo was identified in 1981 and five separate excavations were conducted at different localities between 1981 and 1985 (T1–T5). This is the only known Early Pleistocene site in the Nihewan Basin with 5 m of multiple cultural horizons (Layer A-E). The Donggutuo artifacts have been the subject of two Ph.D. dissertations (Keates 2000, Hou 2000). In both of these studies all artifacts were examined as a single unit independent of stratigraphic provenience.

During 1991 and 1992, the Sino-US collaborative research project conducted two sessions of excavation at Donggutuo T1 and T4, exposing 30 m² at T1 (Schick and Dong 1993). By the end of the 1992 session, the collaborative team excavated stratigraphic levels 12–23 of Layer A/B (top sediments). The excavation reports and artifacts are not available for study. In 1997, an excavation team from the IVPP continued excavation to the bottom of Layer A, recovering 702 artifacts and about 160 faunal remains (Hou 2000).

Between 2000 and 2001, a team from the Royal Ontario Museum and the IVPP carried out two field seasons, excavating one 2 × 3 m unit at T1 and one 2 × 2 m unit at T4 (Fig. 13.2). A total of 2,412 lithic artifacts and faunal remains were recovered from the five layers. The greatest proportion of material came from Layer C, accounting for more than 60% of the total findings. Layer A and Layer B produced fewer artifacts. It is worthy to note that Layer B yielded primarily faunal remains, while Layer D yielded predominantly lithic artifacts. Throughout the layers, artifacts were distributed in clusters. These clusters may represent foci of hominin activity. The clustered artifacts may also represent fluvial activity concentrating artifacts in certain areas. Within Layer C, nearly 800 lithic artifacts were recovered, including cores, various types of flakes, retouch flakes, and waste debris (chips and chunks). The waste debris were oddly under-represented, accounting for less than 20% of the total lithic assemblage at Donggutuo compared to 48%, 33%, and 71% of the lithic assemblages at Goudi, Dachangliang, and Xiaochangliang, respectively. Bipolar cores were the most abundant core type (23% of the total cores). It should be noted that unlike any of the earlier sites, retouched tools, including end scrapers, notches, and points were recovered at Donggutuo. Although detailed study of these materials is still in progress, our preliminary observations suggest that the Donggutuo hominins were engaged in on-site stone tool manufacture.

Dachangliang

Dachangliang was identified during a survey by the ROM-IVPP team in July and August 2000. It is located on a promontory immediately east of Xiaochangliang. Dachangliang is located 160 m from excavation unit A at the Xiaochangliang locality, and about 1,020 m from the Donggutuo T1 excavation . The cultural horizon is a grey-silt, 60 cm in thickness, similar to that found in the excavations at Xiaochangliang. Similar lithostratigraphic horizons suggest that Dachangliang is contemporaneous with Xiaochangliang. This assessment was later supported by the palaeomagnetic analysis of sediments from both localities (Pei 2002; Deng et al. 2006).

A 7 m² area was excavated in 2000 yielding 33 lithic artifacts, 22 fragments of bone identified to the order *Mammalia*, and five pieces of freshwater bivalve mollusk. Most of the fossils and artifacts display light-weathering and low degrees of erosive abrasion. There are no clear clusters of artifacts. Artifacts provenience data suggest an even distribution of artifacts horizontally and vertically within the cultural horizon.

The Dachangliang lithic assemblage includes 4 cores, 18 flakes, and 11 chunks. Three of the flakes display possibly intentional edge modification. The lithic artifacts are made primarily on poor quality quartzite similar to those from Xiaochangliang. All cores exhibit multiple platforms and are usually amorphous in form. Based on core platform orientations and the placement of flake scars, it is suggested that multiple flake removal occurred during several stages of detachment by rotating the core and changing the direction of removal (Fig. 13.3). For example, core

Fig. 13.3 Cores recovered from the Dachangliang site

DCL011D exhibits a total of seven striking platforms and eight final negative scars associated with the platforms. Examination of the relationship between striking platform and flake scars revealed at least five stages of flake removal. Platforms were arranged in parallel and were sometimes opposed. The orientation of flake removals is not indicative of intentional platform preparation. However there does seem to have been an intensity of core reduction that suggests that Dachangliang hominins attempted to obtain as many flakes as they could from the cores. All of the Dachangliang cores appear to be near exhaustion, continuing to detach flakes would be difficult by free-hand percussion due to the small size.

Goudi (Majuangou III)

Located about 1.5 km from Donggutuo village, the site was first excavated in 2001 by archaeologists from the IVPP. Archaeologists from the Hebei Provincial Institute of Archaeology subsequently conducted three additional field sessions in 2001, 2002, and 2003 (Hebei Provincial Institute of Cultural Relics 2002; Xie et al. 2006:16–33). Due to a lack of direct communication in reporting, scientists from the IVPP named the site as Goudi (Wei 2002, 2003), while Hebei province archaeologists referred to the locality as Majuangou III based on the fact that the site was directly below the 1.55 million year old Majuangou site (Hebei Provincial Institute of Cultural Relics 2002).

Faunal remains identified from Goudi by the IVPP include birds (including ostrich), rodents, carnivores, elephantids, *Equus*., rhinocerotids, and cervids. Later excavations by Hebei archaeologists revealed a floor of foot imprints that probably belong to *Mammuthus trogontherii* whose jaw, teeth, and ribs also were recovered. The 2001and 2002 excavations identified an area where natural rocks, hammerstones and stone artifacts were densely integrated with a cluster of *Mammuthus trogontherii* bones. This has been interpreted as a living floor (Xie et al. 2006:34–43).

The first excavation by the ROM-IVPP team exposed a 1 × 10 m area. Over 50 lithic artifacts and more than 600 faunal remains were recovered from clay-silt and sand sediments 50 cm in thickness. Most of the remains were recovered from a concentrated area covering four 1 × 1 m pits at the eastern end of the trench. The cultural remains include flakes that can be refit to cores from adjacent excavation units.

The Goudi lithic assemblage is rather simple, and includes nodules, cores, flakes, and chunks. All of these objects were manufactured using locally abundant poor-quality raw materials, similar to the other sites in this area. Two cores have double-platforms with multiple flake scars. Almost half of the pieces are flakes (24), of which 17 are complete. Four flakes exhibit some degree of intentional edge modification. While waste pieces (chunks) account for 45% of the assemblage, two pieces are identified as nodules with few ambiguous flake removals. Although the lithic assemblage from the ROM-IVPP excavation is small, traces of intentional manufacture, including striking platforms, bulbs of percussion, and edge modification, are clear. One of the cores was reconstructed by refitting three pieces (a core, a flake, and a modified flake) recovered from this 1 × 10 m excavation, indicating on-site activity by early hominins (Fig. 13.4).

Lithic Technology of the Nihewan Hominins

The following discussion on lithic technology is drawn from our preliminary observations on the materials available for study. Our study, focuses on four aspects of stone tool making and use: raw material procurement, core reduction, flake-tool production, and tool utilization. The paleoemagnetic dates of these four sites, suggest that they represent three different phases prior to 1 Ma: Early (Goudi), Middle (Xiaochangliang and Dachangliang), and Later (Donggutuo).

Raw Material Procurement

Sourcing studies of raw materials suggest Nihewan hominins procured abundant locally-available poor quality quartzite in various quality and colors. Fine-grained cherts are locally available, but rare in the core zone of archaeological sites (Pei and Hou 2001; Du 2003). In the Nihewan, chert-like nodules were found usually embedded within quartzite or

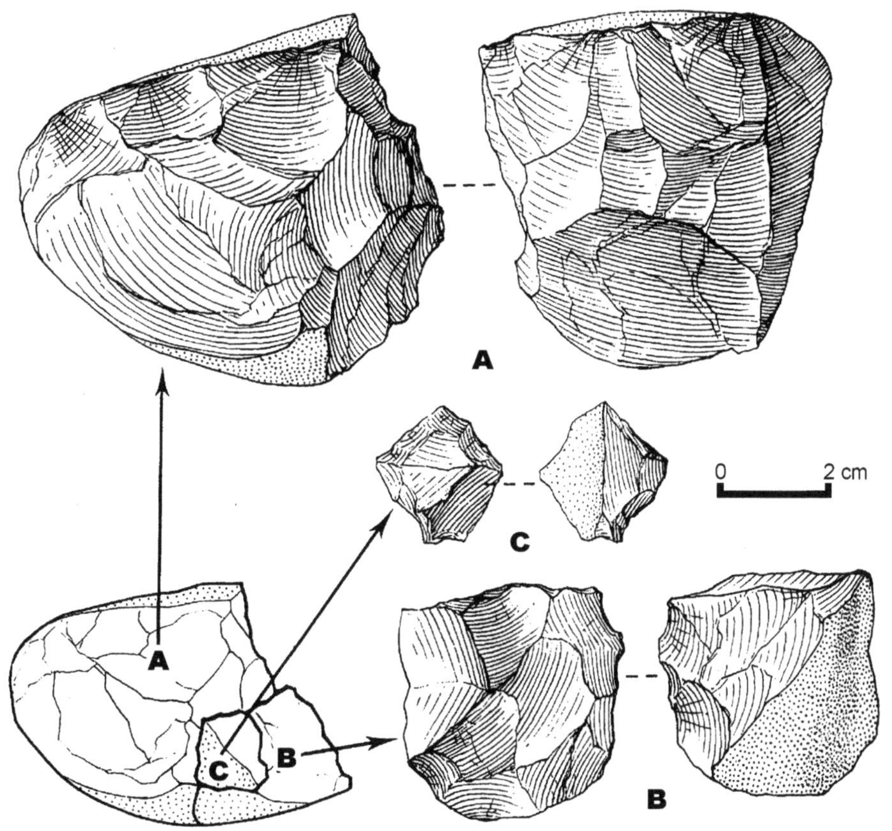

Fig. 13.4 Refitted artifacts from Goudi

other rock types. These kinds of raw materials developed internal fractures and cracks making it difficult to assess the types of raw materials hominins used to make artifacts. Our data suggest that during the early occupations in the Nihewan, hominins did not display an ability to intentionally select high quality nodules from locally rare sources (Shen and Wei 2004; Du 2003).

It is interesting that most of the Nihewan assemblages (Goudi, Dachangliang, and Dongguguo T4 Layer C) have equally high occurrences of chert materials (about 45%), while the rest of the assemblages are comprised of poor quality quartzite. It should be noted that most of the flakes were made on relatively high quality, fine-grained raw material, while the chunks are of poor quality. This may imply that the Nihewan hominins learned to reduce nodules in order to specifically extract the better pieces to further working.

Changes in the core reduction techniques is evident in the middle to later assemblages (Xiaochangliang and Donggutuo) by the emergence of bipolar cores and bipolar flakes. This suggests that two reduction techniques – direct percussion and bipolar percussion – were employed at Xiaochangliang and Donggutuo. Nodules could have been first shattered with a hammer stone using direct percussion. The cores were then continuously selected for further reduction by bipolar percussion. Any chunks unsuitable for direct percussion could have been used for bipolar percussion during the reduction process.

Cores from Xiaochangliang and Donggutuo are relatively small. Average core length, width, and thickness at Xiaochangliang are 53.3 mm (sd = 28), 41.8 mm (sd = 21), and 30.1 mm (sd = 15). Compared to relatively large cores at Goudi and Dachangliang, the data implies that the raw materials were more intensively utilized during the later stages of hominin occupation in the Nihewan.

Core Reduction

Most of the cores from the four assemblages are amorphous in shape. Cores from Goudi and Dachangliang, although rare, are generally angular-shaped multiple-platform cores. Core reduction occurred in multiple stages by rotating and changing the direction of removals, with no clear sign of preparation. Core reduction was primarily through hard-hammer percussion, as hammerstones were recovered at Goudi.

Flake-Tool Production

All assemblages have a large number of small-sized flakes, which is probably related to the nature of poor quality raw materials that are easily shattered into irregular pieces. Examinations of flake attributes suggest no strategic flake removal through core platform preparation. The common features of flakes are non-cortical dorsal surfaces, irregular flake scar patterns on dorsal surfaces, and a relatively high

Fig. 13.5 Lithic artifacts recovered from Donggutuo (*left*) and a closeup of an endscraper (*right*)

number of negative scars on the dorsal surface. All of these attributes suggest that flakes were extensively reduced during the later stage of core reduction.

No retouched tool types were identified in the Goudi and Dachangliang assemblages. Less than 1% of the Xiaochangliang lithic artifacts examined have secondary retouch. Few flakes are identified with intentional modification either from retouch or from use, suggestive of expedient tool use at those early occupation sites. No standardization is evident in the secondary retouch or edge modification of flakes. However, there is a remarkable change with Donggutuo in tool production. Unambiguous forms of end-scrapers, side-scrapers, notches, points, and retouched pieces are identified (Fig. 13.5), although in low numbers.

Tool Utilization

Tool use-patterning can be implied from the results of use-wear analysis. Flake tools that were suitable for direct use in terms of size and shape were randomly selected for microscopic examination for possible use-wear traces based on microfracture scarring, degree of rounding, and polish. The lower-power use-wear method was employed here (Keeley 1980; Odell 1981; Shen 2001). Detailed procedures for microscopic examination of the Nihewan lithic materials are described in previous works (Shen and Chen 2000, 2003; Shen and Wei 2004). Use-wear of flake tools from the four assemblages suggest that the Nihewan hominins engaged primarily in activities related to soft animal substances, followed by woodworking activities.

Our results indicate that 12–24% of the samples selected exhibit unambiguous wear damages caused by direct use (Table 13.2). An additional 6–20% of the samples display ambiguous wear patterns indicating those pieces were possibly used as tools. Our use-wear study suggests that a limited number of tasks were performed. Two used pieces at Dachangliang appear to be related to slicing soft animal substances. At Xiaochangliang, three pieces were used to scrape animal bones, and a different set of three flakes were employed to cut animal meat or hide. An additional flake was used to drill fresh animal bone. At Donggutuo, flake tools were used primarily for scraping on animal substances. Use-wear types related to sawing wood or drilling wood are present, but very limited. Notably, 7 flake tools were used in multiple activities, displaying two or three utilized locations per flake tool.

The use-wear results confirm that unmodified flakes were primarily used at all of these occupations (Fig. 13.6). Although some tool types, including scrapers and notches recovered from Donggutuo, were also found to be employed as scraping tools. Of the 20 pieces that were identified as having positive use wear traces at Donggutou 14 are unmodified flakes. This evidence supports the previous notion that these flakes were used as expedient tools (Keates 2000; Shen and Chen 2003).

Discussion and Summary

The recent investigations in the Nihewan Basin enhance our understanding of earliest hominin behavior in East Asia in two areas. First, paleomagnetic dating has established a more refined chronological framework of the Lower Palaeolithic in the Nihewan. Second, the study of lithic technology has enabled us to interpret early hominin behavior during the Lower Paleolithic in northern China.

The results of recent palaeomagnetic studies support earlier relative dating sequences based on the lithostratigraphic and biostratigraphic evidence (Wei 1997). The magnetostratigraphic dates suggest not only that the Nihewan basin was densely occupied, but also that the hominin occupations lasted continuously for about 1 million years (Shen 2007). However, As Gao et al (2005) argued before, the Nihewan beds containing early hominin artifacts display uneven deposition of sediments across these sites. For example, at Xiaochangliang and Dachangliang, the cultural layer is in a sand-silt sediment, while at Putaoyuan 290 m away, it is gravel. The strata of the five sections analyzed by Zhu and his colleagues differ in thickness and in geological sediments which suggests variation in sedimentation rates (Gao et al. 2005). For example, the Cenjiawan site was dated to 1.1 Ma, based on the stratigraphic distance between the site and a magnetostratigraphic marker horizon, using an average sedimentation rate (Wang 2005; Wang et al. 2006). According to

Table 13.2 Lithic artefacts selected for microscopic examination

	Doudi		Dachangliang		Xiaochangliang		Donggutuo	
Sites	N	%	N	%	N	%	N	%
Artifact examined	25	100	16	100	174	100	88	100
Artifact with no use-wear	15	60	13	81.3	137	78.7	50	57
Artifacts with possible use-wear	4	16	1	6.3	15	8.6	18	20
Aritfacts with positive use-wear	6	24	2	12.5	22	12.6	20	23

Fig 13.6 Flake tools with positive usewear by microscopic examination (*above*). *Lower left*: Artifact XCL98623, 28x, the dorsal side shows a diagnostic woodworking use-wear feature: roll-over scaring on the edge and a series of scars with feathered or stepped termination perpendicular to the rounded edge. *Lower right*: Artifact XCL98272 from Xiaochangliang, 140x. The *ventral side* displays bright polish on the edge, but diffused towards the *centre* in directional fashion. The piece was used for scraping wood

this calculation the site is coeval with Donggutuo. However, the two localites represent drastically different depositional scenarios. In previous studies, Cenjiawan was determined to be an early Middle Pleistocene site based on its stratigraphic position in the Nihewan Formation (Shen and Wei 2004; Wei 2004). Accurate chronostratigraphic control can only be determined by the combination of detailed lithostratigraphy and biostratigraphy in combination with palaeomagnetic studies.

The field investigations over the past decade have yielded abundant materials that allow us to examine the Early Pleistocene lithic technology at Nihewan. Lithic technology of early hominins can be viewed to be a part of the human behavioral system, which directly involved stone tool manufacture and use representing hominin survival strategies. The four lithic assemblages under study represent only a few Nihewan hominin occupations that spanned over half a million years in time. Although our studies of these lithic materials are still in progress, at present our observations suggest a continuous development of lithic technology in this region, indicating some degree of behavioral adaptation to local environments over a long period of time.

Our previous study suggested that the small-sized flake production was probably constrained by the use of poor local raw materials (Shen and Chen 2003). The bipolar technique was probably used to overcome constraints imposed by poor quality materials. Results from our flintknapping experiments indicate that the bipolar method is efficient enough to produce small, workable flakes from these nodules which contain numerous internal fractures. The Donggutuo assemblages display more characteristic bipolar products than the Xiaochangliang assemblages, while the other two early assemblages (Goudi and Dachangliang) have no evidence for the use of the bipolar technique. It is possible that the bipolar technique first appeared in the Nihewan and then spread to other spatio-temporal contexts in China (e.g. Middle Pleistocene). This hypothesis needs further testing.

Evolution of hominin behavior in the Nihewan is represented by a gradual emergence of morphologically-discrete tool types. Retouched tools are not found in the early assemblages, instead appearing initially at the younger Donggutuo site. Standard stone tool-type scrapers, notches, and points, are more common from the late Early Pleistocene assemblages of Cenjiawan and Maliang in the same region (Shen and Wei 2004).

In summary, recent archaeological and chonostratigraphic investigations reveal that the earliest hominins arrived in northern China about 1.7–1.6 Ma. Based on evidence from faunal and pollen studies, the palaeoenvironment during the Early Pleistocene was subtropical, and represents a warmer and moister environment relative to the modern central Northern China habitats (Qiu 2000; Cai and Li 2003). All of these earliest hominin sites are open-air localities near water sources in what is today a hilly environmental setting. The Eastern end of the Nihewan Basin appears to have been heavily populated which thus so far is the only area with such evidence in northern China.

Acknowledgements This work has been supported by Royal Ontario Museum Governor Funds, the Wenner-Gren Foundation for Anthropological Research (Gr.ICRG-41), Chinese Academy of Sciences' Outstanding Overseas-Chinese Scholars Research Fund (No. 2005-1-15), Chinese Academy of Science's Wang-Kuancheng Science Research Fund, as well as Special Fund for Basic Research, Ministry of Science and Technology of China (2001CCA01700). The early draft of this paper was first delivered at the 76th Annual Meeting of the American Association of Physical Anthropologists. March 27–April 1 2007, Philadelphia, PA. The first author thanks Christopher J. Norton for the invitation and for his comments and editorial assistance of this early draft. The authors thank three anonymous reviewers for their constructive comments. We also thank Zhang Xiaoling, Qu Tongli, Chen Hong for their assistance in lithic analysis at the Nihewan Field Station at Donggutuo.

References

Barbour, G. B. (1924). Preliminary observation in Kalgan area. *Bulletin of the Geological Society of China, 3*(3),167–168.

Barbour, G. B. (1925). Geological study of the deposits of the Sankanho Basin. *Bulletin of the Geological Society of China*, *4*(1), 58.

Barbour, G. B., Licent, E., & Teilhard de Chardin, P. (1927). Geological study of the deposits of the Sangkanho Basin. *Bulletin of the Geological Society of China, 6*, 263–278.

Breuil, H. (1935). L'état actuel de nos connaissances sur les industries paléolithiques de Choukoutien. *L'Anthropologie, 45*, 740–746.

Cai, B. Q., & Li, Q. (2003). Early Pleistocene artifacts and palaeoenvironment in the Nihewan Basin. *Sciences in China (Series D), 33*(5), 418–424.

Chen, G. L., Lin, J. L., Li, S. L., Liang, Q. P. (1978). A preliminary paleomagnetic survey of the Nihewan Bed (in Chinese). *Scientia Geologica Sinica*, Issue 3, 247–252.

Chen, Q., Chen,T. M., Yang , Q. (2003). ESR dating of Early Pleistocene archaeological sites. In C. Shen, S.G. Keates (Eds.), *Current research in Chinese Pleistocene archaeology. British archaeological reports international series 1179* (pp. 119–125). Oxford: Archaeopress.

Deng, C. L., Wei, Q., Zhu, R. X., Wang, H. Q., Zhang, R., Ao, H., et al. (2006). Magnetostratigraphic age of the Xiantai Paleolithic site in the Nihewan Basin and implications for early human colonization of Northeast Asia. *Earth and Planetary Science Letters, 244*, 336–348.

Deng, C. L., Xie, F., Liu, C. C., Ao, H., Pan, Y. X., Zhu, R. X. (2007). Magnetochronology of the Feiliang Paleolithic site in the Nihewan Basin and implications for early human adaptability to high northern latitudes in East Asia. *Geophysical Research Letters, 34*(L14301), 1–6.

Dennell, R., & Roebroeks, W. (2005). An Asian perspective on early human dispersal from Africa. *Nature, 438*, 1099–1104.

Du, S. S. (2003). A preliminary study on raw material exploitation in the Middle-Upper Palaeolithic sites in the Nihewan Basin (in Chinese). *Acta Anthropologic Sinica, 22*(2), 121–130.

Gao, X., Wei, Q., Shen, C., Keates, S. (2005). New lights on the earliest hominids in East Asia. *Current Anthropology, 46*(Suppl.), 115–120.

Hebei Provincial Institute of Cultural Relics (2002). Majuanguo site in Nihewan. In State Administration of Cultural Heritage (Eds.), Major Archaeological Discoveries in China in 2001 (in Chinese) (pp. 4–8). House: Cultural Relics Publishing.

Hou, Y. M. (2000). Donggutuo Lithic Industry of the Nihewan Basin, North China (in Chinese). Unpublished Ph.D. dissertation, Institute of Vertebrate Paleontology and Paleoanthropology, Chinese Academy of Sciences.

Jia, L. P. (1985). China's earliest Palaeolithic assemblages. In R. K. Wu & J. W. Olsen (Eds.), *Palaeonanthropology and Palaeolithic Archaeology in the People's Republic of China* (pp. 135–146). New York: Academic.

Keates, S. G. (2000). *Early and Middle Pleistocene Hominid behavior in Northern China. BAR International Series 863*. Oxford: John and Erica Hedges.

Keeley, L. H. (1980). *Experimental Determination of Stone Tool Uses*. Chicago: The University of Chicago Press.

Odell, G. H. (1981). The mechanics of use-breakage of stone tools: some testable hypotheses. *Journal of Field Archaeology, 8*, 197–209.

Olsen, J. W. (2000). China's earliest inhabitants. *Journal of East Asian Archaeology, 2*(1–2), 1–7.

Pei, W. Z. (also known as Pei, W.C.) (1935). Fossil mammals from the Kwangsi Caves. *Bulletin of the Geological Society of China, XIV*(3), 413–425.

Pei, S. W. (2002). The Dachangliang site in the Nihewan Basin (in Chinese). *Acta Anthropologica Sinica, 21*(2), 115–125.

Pei, S. W., & Hou, Y. M. (2001). Preliminary study of raw material exploitation at the Donggutou Site, Nihewan Basin, northern China (in Chinese). *Acta Anthropologica Sinica, 20*(4), 271–281.

Peterson, C. E., Shen, C., Chen, C., Chen, W. Y., Tang, Y. ssssJ. (2003). Taphonomy of an Early Pleistocene archaeofauna from Xiaochangliang, Nihewan Basin, North China. In C. Shen, S.G. Keates (Eds), *Current research in Chinese Pleistocene archaeology. British archaeological reports international series 1179* (pp. 83–98). Oxford: Archaeopress.

Qiang, F., Ma, X. H., & Wu, X. H. (1983). A paleomagnetic study of the Nihewan Bed (in Chinese). *Marine Geology & Quaternary Geology, 3*(3), 17–30.

Qiu, Z. X. (2000). Nihewan fauna and Q/N boundary in China [in Chinese]. *Quaternary Sciences, 20*(2), 142–154.

Pope, G. G. & Keates, S. G. (1994). The Evolution of Human Cognition and Cultural Capacity: a view from the Far East. In R. S. Corruccini and R. L. Ciochon (Eds), *Integrative Paths to the Past: Paleoanthropological Advances in Honor of F. Clark Howell*, (pp. 531–567) Englewood Cliffs: Prentice Hall.

Pope, G. G., An, Z., Keates, S., & Bakken, D. (1990). New discoveries in the Nihewan Basin, Northern China. *The East Asian Tertiary/Quaternary Newsletter, 11*, 68–73.

Schick, K. D., & Dong, Z. (1993). Early Paleolithic of China and eastern Asia. *Evolutionary Anthropology, 2*(1), 22–35.

Schick, K., Toth, N., Wei, Q., Clark, J. D., & Etler, D. (1991). Archaeological perspectives in the Nihewan Basin, China. *Journal of Human Evolution, 21*, 13–26.

Shen, C. (2001). *The Lithic production system of the princess point complex during the transition to agriculture in southwestern Ontario, Canada.* Oxford: BAR International Series 991.

Shen, C. (2007). The palaeolithic cultures of China. In D. M. Pearsall (Ed.), *Encyclopedia of Archaeology* (pp. 570–597). London: Elsevier/Academic.

Shen, C., & Chen, C. (1999). 1998 excavation of Xiaochangliang, an Early-Pleistocene site in northern China. *Current Research in the Pleistocene, 16*, 71–73.

Shen, C., & Chen, C. (2000). A use-wear study of lithic artifacts from Xiaochangliang and hominid activities in Nihewan Basin, Northern China. *Acta Anthropologica Sinica, 19*(Suppl.), 115–121.

Shen, C., & Chen, C. (2003). New evidence of hominid behavior from Xiaochangliang, Northern China: site formation and lithic technology. In C. Shen, S.G. Keates (Eds), *Current research in Chinese Pleistocene archaeology. British archaeological reports international series 1179* (pp. 67–82). Oxford: Archaeopress.

Shen, C., & Wei, Q. (2004). Lithic technological variability of the Middle Pleistocene in the Eastern Nihewan Basin, northern China. *Asian Perspectives, 43*(2), 281–301.

Tang, Y. J., Li, Y., & Chen, W. (1995). Mammalian fossils and the age of Xiaochangliang Paleolithic site of Yangyuan, Hebei (in Chinese). *Vertebrata PalAsiatica, 33*(1), 74–83.

Teilhard de Chardin, P. (1935). Les récents progrÈs de la Préhistoire en Chine. *L'Anthropologie, 45*, 735–740.

Teilhard de Chardin, P., & Piveteau, J. (1930). Les mammiferes fossils de Nihewan (Chine). *Annales de Paleontologie, 19*, 1–154.

Wang, H. Q. (2005). Magnetic Stratigraphy Study on the Nihewan Basin, Northern China (in Chinese). Unpublished Ph.D. dissertation, Graduate University of Chinese Academy of Sciences.

Wang, H. Q., Deng, C. L., Zhu, R. X., Wei, Q., Hou, Y. M., & Boeda, E. (2005). Magnetostratigraphic dating of the Donggutuo and Maliang Paleolithic sites in the Nihewan Basin, North China. *Quaternary Research, 64*(1), 1–11.

Wang, H. Q., Deng, C. L., Zhu, R. X., & Xie, F. (2006). Paleomagnetic dating of the Cenjiawan Palaeolithic site in the Nihewan Basin, North China. *Science in China (Series D), 49*(3), 295–303.

Wei, Q. (1985). The preliminary observations on lithic artifacts recovered from the Dongguto Site (in Chinese). *Acta Anthropologica Sinica, 4*(4), 289–300.

Wei, Q. (1991). Geologic sequence of the archaeological sites in the Nihewan Basin, North China (in Chinese). In Institute of Vertebrate Paleontology and Paleoanthropology, Academia Sinica (Eds.), *Contributions to the XIII INQUA* (pp. 61–73). Beijing: Beijing Scientific and Technological Publishing House.

Wei, Q. (1997). The framework of archaeological geology of the Nihewan Basin (in Chinese). In Y. Tong, Y. Zhang, W. Wu, J. Li, & L. Shi (Eds.), *Evidence for evolution – essays in honor of Prof. Chongchian Yong on the hunhredth anniversary of his birth* (pp. 193–207). Beijing: China Ocean Press.

Wei, Q. (2000). Early Pleistocene paleoliths in China (in Chinese). *Wenwu Chunqiu, 2*, 1–14.

Wei, Q. (2002). Nihewan Basin – cradle of the human culture in the East Asia (in Chinese). *Fossil, 4*, 5–7.

Wei, Q. (2003). Nihewan Basin – Olduvai Gorge of East Asia (in Korean). In Y. Lee (Ed.), *Palaeolithic men's lives and their sites* (pp. 227–254). Seoul: Hakyoun Press.

Wei, Q. (2004). The Palaeolithic cultures in the Nihewan Basin (in Chinese). In Z. E. Lü (Ed.), *Retrospect of Chinese Archaeology in the 20th Century: Palaeolithic* (pp. 84–110). Beijing: Sciences Press.

Wu, R. K., & Dong, X. R. (1985). Homo erectus in China. In R. K. Wu & J. W. Olsen (Eds.), *Palaeonanthropology and Palaeolithic Archaeology in the People's Republic of China* (pp. 79–90). New York: Academic.

Wu, X. Z., Huang, W. W., & Qi, G. Q. (1999). *Palaeolithic sites of China (in Chinese).* Shanghai: Shanghai Science and Education Press.

Xie, F., & Li, J. (1995). The refitting application in the comprehensive analysis of the Cenjiawan lithic artifacts (in Chinese). *Journal of Chinese Antiquity, 1*, 25–38.

Xie, F., Schick, K., Toth, N., & Clarke, D. J. (1994). A refitting study of lithic artifacts recovered from a 1986 excavation at Cenjiawan site (in Chinese). *Journal of Chinese Antiquity, 3*, 86–102.

Xie, F., Li, J., & Liu, L. Q. (2006). *The Nihewan Palaeolithic cultures (in Chinese).* Shijiazhuang: Huashan Wenyi Press.

You, Y. Z., Tang, Y. Y., & Li, Y. (1978). Paleolithic discoveries in the Nihewan formation (in Chinese). *Chinese Quaternary Research, 1*, 1–13.

Yuan, B. Y., Zhu, R. X., Tian, W. L., Cui, J. X., Li, R. Q., Wang, Q., et al. (1996). Age of Nihewan formation and its stratigraphic division and comparison. *Science in China (Series D), 26*, 67–73.

Zhu, R. X., Hoffman, K. A., Potts, R., Deng, C. L., Pan, Y. X., Guo, B., Shi, C. D., Guo, Z. T. Yuan, B. Y., Hou, Y. M., Huang, W. W. (2001). Earliest presence of humans in northeast Asia. *Nature, 413*, 413–417.

Zhu, R. X., An, Z. S., Potts, R., & Hoffman, K. A. (2003). Magnetostratigraphic dating of early humans in China. *Earth Science Reviews, 61*, 341–359.

Zhu, R. X., Potts, R., Xie, F., Hoffman, K. A., Deng, C. L., Shi, C. D., et al. (2004). New evidence on the earliest human presence at high northern latitudes in northeast Asia. *Nature, 431*, 559–562.

Chapter 14
Peopling in the Korean Peninsula

Kidong Bae

Abstract The earliest hominin occupation of the Korean peninsula is likely older than 400 ka. The Chongokni site has recently been dated to 350–300 ka. Komunmoru, Jangsanni, and Jangdongni are likely older than Chongokni. Currently, the oldest hominin fossils in Korea date to the late Middle Pleistocene or early Late Pleistocene and have tentatively been assigned to *Homo erectus* or archaic *H. sapiens*. Unlike many other regions of East Asia the Korean Peninsula records evidence of Pleistocene hominins which employed Acheulean technology.

Keywords Korean peninsula • Chongokni • Komunmoru • Middle Pleistocene • Acheulean • Movius Line • Chongoknian

Introduction

Current evidence from China indicates that initial hominin occupation in northern Asia extends to the Early Pleistocene (~1.6 Ma or older: Zhu et al. 2008). Since the Korean peninsula was never geographically isolated from the Chinese mainland, it has been postulated that the early hominin migrations into northern China could also have extended to the Korean peninsula. However, currently no Early and few early Middle Pleistocene sites have been identified in Korea (Norton 2000; Bae 2001). Current Paleolithic research in Korea is focused on the important questions surrounding the circumstances of the earliest peopling of Korea. This research hopes to understand where the earliest migration of hominins onto the Korean peninsula took place and which hominin species is responsible for this migration. Due to the acidic soils which inhibit bone preservation at most open-air sites, stone artifacts are usually the only hominin traces recovered at Pleistocene localities in Korea. Understanding the nature of these lithic technologies is critical to understanding how early hominins adapted to the Korean environment. The focus of this paper is to present an updated review of the Korean paleoanthropological record since Norton's (2000) synthesis.

Most of the northern part of the peninsula has not been intensively investigated for paleoanthropological traces. Some cave sites have yielded hominin fossils and stone industries from the Late Paleolithic. Due to an improved socio-economic situation and a growing interest in paleoanthropology, more than 100 Paleolithic sites have been identified in the southern part of the peninsula (Norton 2000, 2007; Sohn et al. 2002; Fig. 14.1). Since few absolute dates are available and many sites are considered to have been secondarily deposited, it is difficult to construct an absolute chronology for many of these sites and associated materials (Bae 2002a; Table 14.1). The age of many of these sites unfortunately falls between the limits of most dating methods. Fortunately, the recent application of tephrochronological studies provides an invaluable methodology for developing a confident chronostratigraphy of Early Paleolithic sites in Korea.

The Oldest Hominin Fossils from the Korean Peninsula

The earliest hominin fossils were excavated from limestone caves in central South Korea and the Pyongyang area in North Korea (Norton 2000; Park and Lee 2004). All of the hominin fossils in South Korea have been dated to the Late Pleistocene. Some hominin fossils found in North Korea are considered to be Middle Pleistocene in age. The North Korean hominin fossil material is discussed here (see Norton 2000 for discussion of the South Korean hominin fossil record).

Two molars from Dokchon are considered to be archaic *Homo sapiens* (Institute of Archaeology 1978). Unfortunately no stone artifacts were found in association with these specimens, and no absolute dates are available for this site.

K. Bae (✉)
Department of Anthropology, Hanyang University, Ansan, Kyunggi-do, 425-791, South Korea
e-mail: Bkd5374@gmail.com

C.J. Norton and D.R. Braun (eds.), *Asian Paleoanthropology: From Africa to China and Beyond*, Vertebrate Paleobiology and Paleoanthropology, DOI 10.1007/978-90-481-9094-2_14,

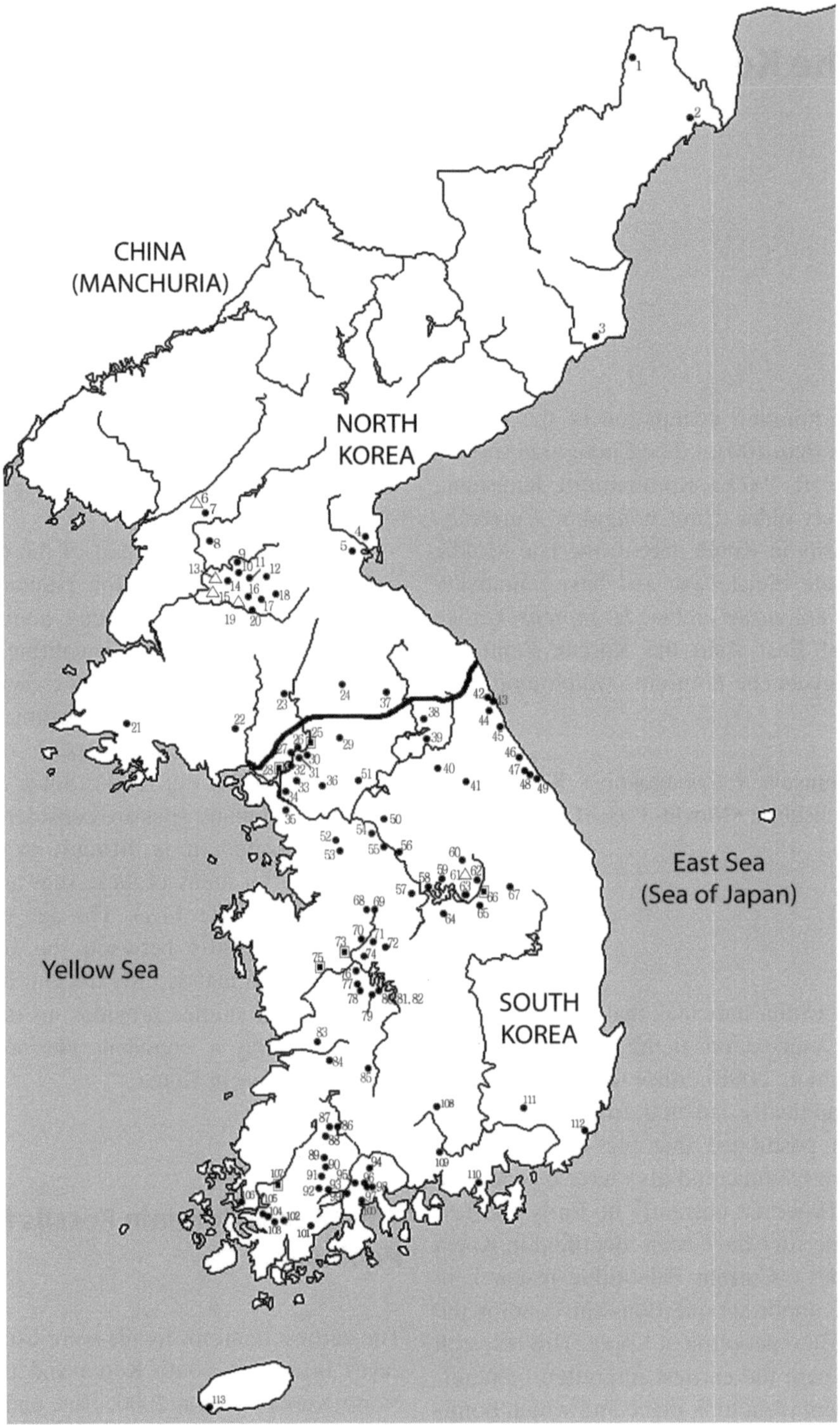

Fig. 14.1 Primary Paleolithic sites from the Korean Peninsula

Legend of Paleolithic Sites in Korea

No.	Sites	Notes
1	Donggwanjin	Paleolithic Site
2	Gulpori	Paleolithic Site
3	Jangdeongni	Paleolithic Site
4	Dajeonni	Paleolithic Site
5	Guljaedeok	Paleolithic Site
6	Dokchon-seungrisan	Hominid Fossil Site

Legend of Paleolithic Sites in Korea (continued)

No.	Sites	Notes
7	Punggongni	Paleolithic Site
8	Miljeonni	Paleolithic Site
9	Geumpyeongni	Paleolithic Site
10	Seungho III cave	Paleolithic Site
11	Maeri	Paleolithic Site
12	Heungnyeong	Paleolithic Site

Legend of Paleolithic Sites in Korea (continued)

No.	Sites	Notes
13	Mandalri	Hominid Fossil Site
14	Hwacheonni	Paleolithic Site
15	Ryokpori	Hominid Fossil Site
16	Komonmoru	Paleolithic Site
17	Cheonjaeam	Paleolithic Site
18	Bancheonni	Paleolithic Site
19	Ryonggokni	Hominid Fossil Site
20	Geumcheon cave	Paleolithic Site
21	Daetanni	Paleolithic Site
22	Haesang	Paleolithic Site
23	Hoengsanni	Paleolithic Site
24	Jangheungni	Paleolithic Site
25	Chongokni	Earliest Paleolithic Site
26	Namgyeri	Paleolithic Site
27	Wondangni	Paleolithic Site
28	Jangsanni	Earliest Paleolithic Site
29	Hwadaeri Shimteo	Paleolithic Site
30	Gawolni	Paleolithic Site
31	Juwolni	Paleolithic Site
32	Geumpari	Paleolithic Site
33	Unjeong	Paleolithic Site
34	Gawaji	Paleolithic Site
35	Janggi	Paleolithic Site
36	Minrakdong	Paleolithic Site
37	Geoduri	Paleolithic Site
38	Sangmuryongni	Paleolithic Site
39	Budongni	Paleolithic Site
40	Hahwagyeri	Paleolithic Site
41	Hyeoncheonni	Paleolithic Site
42	Guhodong	Paleolithic Site
43	Gumidong	Paleolithic Site
44	Jusuri	Paleolithic Site
45	Simgokni	Paleolithic Site
46	Gigok	Paleolithic Site
47	Nobong	Paleolithic Site
48	Balhandong	Paleolithic Site
49	Bukpyeongdong	Paleolithic Site
50	Andeonkni	Paleolithic Site
51	Hopyeongdong	Paleolithic Site
52	Pyeongchangni	Paleolithic Site
53	Samri	Paleolithic Site
54	Byeongsanni	Paleolithic Site
55	Yeonyangni	Paleolithic Site
56	Keungilga	Paleolithic Site
57	Ochangri	Paleolithic Site
58	Myeongori	Paleolithic Site
59	Sagiri	Paleolithic Site
60	Jeommal cave	Paleolithic Site
61	Sangsiri	Hominid Fossil Site
62	Gunang cave	Paleolithic Site
63	Suyanggae	Paleolithic Site

Legend of Paleolithic Sites in Korea (continued)

No.	Sites	Notes
64	Sangsi-geuneul	Paleolithic Site
65	Dodamni	Paleolithic Site
66	Keum cave	Earliest Paleolithic Site
67	Yeondang Ssang cave	Paleolithic Site
68	Janggwanni	Paleolithic Site
69	Songduri	Paleolithic Site
70	Sorori	Paleolithic Site
71	Yulyangdong	Paleolithic Site
72	Durubong	Paleolithic Site
73	Mansuri	Earliest Paleolithic Site
74	Saem cave	Paleolithic Site
75	Sokchangni	Earliest Paleolithic Site
76	Nosanni	Paleolithic Site
77	Daejeongdong	Paleolithic Site
78	Gwanpyeongdong	Paleolithic Site
79	Gujeukdong	Paleolithic Site
80	Dunsan	Paleolithic Site
81	Yongsandong	Paleolithic Site
82	Noeundong	Paleolithic Site
83	Sinmak	Paleolithic Site
84	Sageunni	Paleolithic Site
85	Chimgokni	Paleolithic Site
86	Jewolni	Paleolithic Site
87	Songjeonni	Paleolithic Site
88	Okgwa	Paleolithic Site
89	Hwagok	Paleolithic Site
90	Jukongni	Paleolithic Site
91	Chipyeongdong	Paleolithic Site
92	Daejeonni	Paleolithic Site
93	Juksanni	Paleolithic Site
94	Juknaeri	Paleolithic Site
95	Geumcheon cave	Paleolithic Site
96	Usanni Gokcheon	Paleolithic Site
97	Sinpyeongni Geumpyeong	Paleolithic Site
98	Wolpyeong	Paleolithic Site
99	Okcheon	Paleolithic Site
100	Deoksanni	Paleolithic Site
101	Sanging	Paleolithic Site
102	Yongho	Paleolithic Site
103	Dangga, Chongok	Paleolithic Site
104	Dasanni	Paleolithic Site
105	Piseori	Paleolithic Site
106	Danghasan	Paleolithic Site
107	Jangdongni	Earliest Paleolithic Site
108	Imbulri	Paleolithic Site
109	Naechonni	Paleolithic Site
110	Igeumdong	Paleolithic Site
111	Goryeri	Paleolithic Site
112	Haeundae	Paleolithic Site
113	Gosanri	Paleolithic Site

Table 14.1 List of radiometric dates for Paleolithic sites believed to be older than 40 ka in Korea

Site Name	Sample Name	Layer	Age	Method	Reference
Chongokni		Basalt bed	<0.27 Ma	K-Ar	Takayanagi (1983)
		Basalt bed	0.6 ± 0.2 Ma	K-Ar	Yi (1983)
		Basalt bed	2.9 ± 0.3 Ma	K-Ar	Yi (1983)
		Basalt bed	0.4 ± 0.1 Ma	K-Ar	Yi (1983)
		Basalt bed	0.6231 ± 0.018 Ma	K-Ar	Deino (1988)
		Burnt Silt	0.51 ± 0.07 Ma	FT	Danhara et al. (2002)
		Basalt bed	0.49 ± 0.05 Ma	K-Ar	
		Loess/Soil	95,000 BP	Tephra(K-Tz)	
		Sediment	>200,000	TT-OSL	Kim et al. (2008)
		Sediment	>160,000	OSL	Naruse et al. (2003)
Chatanni		Basalt bed	0.16 ± 0.05 Ma	K-Ar	Danhara et al. (2002)
Jangdongri		L1S1	81,000 ± 2,000 BP	OSL	Naruse et al. (2006)
		S1L1	98,000 ± 3,000 BP	OSL	
		S1L2	134,000 ± 4,000 BP	OSL	
		S1L3	137,000 ± 3,000 BP	OSL	
		L2	148,000 ± 6,000 BP	OSL	
		S2-1	225,000 ± 6,000 BP	OSL	
		S2L1	156,000 ± 6,000 BP	OSL	
		S3L1	194,000 ± 11,000 BP	OSL	
		L4	187,000 ± 5,000 BP	OSL	
		S4	193,000 ± 6,000 BP	OSL	
Kum cave		Culture 3	107,450 BP	ESR	Sohn (1972)
		Culture 2	185,870 BP	ESR	
Mansuri	OS1	Culture 3	92,000 ± 3,000 BP	OSL	Park and Hong (2007)
	OS2	Culture 4	95,000 ± 4,000 BP	OSL	
	OS3	Culture 5	103,000 ± 8,000 BP	OSL	
	OS6	Culture 3	104,000 ± 7,000 BP	OSL	
	MS1		44,000 ± 4,000 BP	OSL	
	MS2		68,000 ± 4,000 BP	OSL	
	OM1 (250–125 μm)	Culture 3	51,000 ± 4,000 BP	OSL	
	OM2 (250–125 μm)	Culture 4	53,000 ± 3,000 BP	OSL	
	OM2 (125–63 μm)	Culture 4	52,000 ± 4,000 BP	OSL	
	OM3 (250–125 μm)		49,000 ± 3,000 BP	OSL	
	OM3 (125–63 μm)		49,000 ± 4,000 BP	OSL	
	OM4 (250–125 μm)		44,000 ± 4,000 BP	OSL	
Songduri		Lower	>48,000 BP	AMS	Chungwon Munhwajae Yeonguwon (2006)
Naeheungdong			>40,800 BP	AMS	Chungcheong Munhwa Yeonguwon (2003)
Bongmyoungdong		Culture 4	>49,860~48,450 BP	AMS	Yi and Hong (1999)
Noeundong		Culture 3	>54,720 BP	AMS	Han (2003)

Several fragments of a hominin juvenile male's skull were collected from sediments of Middle Pleistocene age from Daehyundong in Pyongyang, North Korea (Kim et al. 1985). The hominin fossils were assigned to archaic *Homo sapiens* on the basis of slightly raised brow ridges and a receding forehead. Even though no absolute dates exist for this site the associated fauna suggest deposition occurred during the early part of the Middle Pleistocene (Chang 1989, 1990).

The hominin fossils from Ryonggok cave were originally assigned to *Homo erectus* on the basis of thermoluminescence (TL) dates of 530–450 ka (Lab of Human Evolution 1995b). However, more recent dates, using a combination of uranium series and TL methods, range between 111 ± 10 and 44.3 ± 2 ka for cultural layer 2 from which the hominin fossils are associated (Lab of Human Evolution 1995). The hominin fossils were reclassified as *Homo sapiens* on the basis of these more recent dates and the presence of derived features. Similar hominin fossils were found in Rangjeong cave in Hwanghaedo Province and are penecontemporaneous with the Ryonggok materials on the basis of faunal correlations (Lab of Human Evolution 1995).

The taxonomic affinity of the earliest hominins in the Korean peninsula is one of the most pressing issues in Korean paleoanthropology. *Homo erectus* fossils have yet to be clearly reported east of Zhoukoudian in eastern Asia (but the Daehyundong materials need further investigation [see Norton 2000]). However, the karstic caves of northwestern Korea may provide the best evidence for *H. erectus* fossils on the Korean peninsula, particularly considering the many *H. sapiens* localities in the area.

The Earliest Archaeological Sites

Currently, the oldest stone industry on the Korean peninsula is the Komunmoru lithic assemblage. Komunmoru was discovered in the mid-1960s and was claimed to be the oldest hominin occupation in the country based on the associated faunal assemblage, which is similar to Zhoukoudian Locality 1 (Institute of Archaeology 1969). The faunal assemblage is dominated by extinct species including *Equus sangwonensis*, *Megaloceros sangwonensis* Kimsingyu, *Macaca* sp. *Dicerorhinus kirchbergensis* Jaeger, Elephantidae *gen.* sp., *Bubalus* sp., *Cervus nippon grayi*. Other paleoanthropological sites (e.g., Dokchon cave site, Yonggok cave site), on the Korean peninsula have much lower percentages of extinct forms. Currently, no absolute dates exist for this locality. Several specimens present in the cave were identified as sub-tropical species indicating a warmer Middle Pleistocene age (Norton 2000). Only four stone artifacts have been reported, though there may be more (Institute of Archaeology 1977). These include an axe-form, a rectangular shaped piece and two irregular forms. However, these fractured stones representing hominin activity must be questioned because most of the sediments in Komunmoru represent alluvial deposition; therefore post-depositional processes currently cannot be discounted. Most of these deposits represent gravels and sands formed by the Sangwon river channel, which is presently about 20 m below the cave.

The stone industries from the lower part of Sokchangni have long been claimed to be one of the oldest in the peninsula (Sohn 1972, 1988, 1990, 1993). The site is situated on the lowest terrace of the Keum river in central Korea. A sequence of 27 stratigraphic layers was identified, with the lower part of the stratigraphic column thought to have been deposited during the Early Pleistocene. An AMS date of 50 ka was obtained on a charcoal sample collected from Layer 10 (Sohn 1993). However, the presence of alluvial deposits on the first terrace suggests Sokchangni may consist of reworked sediments from an earlier period in the geomorphological history of the river system; thus, the antiquity of the site cannot currently be confirmed.

The stone artifacts from the Keum cave site in the upper Namhan river basin are claimed to be of early Middle Pleistocene age (Sohn 1985, 1990). Paleolithic to Bronze Age cultural materials were recovered in seven stratigraphic layers; with Paleolithic industries (from Lower Paleolithic to Mesolithic)[1] excavated from the bottom 5 layers. The artifacts from the lowest two layers consist of heavy duty tools, such as proto-handaxes and choppers. An electron spin resonance (ESR) date from a rhino tooth from Level 6 suggests a late Middle Pleistocene age (0.18 Ma). However, the excavators assert that this layer should be 0.45–0.35 Ma and the lowest cultural layer to be early Middle Pleistocene (0.7–0.6 Ma: Sohn 1985, 1988). Abundant faunal remains were found in Level 2, including rhinoceros, bear, macaque, and horse. Even though the stone industry and faunal assemblage from the lower layers may be relatively old, it is unlikely that the cave was inhabited before 200,000 BP. The cave is currently just above the modern day river bank. Considering the geological development of the river system in this region, the cave was quite likely inundated for much of the past. Regardless of the age, the archaeological collection is very important because few sites in Korea contain lithics and faunal materials in spatial association. The site formation processes and nature of hominin-carnivore interactions are currently poorly understood at Keum Cave; a situation similar to many other cave localities in Korea (Norton 2000) (Table 14.1).

The Current State of Research on the Oldest Industries in Korea

Due to increasing urban development since the 1970s, more than 150 Paleolithic localities have been discovered in Korea. The vast majority of these sites date to the Late Pleistocene. The few sites that are considered to derive from Middle Pleistocene sediments base these age determination on geomorphic observations of the surrounding sediments. New applications of various dating methods indicate the potential for developing an in depth chronological sequence from age determinations at these localities.

The Chongokni stone industry, first found in 1978, is often regarded as the oldest lithic technology in the Korean peninsula (Chung 1978; Kim and Chung 1979). The stone industry includes many different handaxe forms. The discovery of this site initiated a new discussion on the nature of Movius' hypothesis of a behavioral dichotomy between eastern Asian and European/African stone artifact traditions during the Lower Paleolithic (see Norton et al. 2006 and references therein). In the early stages of research at the Chongokni site, the presence of Acheulean forms was the basis for the assumed Pleistocene

[1] The terms of 'Lower', 'Middle' and 'Upper Paleolithic' are often used for description of stone industries in Korea, however 'Early Paleolithic' is sometimes used for the Lower and Middle Paleolithic because it is very difficult to find any morphological difference in stone industries of these ages.

age of this locality (Chung 1981). Based on the morphology of these artifacts the site was suggested to be from somewhere in the Middle Pleistocene. Early attempts to date the locality using radiometric methods have proven difficult. The earliest dating studies suffered from methodological inconsistencies. For instance, age estimations based on the TL method cast doubt on the Middle Pleistocene age of Chongokni (e.g., Yi 1989, 1996). However, contradictory age estimations using the optically stimulated luminescence (OSL) obtained from silt and sandy deposits at various localities in the Hantan-Imjin river basin suggested that the earlier attempts to date Chongokni were incorrect. Initial attempts to provide a date for this locality resulted in a wide range of ages from earlier than 200 to 20 ka (Bae 2002a and b, 2003).

Most recently, research carried out by a Korean-Japanese team has determined a new estimation of the age of the Chongokni sediments. The Chongok basalt forming the bedrock of the site was dated to 0.5 Ma using two different dating methods, K/Ar and fission track dating (Danhara et al. 2002). Fission track dating was used to date the lacustrine silt that was burned by the emplacement of the underlying basalt. This method indicated a date of 0.49 Ma, while the basalt itself was dated to 0.51 Ma. These dates represent a lower limit for the age of the locality.

Overlying the Chongok basalt are two different types of sediment. Lacustrine and fluvial sediments are found directly above the basalt surface. Subsequently a large accumulation of aeolian loess was deposited overlying the lacustrine and fluvial sediments. The lacustrine and fluvial deposits were formed during the time when the ancient river channel was scouring the top of the basalt. However loess deposition, which is believed to be geochemically similar to Chinese loess (Hayashida 2006), continues into the present. Two different types of tephra originating from pyroclastic eruptions from the Japanese archipelago were identified in the upper part of the aeolian deposit. These two ash lenses were separated by 70 cm and have dates of 25 ka (AT tephra) and 95 ka (Ktz tephra) (for detailed tephrochronology discussion see Machida 1999). This allows for the construction of a relatively detailed sedimentation rate for the upper part of the loess deposit. The lowermost artifact horizon at Chongokni is found 2.5 m below the Ktz tephra, and assuming a constant aeolian deposition rate, the lowest cultural layer should date to 350 ka. One OSL date of 160 ka was obtained below the level of the Ktz horizon in the same geological trench as the tephra used for K/Ar dating methods (Matsufuji et al. 2005). The OSL date has been turned out matching well with new TT-OSL dates (Kim et al. 2008). Alternating loess-paleosol deposition at Chongokni is believed to represent climatic fluctuations as has been shown in northern China (Naruse et al. 2003).

The earliest occupation of Chongokni is believed to be older than the advent of loess sedimentation at the site. Stone artifacts were excavated in the fluvial deposits which are overlain by the loess deposit. These artifacts should therefore be considered much older than the artifacts found in the loess deposit. Importantly, several handaxes were excavated in situ from a sandy clay layer overlain by this same loess deposit (Kim and Bae 1983). Based on this new data, the appearance of handaxes predates the 350 ka date for the earliest occupation of Chongokni. It appears likely that the hominin occupation of Chongokni is associated with Acheulean forms well before the Late Pleistocene

In addition to Chongokni several localities in the Hantan-Imjin river basin are claimed to date from the Middle Pleistocene. For instance, the Jangsanni site near Munsan city in the lower reach of the Imjin river basin is thought to be Early Pleistocene or early Middle Pleistocene (Yi 2002). The stone industry from Jangsanni includes a handaxe found in a colluvium deposit on the top of a fluvial terrace. This terrace is believed to be the third terrace from the present channel bed, which is situated about 40 m higher than the present river channel. Age estimation was based on the geomorphological development of the river terraces in the lower basin of the Imjin River. It is believed that the terrace was formed before the basalt flow in the basin, possibly earlier than 0.5 or 0.6 Ma (Yi 2002). The age of Jangsanni is still under review, and further analyses must be conducted to determine if this locality is contemporary or older than Chongokni.

Aso 4, Ktz, and AT tephra were recently identified in the upper section of a long depositional sequence from the Mansuri site, west of Chongju city in the central part of the Korean peninsula (Danhara 2008). Aso 4 and Ktz tephras are dated to 90 and 95 ka, respectively. It is not clear how much older the archaeological deposits are that lie below the level of these tephras. It is possible that the lowest level of artifact horizons could be somewhere in the Middle Pleistocene based on stratigraphic position of the two tephras (Bae 2007; Bae et al. 2006). However, some OSL dates obtained from silty and sandy deposits under the colluvium from which most of artifacts were excavated indicate a Late Pleistocene age, with most ranging between 68 and 44 ka, and some between 103 and 44 ka. The latter OSL dates were obtained from the fine sediments in the upper part of the stratigraphy (Yi 2007; Park and Hong 2007). Another proposal has been suggested that all of the sediment at the site could not be earlier than 75 ka when the second terrace of the Miho river was formed (Kim 2007). Mansuri is an important locality that warrants further study.

The Jangdongni site in the lower basin of the Youngsan river in the southwestern part of Korea has also been suggested to be of Middle Pleistocene age. Although archaeological excavation was limited to the upper part of the sequence, extensive geological studies were done on sediments below the excavated horizon (Lee et al. 2003; Lee et al. 2006). The fine sediment of the upper part of the section is believed to be loess/paleosol while the lower part is

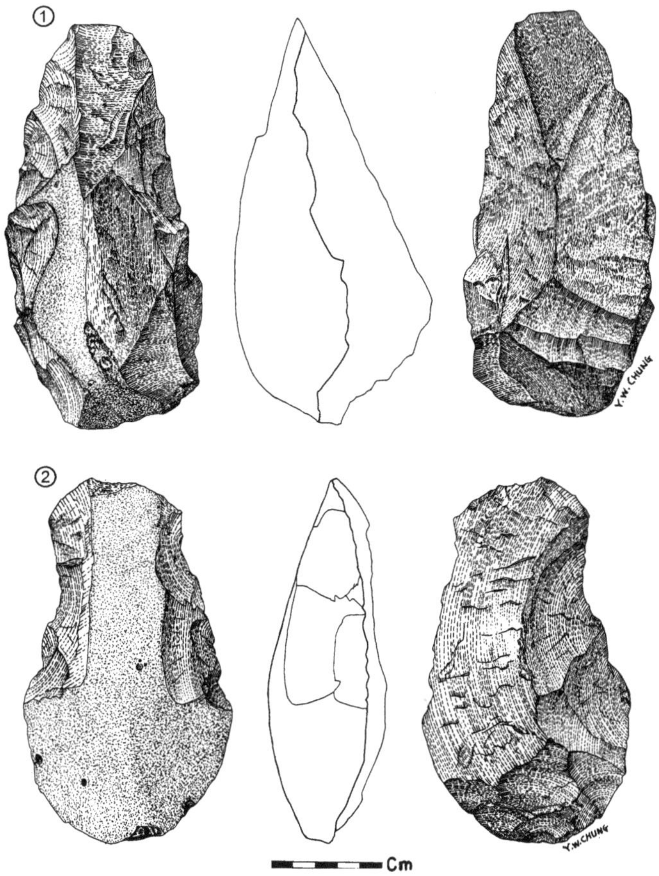

Fig. 14.2 Handaxe and cleaver from Chongokni

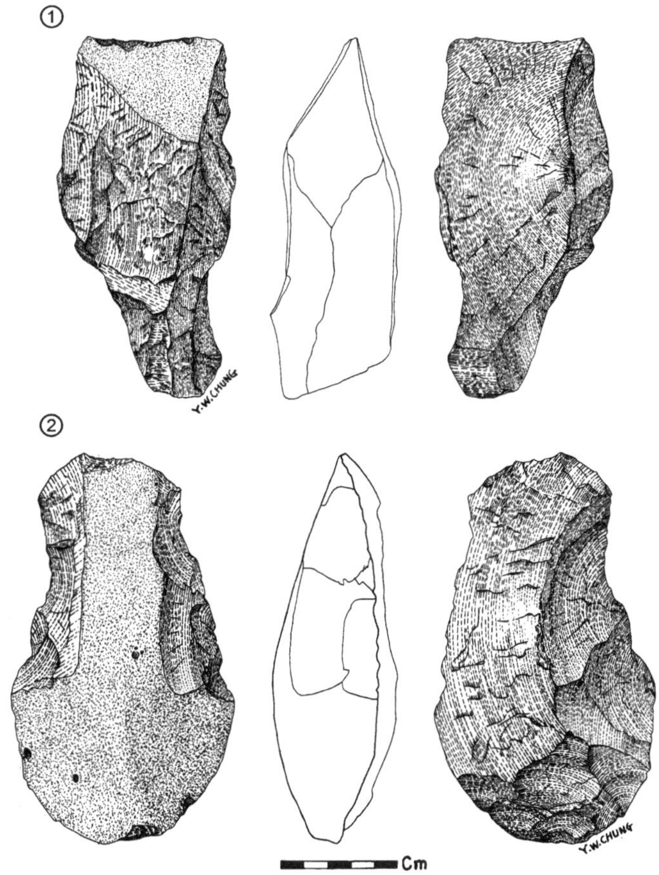

Fig. 14.3 Cleaver from Chongokni

alluvium. OSL dates on aeolian loess indicate the bottom of the upper strata is older than 220 ka. Therefore, Jangdongni must be from the Middle to Late Pleistocene (Naruse et al. 2006; Lee et al. 2006). Although no excavation has been carried out in the lowest deposits it should not be surprising if Early Paleolithic artifacts are present at the site.

Although few sites from the Korean peninsula have radiometric dates, current evidence clearly indicates that hominin occupation began sometime during the Middle Pleistocene. Extensive research of the major river basins and limestone caves, especially in the northwestern part of Korea, will surely provide some new evidence for the earliest hominin presence on the Korean peninsula. The unification of the Korean peninsula will be a great catalyst to continuing this research endeavor (Figs. 14.2 and 14.3).

Old Fashioned Artifacts in New Contexts

Developing a chronology of stone industries in Korea is particularly difficult. This is especially true for the assemblages that lack any evidence of a blade or micro-blade technology. Even in those sites that are dominated by laminar industries, crude and expedient tool types appear in association with more developed tool types. This may indicate conservatism in the early stages of stone tool production. However, in some cases, crudely shaped tools may have been derived from earlier preexisting geological contexts. Other problems include finding a few isolated crude stone artifacts in colluvium deposits on hill slopes are believed to be much older than the age of the sediment surrounding the artifacts. Furthermore, lithics that were surface collected (e.g. Lee 2002) or were found in gravel layers at the bottom of river courses (e.g. Yi and Kong 2004; Jeon and Hyun 2005) are sometimes considered to be remnants of older cultural horizons. Generations of erosional and depositional events have caused the modern day juxtaposition of older artifacts with younger geomorphological features. For instance, at the site of Bongmyoung-dong (Yi and Hong 1999), several Early Paleolithic artifacts may have been eroded and subsequently redeposited in sediments of a younger age. Similar archaeological contexts have been observed at recently excavated Paleolithic sites in Kimpo and Paju, central part of the peninsula. This type of post-depositional history would prove very difficult to determine, even utilizing modern excavation techniques. It is very difficult to determine the amount of time that elapsed between the initial erosion of the artifact and its subsequent deposition in younger deposits. It is generally assumed that Early Paleolithic sites have undergone several episodes of post-depositional modification by erosional processes during the course of the Pleistocene (Bae 2000). This may be the reason for poor preservation of Early Paleolithic sites on the Korean peninsula. The exception is the group of localities in the Hantan-Imjin river basin where volcanic activities modified the channel systems in the basin during the Middle Pleistocene.

What Is the Age of the Earliest Occupation and What Types of Stone Industries Were Associated with This Earliest Dispersal?

Multiple lines of evidence presented here indicate a Middle Pleistocene age for the earliest hominin presence on the Korean peninsula. The evidence from Chongokni suggests that hominins appeared possibly as early as 400 ka in the basin. Current geological observations and preliminary analyses, suggest several other localities in the peninsula, including Jangsanni and Jangdongni, could predate Chongokni. In addition, Komunmoru in Pyongyang and Keum cave are also considered representative of early evidence of hominins on the peninsula. The Komunmoru site may be the oldest locality yet known. The fauna associated with this locality suggest the site may be much older than the sites in the Hantan-Imjin river basin. Yet, there is still some doubt about the nature of the stone artifacts from Komunmoru. Considering the early age of hominin occupation in northern China there is still the possibility that hominins appeared in the Korean peninsula during the Early Pleistocene. It is quite possible that hominins dispersed along the coastal plain of the current and ancient Yellow Sea into the southern part of the Korean peninsula shortly after their appearance in northeast China. Extensive erosion on the Korean peninsula during the Early Pleistocene would have removed the deposits that would have recorded this initial occupation.

Current evidence indicates that hominins reached the Japanese archipelago much later than the Korean peninsula. After the dust settled from Fujimura's fraud of Early Paleolithic industries in Japan, the Kanedori site in Iwate Prefecture, northern Japan is often considered the oldest site. Stone artifacts from Layer IV of Kanedori possibly date back to the early-middle part of the Late Pleistocene on the basis of tephrachronology (Kikuchi and Yoshiro 2003). The first hominins to have reached the Japanese archipelago is assumed to have migrated from China or Korea during OIS 6 via exposed continental shelves in the Yellow Sea basin and the Korean Strait (Matsufuji 2002).

The discovery of Acheulean-like handaxe forms at the Chongokni site was initially regarded as an exception to the Movius Line (see Norton et al. 2006). However, it would not be surprising if more localities with handaxes are discovered. Handaxe, cleaver and proto-handaxe forms have been recovered from more than 20 separate localities on the Korean peninsula. Early Paleolithic hominins on the Korean peninsula clearly had the ability to produce Acheulean-like bifacial implements. Assemblages that are primarily core and flake tools with a few handaxes and cleavers are referred to as 'Chongoknian' (Bae 1994). Vein quartz and quartzite are the most common raw material for ancient toolmakers during the Pleistocene in Korea. These materials are not suitable for producing typical Acheulean handaxes, because of the numerous joint surfaces in these types of rock. Considering that handaxes have recently been found in China (Hou et al. 2000; Huang 1989, 1993; Li 2002), it is quite likely that hominins dispersed into the Korean peninsula with the ability to manufacture bifacial implements sometime during the Middle Pleistocene.

Acknowledgements I would like to thank the editors, Christopher Norton and David Braun, for inviting me to contribute this paper to the *Asian Paleoanthropology* volume. I greatly appreciate the comments from Tim Reynolds and two anonymous reviewers that served to strengthen the quality of this manuscript. I would like to also express my sincere thanks to Christopher Norton and David Braun for their comments and strong copyediting of earlier versions of this paper. Thanks also to Sohee Hwang and Jonghyun Park for help with the table and figure. I take full responsibility for any errors that may be present here.

References

Bae, K. D. (1994). Paleolithic tradition of East Asia: Chongoknian, presence of few bifaces in Paleolithic stone industries. In Paleolithic Culture of East Asia. National Research Institute of Cultural Properties. Seoul, pp. 193–211.

Bae, K. D. (2000). Transformation processes of Paleolithic sites in Korea. *Journal of Korean Paleolithic Archaeology Society, 1*, 45–56.

Bae, K. D. (2001). Lower and Middle Paleolithic industries in the Korean Peninsula. *Paleolithic Archaeology, 62*, 1–10 (in Japanese).

Bae, K. D. (2002a). Chongokni Paleolithic site: current understandings. In K. D. Bae, K. D., J. C. Lee (Eds.), *Paleolithic archaeology in Northeast Asia* (pp. 55–75). Yeoncheon County and the Institute of Cultural Properties Seoul.

Bae, K. D. (2002b). Radiocarbon dates from Paleolithic sites in Korea. *Radiocarbon, 44*(2), 473–476.

Bae, K. D. (2003). Review of chronological researches of the Chongokni site. In K. D. Bae & J. C. Lee (Eds.), *Paleolithic archaeology in Northeast Asia* (pp. 13–20). Yeoncheon County, Korea: Yeoncheon County and the Institute of Cultural Properties Seoul.

Bae, K. D. (2007). Stone industry of the Mansuri site and provision of Korean Paleolithic research. In Proceedings of Prehistoric Seminar of the Mansuri Paleolithic Site, Cheongwon, pp. 47–58.

Bae, K. D., Lee, H. Y., & Kim, K. R. (2006). Report of excavation of the locality 4, Mansuri, Cheongwon. *Hankukgogohak (Korean Archaeology), 4*, 92–107.

Chang, W. J. (1989). *Origin of Korean People (Josunsaram uy Giwon)*. Pyoungyang: Social Science Press.

Chang, W. J. (1990). Several anthropological problems of the Ryokpo Man. *Chosungogoyonku (Study of Chosun Archaeology), 74*, 22–27.

Chung, Y. H. (1978). New find of Acheulean handaxe bifaces at the Chongokni site. *Youngdaemunhwa, 11*, 5–34.

Chung, Y. H. (1981). Chongokni site, *Study of Paleolithic Culture in Korea*. Seoul; Academy of Korean Studies, pp 19–96.

Chungcheong Munhwa Yeonguwon. (2003) *The Report of excavation of Naeheungdong Paleolithic Site in Kunsan*. Cheonju.

Chungwon Munhwajae Yeonguwon (2006). *The Report of excavation of Songduri Paleolithic Site in Jincheon*. Cheongju.

Danhara, T. (2008). Volcanic Glasses Found in the Sediments of Paleolithic Sites in Korea and China, and their Comparison with Wide-spread Tephras, In Matsufuji (Ed.), *Basic Studies of the Paleolithic Chronology and Paleoenvironmental Changes in East Asia*. pp 41–52.

Danhara, T., Bae, K. D., Okada, T., Matsufuji, K., & Hwang, S. H. (2002). What is the real age of the Chongokni Paleolithic site? In K. D. Bae & J. C. Lee (Eds.), *Paleolithic archaeology in Northeast Asia* (pp. 77–116). Seoul: Yeoncheon County and the Institute of Cultural Properties.

Deino, A. (1988). K/Ar Date obtained from the Chongok Basalt, *The Significance of the Chongokni Stone Industry in the Tradition of the Paleolithic Culture in East Asia*, p. 459.

Han C. K. (2003). *The Noeundong Site, The Construction area of Daejun Worldcup stadium*, Central Museum, Hannam University. Taejeon.

Hayashida, A. (2006). Magnetic properties of the quaternary sediments at the Chongokni Paleolithic site: comparison with Chinese loess and paleosol. In *Proceeding of Geological Research and New Age Dates of the Chongokni Paleolithic Site*. Institute of Cultural Properties, Hanyang University, pp. 11–12.

Hou, Y. M., Potts, R., Yuan, B. Y., Guo, Z. T., Deino, A., Wang, W., et al. (2000). Mid-Pleistocene Acheulean-like stone technology of the Bose Basin, South China. *Science, 287*, 1622–1626.

Huang, W. W. (1989). Bifaces in China. *Human Evolution, 4*, 87–92.

Huang, W. W. (1993). Typology of the heavy-duty implements in the East and Southeast Asian Lower Paleolithic and a comment on the Movius typological system. *Acta Anthropologica Sinica, 12*, 297–304 (in Chinese).

Institute of Archaeology (1969). *Report of excavation of the Kommonmoru cave site, Sangwon, Pyoungyang*. Collection of Papers of Archaeology and Ethnography, vol. 1.

Institute of Archaeology. (1977). *Paleolithic Age in Korea*. Pyoungyang: Baekguasajeonchulpansa.

Institute of Archaeology (1978). Report of the excavation of the Seungnisan cave, Dokchon, Pyoungyang. Science Encyclopedia Press.

Jeon, B. H., & Hyun, N. J. (2005). Preliminary report of the excavation of the Paleolithic localities in the Janggi development area. *Journal of Korean Paleolithic Archaeology Society, 11*, 45–53.

Kikuchi, K., & Yoshiro, T. (2003). An introduction of the Kanedori site, Japan. In K. D. Bae & J. C. Lee (Eds.), *Geological Formation of the Chongokni site and Paleolithic Archaeology in East Asia* (pp. 163–166). Seoul: Yeonchon County and Institute of Cultural Properties, Hanyang University.

Kim, J. Y. (2007). Quaternary geology of the Mansuri site, Cheongwon. In Proceedings of Prehistoric Seminar of the Mansuri Paleolithic Site, Cheongwon, pp. 59–76.

Kim, W. Y., & Chung, Y. H. (1979). Preliminary report of Acheulean handaxe bifaces from the Chongokni site, Jindanhakbo, vol. 46–47.

Kim, W. Y., & Bae, K. D. (1983). *Report of excavation, 1979–1983*. Chongokni: Bureau of Cultural Properties.

Kim, S. K., Kim, K. K., Baek, K. H., Chang, W. J., & Seo, K. T. (1985). *Report of excavation of Daehyundong cave, Ryokpoku. Excavation report of caves sites in Pyoungyang area*. Pyoungyang: Science Encyclopedia Press.

Kim, J. C., Duller, G. A. T., Roberts, H. M., Wintle, A. G. Lee, Y. I., Yi, S. (2008). Thermally transferred optically stimulated luminescence dating of fine-grained quartz(4-11 micron) in Jeongokri(Chongokni), Korea. In S. Yi (Ed.), *Diversity and variability in the East Asian Paleolithic toward an improved understanding* (pp. 95–100). Seoul: Department of Archaeology and Art History, Seoul National University.

Lab of Human Evolution. (1995a). *Cave sites in Northwest Chosun (Korea)*. Pyoungyang: Kimilsung University Press.

Lab of Human Evolution. (1995b). *Younggok Cave site*. Pyoungyang: Kimilsung University Press.

Lee, H. J. (2002). Preliminary report of the Dangga Paleolithic site, Naju, Korea. In K. D. Bae & J. C. Lee (Eds.), *Paleolithic Archaeology in Northeast Asia* (pp. 173–180). Seoul: Yeoncheon County and the Institute of Cultural Properties Hanyang University.

Lee, H. J., Kim, J. B., Kim, J. Y., Yang, D. Y., Hong, M. Y., & Lee, H. Y. (2003). New Paleolithic finds in the Lower basin of the Youngsan river. *Journal of Korean Paleolithic Archaeology Society, 8*, 67–98.

Lee, H. J., Kim, J. B., Chung, C. W., Im, H. S., Lee, H. Y., Derevianko, A. P., et al. (2006). *The Paleolithic archaeology and quaternary geology in the Yongsan river region*. Seoul: Hakyeon.

Li, C. R. (2002). The Handaxes unearthed in China. In K. D. Bae & J. C. Lee (Eds.), *Paleolithic archaeology in Northeast Asia* (pp. 29–39). Seoul: Yeoncheon County and Institute of Cultural Properties.

Machida, H. (1999). Quaternary widespread tephra catalog in and around Japan: recent progress. *Quaternary Research, 38*, 194–201.

Matsufuji, K. (2002). The most ancient industry in Japan – when immigrated hominid to Japan? In K. D. Bae & J. C. Lee (Eds.), *Paleolithic Archaeology in Northeast Asia* (pp. 41–51). Seoul: Yeoncheon County and the Institute of Cultural Properties, Hanyang University.

Matsufuji, K., Bae, K. D., Danhara, T., Naruse, T., Hayashida, A., Yu, K., et al. (2005). New progress of studies at the Chongokni Paleolithic site, Korea: Korea-Japan cooperative project in 2001–2004. *Paleolithic Archaeology, 66*, 1–16.

Naruse, T., Bae, K. D., Yu, K., Matsufuji, K., Danhara, T., Hayashita, A., et al. (2003). Loess-Paleosol sequence in the Chongokni Paleolithic site. In K. D. Bae & J. C. Lee (Eds.), *Geological formation of the Chongokni site and Paleolithic archaeology in East Asia* (pp. 143–156). Seoul: Yeonchon County and Institute of Cultural Properties, Hanyang University.

Naruse, T., Matsufuji, K., Lee, H. J., Danhara, T., Hayashida, A., Kim, C. B., et al. (2006). Preliminary report of the loess-paleosol stratigraphy in the Jangdongni site, Korea. In H. J. Lee, J. B. Kim, C. W. Chung, H. S. Im, H. Y. Lee, A. P. Derevianko, M. I. Deregacheva, & N. S. Bolikhovskaya (Eds.), *The Paleolithic Archaeology and Quaternary Geology in the Yongsan river region* (pp. 269–290). Seoul: Hakyeon.

Norton, C. J. (2000). The current state of Korean paleoanthropology. *Journal of Human Evolution, 38*, 803–825.

Norton, C. J. (2007). Sedentism, territorial circumscription, and the increased use of plant domesticates across Neolithic-Bronze Age Korea. *Asian Perspectives, 46*(1), 133–165.

Norton, C. J., Bae, K., Harris, J. W. K., & Lee, H. Y. (2006). Middle Pleistocene handaxes from the Korean Peninsula. *Journal of Human Evolution, 51*, 527–536.

Park, S. J., & Lee, E. K. (2004). Fossil men in Korea. *Journal of Korean Paleolithic Archaeology Society, 7*, 41–51.

Park, J. H., & Hong, D. K. (2007). OSL dates of the Mansuri site, Cheongwon. Proceedings of Prehistoric Seminar of the Mansuri Paleolithic Site, Cheongwon, pp. 78–85.

Sohn, P. K. (1972). Lower and Middle Paleolithic cultural layers at the Sokchangni site. *Hankuksayongu (Study of Korean History), 7*, 1–58.

Sohn, P. K. (1985). Report of excavation of the Dodam Keumgul site. In *Report of extended excavations in the Chungju Dam construction area* (pp. 1–110). Chongju: Chungbuk University Museum.

Sohn, P. K. (1988). *Guidebook of study of Paleolithic archaeology in Korea*. Yonsei University Press.

Sohn, P. K. (1990). *Paleolithic site in Korea and Manchuria*. Seoul: Korean Institute of Prehistory.

Sohn, P. K. (1993). *Sokchangni Prehistoric Site (Sokchangni Sonsa Yujok)*. Seoul: Dong A press.

Sohn, P. K., et al. (2002). *Paleolithic Culture in Korea*. Seoul: Yonsei University Press.

Takayanagi, M. (1983). K/Ar date of basalt from the Chongokni Paleolithic site, *Chongokni*. Seoul; Research Institute of Cultural Properties, pp. 586–588.

Yi, S. B. (1984). Geoarchaeological Observations of Chongokni, Korea. *Korea Journal 24*(9), 5–15.

Yi, S. B. (1989). *Northeast Asian Paleolithic Research*. Seoul: Seoul National University Press (in Korean).

Yi, S. B. (1996). On the age of the Paleolithic sites from the Imjingang Basin. *Journal of the Korean Archaeological Society, 34*, 135–159 (in Korean).

Yi, S. B. (2002). Preliminary report of the Jangsanni site – Paleolithic site of the Lower or early Middle Pleistocene. In K. D. Bae, J. C. Lee (Eds.), *Paleolithic archaeology in Northeast Asia* (pp. 199–206). Seoul, Korea: Yeoncheon County and the Institute of Cultural Properties, Hanyang University.

Yi, Y. J. (2007). Overview of the Excavation of the Mansuri Paleolithic Site. In Proceedings of Prehistoric Seminar of the Mansuri Paleolithic Site, Cheongwon, pp. 13–43.

Yi, Y. J., & Hong, M. Y. (1999). Preliminary report of the excavation of the Bongmyoungdong Paleolithic site. Annual Review 8, The Chungbuk National University Museum, pp. 103–112.

Yi, Y. J., & Kong, S. J. (2004). Paleolithic Culture of the Songduri site, Jinchon, Korea. *Paleolithic Archaeology in Korea, 10*, 47–49.

Zhu, R. X., Potts, R., Pan, Y. X., Yao, H. T., Lu, L. Q., Zhao, X., et al. (2008). Early evidence of the genus *Homo* in East Asia, *Journal of Human Evolution, 55*(6), 1075–1085.

Chapter 15
When Were the Earliest Hominin Migrations to the Japanese Islands?

Kazuto Matsufuji

Abstract In the Japanese islands, no hominin fossil beyond 35 ka has been recovered thus far. Indirect evidence of hominin migration to Honshu before 35 ka is known from only stone artifacts. The earliest stone industries were recovered stratigraphically from two different layers at the Kanedori site in northeast Japan. Both industries are made of hornfels as local raw material. Kanedori IV with large flakes and an irregular biface is dated c. 80 ka by tephrochronology. Kanedori III with small flake tools and a large biface is estimated 67 ka by fission track dating. Such a small flake tool tradition lasted until the appearance of blade technology around 35 ka. The typological and chronological study of Kanedori IV and III suggests that the earliest migration to the Japanese islands was from northeast China.

Keywords Tephrochronology • Nojiriko-Tategahana • Kanedori • Small flake industry • Edge-polished axe • Blade technology • Happusan II

Introduction

Currently, archaeological evidence suggests that the earliest hominin occupation of the Japanese archipelago occurred during Marine Isotope Stage (MIS) 5 (127–71 ka). Since the Japanese archipelago is surrounded by deep sea, hominin migrations could have occurred when sea levels were low enough to permit crossings. The other possibility is early hominins used some type of boating/rafting technology. This is likely the reason why the earliest occupation of Japan occurred relatively late in human prehistory.

More than 5,617 Paleolithic sites have been identified in Japan through systematic surveys and excavations (JPRA Data Base Committee 2007). The majority of these sites are assigned to the backed knife and microblade culture periods (30–12 ka), with very few purportedly older than 30 ka. Most of these open-air sites are located on river terraces. Very few archaeological traces have been found in limestone caves and rock-shelters indicating these areas were rarely utilized as base camps and/or burial places during the Late Paleolithic period. Burial pits and pit dwellings are also very scarce in Japanese open-air sites (Matsufuji 1999) (Fig. 15.1).

Tephrochronology is one of the most reliable methods for reconstructing Paleolithic chronology. Certain tephra has a wide distribution, some reaching hundreds of kilometers (Machida and Arai 2003). For instance, both Aira-Tanzawa tephra (AT: 26–29 ka) and Kikai-Tozurahara tephra (K-Tz: 95 ka) from South Kyushu have been discovered in the Shandong Peninsula in North China and the Chongokni site in the Korean Peninsula. Tephra is not restricted to only Late Pleistocene sequences. For instance, the Baektusan Volcano in North Korea erupted several times during the Middle Pleistocene. Currently the oldest tephra has been dated to c. 450 ka, found in several boring cores under the Japan Sea bottom (Shirai 2001). In Japan, tephra is used to determine chronological and stratigraphic position of sites and associated materials.

In this paper, I review the evidence of the earliest hominin occupations in Japan, in light of the vertebrate paleontology, hominin fossils, and particularly, the archaeological records. I then introduce Kanedori, which may prove to be the earliest site in Japan.

Vertebrate Paleontology

Many taxa (e.g., *Stegodon orientalis*, *Palaeoloxodon naumanni*, *Cervus katokiyomai*, *Sinomegaceros yabei*, *Elaphurus menziesianus*, *Bubalus* cf. *teilhardii*, *Panthera tigris*) appear for the first time in Japan during the Middle Pleistocene (Otsuka 1987). The only way these taxa could have migrated to Japan is via a land bridge. Konishi and Yoshikawa (1999) offered an interesting hypothesis on the

K. Matsufuji (✉)
Archaeology Lab, Faculty of Letters, Doshisha University, Kyoto, Japan, 602-8580
e-mail: kmatsufu@mail.doshisha.ac.jp

C.J. Norton and D.R. Braun (eds.), *Asian Paleoanthropology: From Africa to China and Beyond*, Vertebrate Paleobiology and Paleoanthropology, DOI 10.1007/978-90-481-9094-2_15,

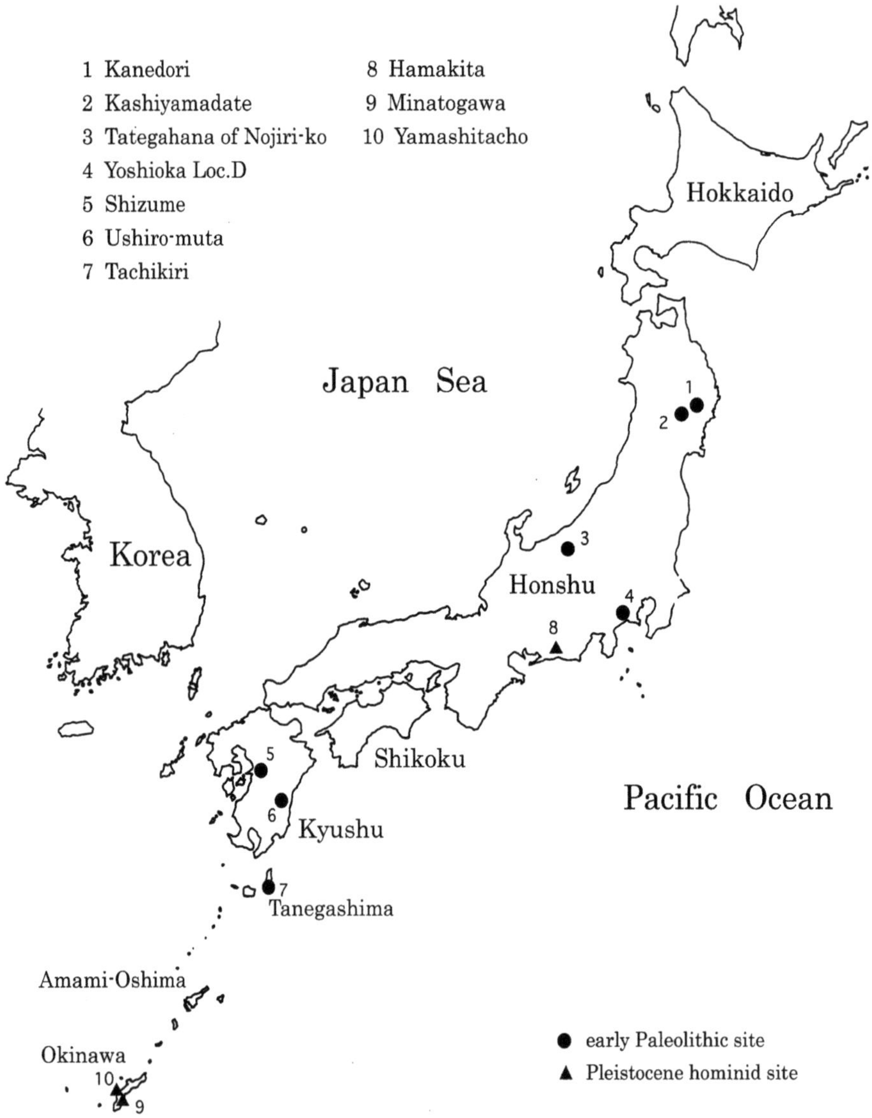

Fig. 15.1 Referred Paleolithic sites and Pleistocene hominid sites in Japanese Archipelago

timing of these land bridges. They reconstructed the stratigraphic position of Japanese Quaternary proboscidean species using tephrochronology and marine isotope stages to build their model. According to Konishi and Yoshikawa (1999) *S. orientalis* migrated to the archipelago through the dried continental shelf from southern China during MIS stage 16 (c. 630 ka) and *P. naumanni* migrated to the islands via the Tsushima land bridge from northern China during MIS 12 (c. 430 ka). Kawamura (2001) has also suggested that most of these temperate-forest type faunas on the Honshu, Shikoku and Kyushu islands appeared only during the middle Middle Pleistocene. Kawamura (2001) suggests that the formation of the land bridges only occurred a few times during the Pleistocene and the durations were probably too short to have facilitated major migrations of these megafaunas.

Pleistocene Hominin Fossils in Japan

The highly acidic soil hinders preservation of organic remains in mainland Japan. The majority of the hominin fossils have been found in the Ryukyu Islands (Table 15.1). Most of the Ryuku Islands consists of coral reef that contains calcium carbonate suitable for fossilization. This can at least partially explain the more abundant hominin fossils in this region compared to the main islands in Japan. Nevertheless, no lithic artifacts have been discovered in association with hominin bones anywhere in the Japanese Archipelago thus far.

Modern *Homo sapiens* fossils from Minatogawa, Oyama cave, Yamashitacho cave, Kadabaru cave, Shimojibaru cave, Pinza abu in the Ryukyu Islands, and Akashi, Kuuzu, Ushikawa, Mikkabi, Hamakita in Honshu and Hijiridaki in Kyushu were

Table 15.1 Pleistocene *Homo sapiens* localities on the Japanese Archipelago

Site name	Site	Prefecture	Discovery	Hominin remains	Vertebrate Paleontology	^{14}C age	References
Hamakita	Quarry	Shizuoka	1960–1962	Skull fragments, clavicle	*Homo* bed; *Meles leuculus*, *Cervus nippon*?	c. 14,000 BP (upper)	Baba (2001a), Hasegawa (1980)
				Humerus, ulna, innominate, tibia	*Lepus timidus*? Felis bed; *Felis* cf. *pardus*, *Meles leuculus*, *Cervus* sp., *Macaca fuscata*, etc.	c. 18,000 BP (lower)	Narasaki et al. (2000)
Iwato	Burial pit	Oita	1979	Canine tooth, incisor			
Yamashitacho Cave No.1	Cave	Okinawa	1968	Femur and tibia (child)	*Cervus astylodon*	32,000 ±1,000 BP	Hasegawa (1980)
Minatogawa	Fissure	Okinawa	1970–1971	Skeltons (5 individuals at least)	*Rhinolophus cornutus*?, *Tokudaia osimensis*	18,250 ±6500 BP	Suzuki and Hanihara (1982)
					Diplothrix legata, Sus leucomystax	16,600 ±300 BP	Narasaki et al. (2000)
					Cervus astylodon		Baba (2001a)
					Muntjacinae gen. et sp. indet., etc.		
				Talus, metatarsal	*Sus leucomystax*, etc.		Baba (2001a)
Shimojibaru	Cave	Okinawa	1983, 1986	Infant skeleton, mandible,	*Cervus astylodon*, Muntjacinae gen. et sp. indet.	32,000 ±1,000 BP	Narasaki et al. (2000)
				Right humerus, right femur			
Pinza-abu	Cave	Okinawa	1979–1983	Parietal, occipital bone,	*Trimeresurus flavoviridis, Japalura* sp.	25,800 ±900 BP	Narasaki et al. (2000)
				Lumbar vertebra, hand bones,	*Testudo sp., Caprcolus miyakoensis*	26,800 ±1,300 BP	Baba (2001a)
				Isolated teeth	*Diplothrix legata,* etc.		

originally considered to be of Pleistocene origin. Most of these human fossils are represented by fragmented crania and post-crania, except Minatogawa No.1 which preserves an almost complete skeleton. The original innominate from Akashi was lost in an air raid during World War II. On the basis of morphological reexamination and direct AMS ^{14}C dating on the specimens, Kuuzu, Ushikawa, Mikkabi, and Hijiridaki are no longer considered Pleistocene modern *Homo sapiens* (Baba 2001a; Matsu'ura and Kondo 2001).

Currently, only the Hamakita hominin is considered to be of Pleistocene origin on the Japanese main islands (Baba 2001c). The fragmented human fossils were found in a limestone quarry in Hamakita City, Shizuoka Prefecture in Honshu Island. Two mammalian fossil beds identified between 1960 and 1962 without any cultural remains. The lower feline bed includes human tibia, while the upper hominin bed contains skull fragments, a clavicle, humerus, ulna, and an innominate bone (Baba 2001a). The Hamakita materials are estimated to be c. 18 ka for the lower feline bed and c. 14 ka for the upper hominin bed by ^{14}C dating (Matsu'ura and Kondo 2001). The age estimates correspond to the backed knife and microblade cultural periods in Japan.

Pleistocene *Homo sapiens* fossils are better represented on the Ryukyu (Okinawan) Islands. The Minatogawa hominins were excavated from a limestone fissure in a quarry in Yaese Town (former Gushigami Village) in Okinawa. The Minatogawa fossils represent at least five individuals. Minatogawa hominin No.1 is the most complete Pleistocene skeleton in East Asia. The charcoals associated with the hominin were dated 16 and 18 ka by ^{14}C (Suzuki and Hanihara 1982). Suzuki suggested Minatogawa Man is similar to Liujiang Man in South China (Suzuki and Hanihara 1982). On the other hand, Baba (2001b) indicates that Minatogawa is morphologically more similar to Wadjak No.1 from Java, Indonesia, rather than Liujiang Man. It is probable that Minatogawa hominins came from southern China or Southeast Asia during the Last Glacial Maximum, crossing over the Kerama Channel (more than 1,000 m in depth) by raft or canoe from island to island.

Lithic Industries Older than c. 30 ka in Japan

In the 1970s there was a major debate on whether an earlier Paleolithic than blade and microblade technologies exists in Japan. Professor Chosuke Serizawa insisted vein quartz or quartzite from Sozudai in Kyushu, and Hoshino and other localities in central Japan were early Paleolithic artifacts, but most Japanese researchers did not accept these. In the case of the latter sites, the quartzite materials were considered to be geofacts.

In the 1960s Tategahana, Nojiriko in central Japan, was reported as a *Paleoloxodon naumanni* and *Sinomegaceros yabei* kill site with many small chipped stones made of various raw materials and bone/tusk tools made of *Paleoloxodon* and *Sinomegaceros*. The fossils and artifacts were identified from several different horizons of fluvio-lacustrine sediments. However, some researchers doubt the artifactual nature of the materials. Recently Yoshikatsu NAKAMURA has carefully reexamined all of the stone materials from four fluvio-lacustrine units, which were chronostratified by tephras and 45 AMS ^{14}C dates. As a result, ten artifacts were identified as true ones, and classified into two side-scrapers, three utilized flakes, one retouched flake, two flakes and two cores produced on non-porphyritic andesite, chert, jasper, chalcedony and obsidian (Anthropology and Archaeology Research Group for Nojiri-ko Excavation 2006). Based on AMS ^{14}C dating on molars, antlers and wood from the ten stratified layers, the calibrated age range is between 47 ka to 31 ka (Geology Research Group for the Nojiri-ko Excavation 2004).

Stone artifacts underlying the Aira-Tanzawa (AT) volcanic ash (26–29 ka: Machida and Arai 2003) have been excavated from the Tsuchihama-Yaya site in the Amami Islands, and the Tachikiri site in Tanegashima Island in southern Kyushu. Edge-polished axes, similar to ones from the Japanese mainland, were excavated from both sites. Traces of hearths (burnt soil) and fire cracked cobbles just under Tane-IV volcanic ash layer (Tn-IV; c. 30 ka) from the Kikai caldera were also found at Tachikiri. Many grinding stones and stone plates, which were probably used to prepare plant foods, were unearthed at Tachikiri. Such tools were also found under AT at the Tanukidani site in South Kyushu Island (Kizaki 1987).

At present, about 50 sites are estimated to be between 35–30 ka from Honshu to Kyushu. The associated lithic industries are characterized by thin edge-polished axes and so-called trapezoids made on small irregular flake without any blade technology. At the same time, several blade industries appeared in the Kanto Plain of Honshu and the surrounding areas (Matsufuji 2004).

In the central mountainous region of Honshu in Nagano Prefecture the Happusan II atelier site has been identified (Sudoh 1999). A large number of stone artifacts, primarily produced on locally-available black andesite, with five concentrations were unearthed in the layer Xb under AT and Yt-Pm4 tephras. Five AMS ^{14}C ages derived from associated charcoal date the site to 32–31 ka (36,000–35,000 cal. BP). The lithic assemblage is composed of 21 knives, 15 end-scrapers, 44 side-scrapers, 1 edge-polished axe, 67 blades and flakes with minute scars, 289 blades, 859 flakes, 4,454 chips and 43 cores.

Due to the abundance of refitted materials, the blade manufacturing technology is well understood. Initially, a large andesite pebble (50 cm in maximum length) was divided into several parts for core blanks. Next, a single striking platform was formed by hard hammer percussion. During the third step, blades were continuously detached from the narrow face of the core blank without any preliminary preparation of core or crested blade removal. These cores have the primary flaking faces divided on both sides. This core reduction technique is referred to as "Bisan-Seto type blade technology" (Satoh 1987), named after a channel situated in the central part of the Japan Inland Sea where the technology was first identified.

Another blade production methodology is also known in Happusan II, which is a blade technology with crested flakes. This technology is characterized by distinctive marks on the dorsal side of the knife-blades. Finished knives of Happusan II have, in the most cases, a Y-shaped ridge on the dorsal side, keeping their striking platforms. These are very similar to knives from the Tachikawa loam layer X of the Takaido-Higashi site in Tokyo Metropolis (Oda et al. 1977). Similar cores and knives were also excavated from several sites in Northeast Japan. Most of these have proto-knives with slight retouch around the platform on the dorsal and pointed tip. These are considered to be incipient backed knives based on techno-morphology and geochronology (Matsufuji 1986).

The Earliest Site in Japan: Kanedori

The Kanedori site (39°23′13″ north, 141°20′36″ east), situated in Tono city in Northeast Japan, was discovered by Yoshio Takeda, an amateur archaeologist in 1984. Kyoichi Kikuchi (1986) and his research team conducted a salvage archaeology excavation in 1985. In 2003 and 2004 the Education School Board of Miyamori Village (present Tono City) excavated around the periphery of the site again (Kuroda 2005). The excavations in 2003 and 2004 led to the recovery of only one retouched flake and ten chips.

Kanedori is situated in a small basin within the Kitakami Mountainous region. The site is located on the middle terrace along the Yuya River, a branch of the Tatsusobe River. Kanedori is 245 m above sea level and 25 m above the present Yuya River. The stratigraphy is as follows (Fig. 15.2):

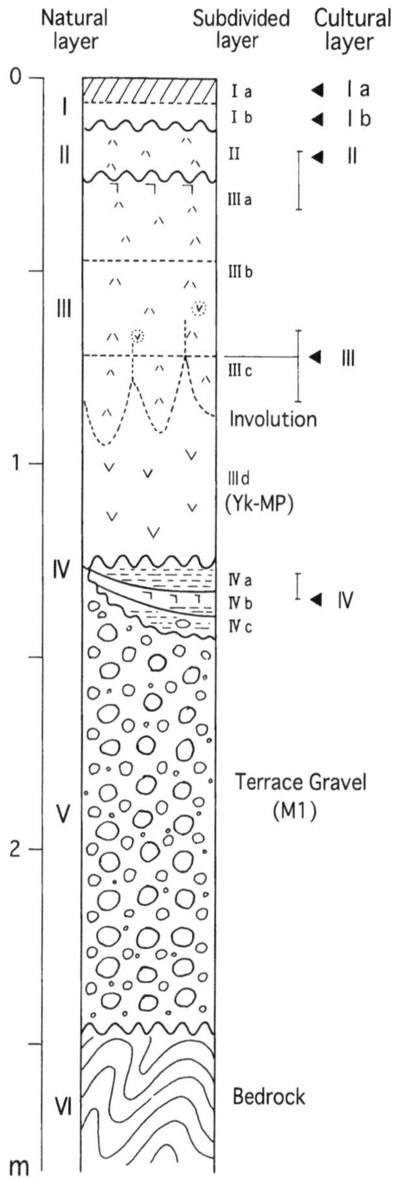

Fig. 15.2 Stratification of Kanedori site (After Kikuchi 1986)

Layer I; Holocene black soil, 10 cm in thickness, contains Jomon pottery.
Layer II; Light yellow colored silt, 10–15 cm. ^{14}C dating is 23,580 ± 450 B.P. (Gak-13090).
Layer III; Bright orange colored silt, 90–120 cm. This is divided into four sub-layers. The lowest sub-layer IIId is Murasakino pumice (MP) or Yakeishi-Murasakino pumice (Yk-MP). This tephra originated from the Yakeishi Volcano in southern Iwate Prefecture. The top of this layer shows a wave structure (involution) derived from a periglacial phenomenon widely observable in the Kitakami Mountainous area, which probably occurred under the periglacial conditions of MIS 4.
Layer IV; Light yellow to light yellow-gray colored clay, 40–50 cm. This layer of fluvial sediment is divided into three sub-layers (IVa–IVc).

Layer V: Weathered reddish gravel with mud matrix, 110–120 cm.
Layer VI; Slate bedrock.

Lithic Industries from the Kanedori Site

Jomon artifacts were identified in Layer I. Paleolithic artifacts yielded from Layers II, III and IV, which were named cultural layers II, III and IV. Cultural layer II, yielding only 2 flakes, belongs to the Late Paleolithic based on an associated ^{14}C date of 23,580 ± 450 BP. Cultural layer III yielded 1 bifacial tool with wide cortex on one side (Fig. 15.3-1), 1 large discoidal core (Fig. 15.3-2), 1 chopper, 6 retouched and/or utilized flakes (Fig. 15.4-1, 2, 3, 4, 9, 10), 13 flakes (Fig. 15.4-6, 7, 8), 5 chips, 3 burnt pebbles, and 500 charcoal particles. Hornfels was the raw material for the large cores and flakes, while the small flakes and scrapers were produced on siliceous shale. Hornfels is a local raw material, abundantly present on the Yuya riverbed and in the terrace gravel under the site. However, the closest siliceous shale source is in the Ou Mountains, about 50 km from Kanedori. Seven large flakes produced on hornfels were recovered in a pit from the upper part of the Jomon period layer. Four of these flakes were refit.

Layer IV which is fluvial silt without any cobbles contains the oldest industry in this site. Although the lithic artifacts were unearthed from the 2 subdivided layers of IVa and IVb, the top of layer IVb with sun-cracks and abundant charcoal particles is considered to be the living floor. The lithic assemblage is comprised of one biface (Fig. 15.5-1), two choppers (Fig. 15.5-2), one chopping-tool (Fig. 15.6-3), 1 pointed tool, two flakes (Fig. 15.5-3), and one retouched flake (Fig. 15.5-4 and Fig. 15.6-2). Two of these lithics (Fig. 15.6-2 and Fig. 15.6-3) were refit (Fig. 15.6-1). Local hornfels river cobbles were used to produce the large core and flake tools. A non-local chert was used for producing the small flakes or retouched flakes (Terui 2005).

Even though the Kanedori assemblage is small, intrasite variation is present. The Kanedori III industry is different from Kanedori IV. The former exhibits small flakes produced on siliceous shale, but the latter's small flakes were made from chert. Kanedori IV industry is comprised of crude and large tools except for two small flakes. This industry belongs to the broader East Asian core and flake tool tradition. Kanedori III can be considered a flake industry, though a massive handaxe-like tool

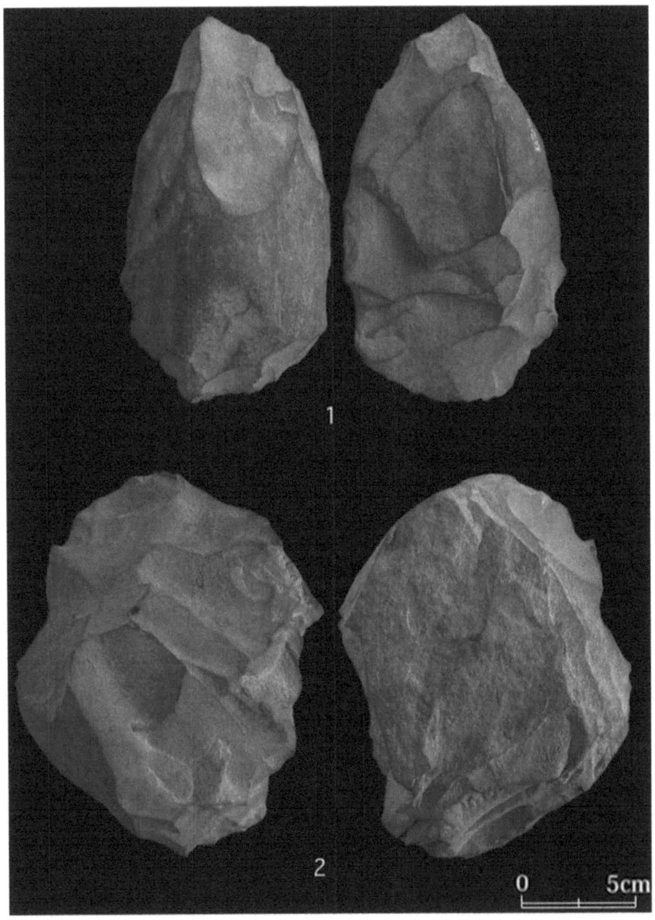

Fig. 15.3 Cultural Layer III of Kanedori site (After Kuroda 2005)

Fig. 15.4 Cultural Layer III of Kanedori site (After Kuroda 2005)

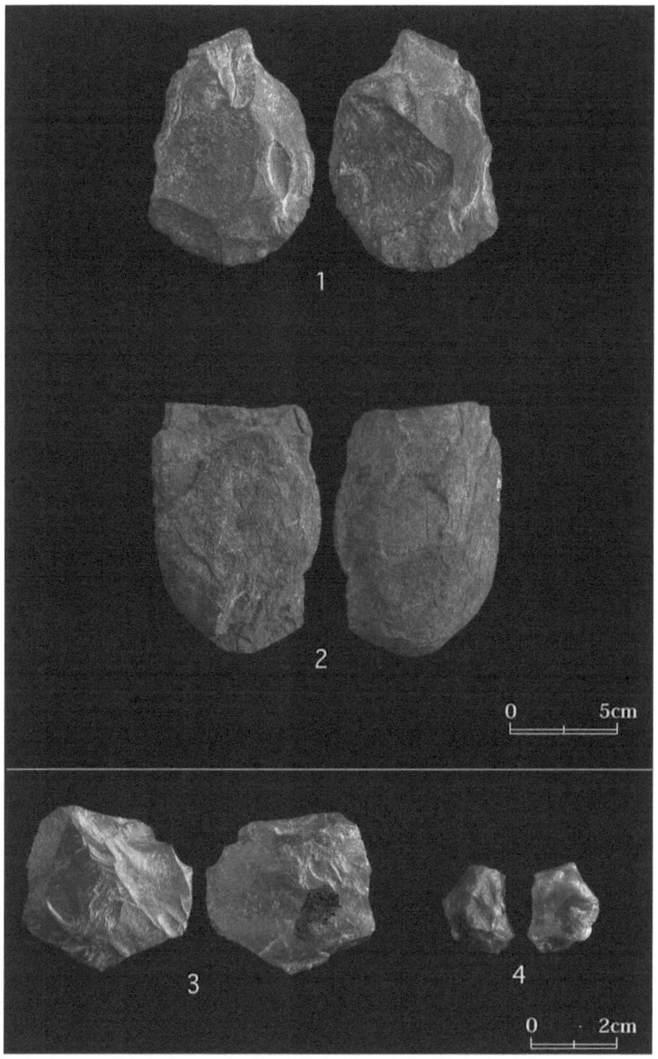

Fig. 15.5 Cultural Layer IV of Kanedori site (After Kuroda 2005)

Fig. 15.6 Cultural Layer IV of Kanedori site (After Kuroda 2005)

and a large discoidal core are present. This assemblage is very similar to later flake tool industries. For instance, the Kanedori materials are similar to lithics from the lowest stratum of the Tachikawa loam in the Kanto Plain. The massive handaxe-like tool from Kanedori III may be considered a prototype of the latter edge polished axe, which is common in Japan around 30 ka.

Tephrochronology at the Kanedori Site

Tephrochronological research has played an important role in determining the date of the Kanedori deposits. Many Pleistocene volcanic ashes have been identified in the lower Kitakami river basin, Northeast Japan (Fig. 15.7; Watanabe 1991). In 2001 we took seven block samples at 5 cm intervals from layer IV for tephra analysis conducted in the Kyoto Fission-Track Laboratory. In 2002 we systematically collected a series of samples on the same profile again.

Several kinds of volcanic glasses from different volcanoes were found in layer IVb (Danhara 2002). These volcanic glasses were identified as Aso-4 (c. 85–90 ka), Kitahara (Kth) and Atago (84 ± 12 ka by FT). In the second and third excavations, Narugo-Nisaka (c. 90 ka) and Toya (c. 112–115 ka) volcanic glasses were also identified in the same horizon (Soda 2005). We believe all the volcanic glasses are of secondary deposition, since a distinctive tephra bed is not visible in the profile. Their tephras are dated between 84 and 115 ka. It is important to bear in mind that Layer IVb contains no younger tephra than Aso-4. Accordingly, the artifacts from Layer IVb were left just after the Aso-4 deposition. The upper age limit of Layer IVb is determined by Yk-MP tephra, but Yk-MP does not have a sufficient amount of zircon to measure by fission track. However Yk-YP tephra is directly superimposed on Yk-MP tephra. The fission-track age of Yk-YP from another location has been recently measured to be 82 ± 19 ka (Watanabe et al. 2005).

Due to the periglacial structure at the top of layer IIId, it may be possible Kanedori III is younger than MIS 4. A few volcanic glasses comparable to Iwate-Oide (Iw-Od) volcanic

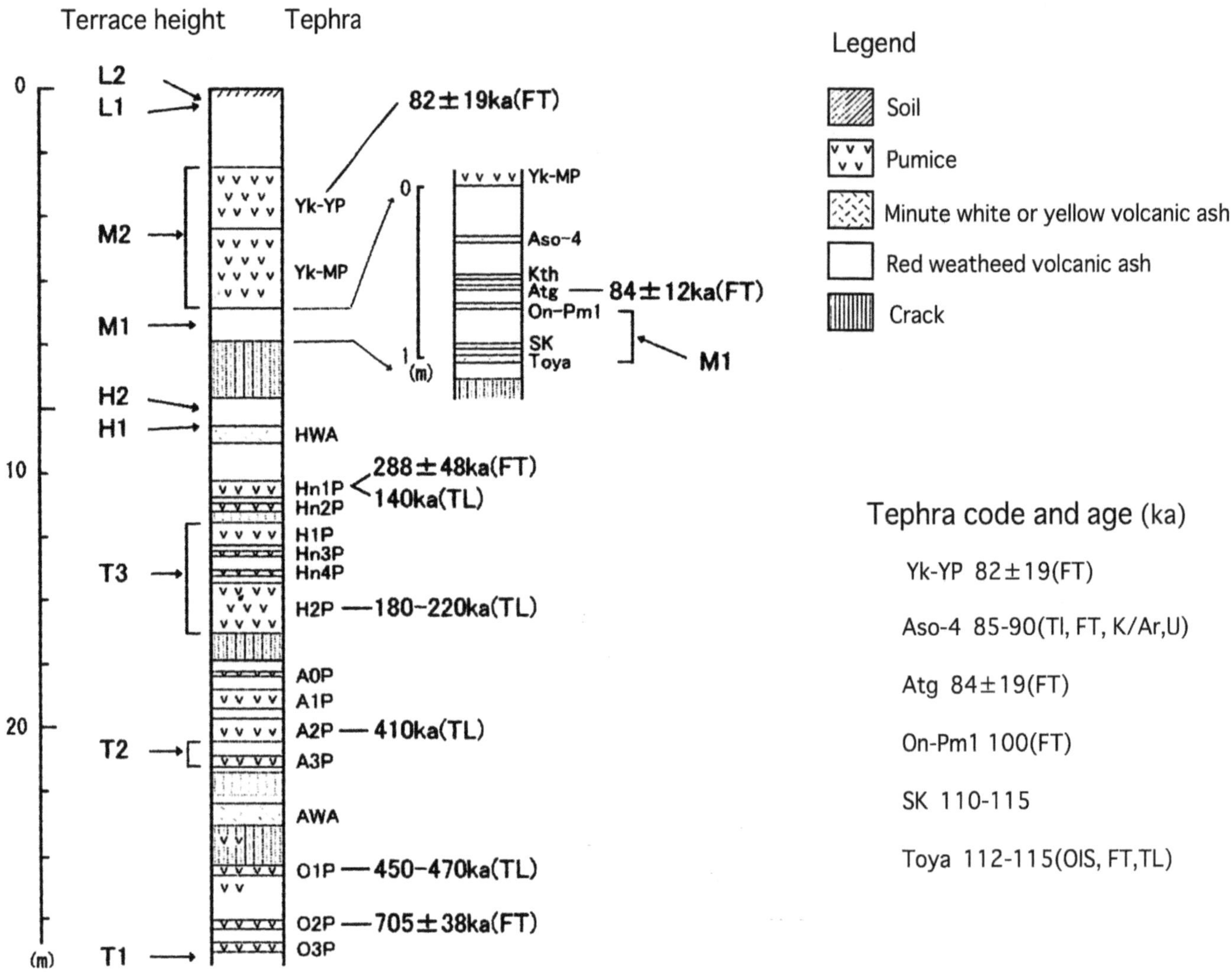

Fig. 15.7 Tephrochronology in Kitakami Lowland (After Watanabe et al. 2005)

ash was found in the same horizon as the Kanedori cultural layer III (Danhara 2002). This tephra originates from the Iwate Volcano near Morioka City, and the age was estimated to be 50–35 ka by Machida and Arai (2003). A fission track age of the lower unit was recently measured to be 67 ± 7 ka (Itoh et al. 2007). The latter age suggests the lower limit of the Iwate-Oide tephra group. Therefore, cultural layer III will be younger than the fission track age. I believe the lowest Kanedori cultural layer to be older than 35 ka on the base of typological examination for the industry.

Discussion

Both Kanedori III and IV industries were produced on locally available hornfels and non-local siliceous shale and chert. Due to small sample size it is quite possible that both industries belong to the same cultural tradition that relied primarily on hornfels. In the case of the Kanedori III industry, large hornfels cores are very characteristic with no evidence of retouched flakes.

The Kanedori III industry shows centripetal flake detachments and technology of braking flake intentionally. This latter technology is commonly found in the lowest layers of the Tachikawa loam, Kanto district that date to c. 35 ka (Shiraishi 2002). For instance, Layer B5 industry of Locality D at the Yoshioka site in the Sagamino Upland is represented by slight retouched flakes and a few scrapers on irregular flakes without large core tools (Shiraishi and Kato 1996). The Yoshioka lithics are basically a small flake tool industry made on chert. A similar lithic industry as that of Yoshioka Locality D was also recovered at the Shizume site in Kyushu Island. The assemblage consists of a few retouched or utilized flakes, irregular flakes, discoidal cores, cubic cores made of chert, and hammerstones. It is likely that such an industry diffused before the appearance of the trapezoid and edge polished axe industries in the Japanese main islands (Matsufuji 2004).

Conclusions

Marine isotope stage 5 is from 127 to 71 ka, which is referred to as the last interglacial and represents a period of global warming. I argued elsewhere that there is no evidence that East and South Asian hominins made ocean crossings during stage 5 (Matsufuji 1999). Accordingly, East Asian hominins would have migrated to the Japanese Islands through the last land bridge that appeared during the cold period of MIS 6 which lasted from 186 to 127 ka (Matsufuji 2002). It seems MIS 6 was as cold as MIS 12 and 16 (Shackleton 1995). The peak of this glacial period was between 135 and 155 ka.

The Kanedori III industry is estimated to be 50–35 ka based on tephrochronology and typology. It is basically a small flake industry with a large biface. However, both trapezoids and edge polished axes are absent in the assemblage. Therefore, we should consider that the occupation of Kanedori III preceded the introduction of trapezoid and edge-polished axe industries. It is probably contemporaneous with the stone artifacts from the lower fluvio-lacustrine sediments of the Tategahana site of Lake Nojiri, and Layer B5 industry of Locality D at the Yoshioka site.

The Kanedori IV lithic industry is the oldest in the Japanese Archipelago thus far. Kanedori IV hominins used hornfels as local raw material. Hornfels was also used in several sites such as Nakazanya (Kidder and Oda 1975) in Tokyo Metropolis, Takesa-Nakahara (Otake 2005) in the midland and Ushiromuta (Tachibana et al. 2002) in South Kyushu. It is certain that these industries are older than the introduction of the edge-polished axe and trapezoid industries (c. 30–35 ka) as well as Kanedori III, because they do not contain such standardized tools.

There is a remarkable flake-tool industry with a rough hornfels biface from the lowest layer at Xianrendong cave in Jilin Province in Northeast China (Chen and Li 1994). Its uranium-series age is 160 ka. In addition another small industry was recovered with *Mammuthus, Equus*, and *Bos* fossils from the reddish brown clay on the second terrace at Xinxiang Brickfield (Chen and Cheng 1996). It consists of a rough handaxe of hornfels, cores and a chopping-tool of siliceous limestone. Its uranium-series age is 62 ± 0.6 ka. We can guess that flake-tool industries of hornfels were distributed over Northeast China from c. 160 to c. 62 ka. This data leads to an interesting suggestion to look for the homeland of Kanedori Man around the Japanese Archipelago. We should pay attention to the connection through the northern route during MIS 6 between Northeast China and Japan after this. In this way, we may be able to determine by which route hominins initially migrated to Japan.

Acknowledgements I would like to thank the editors, Christopher Norton and David Braun for inviting me to contribute this paper to the volume. The comments by Christopher Norton, Peter Bleed, Alison Brooks, and others greatly strengthened the quality of this manuscript. They are much appreciated for taking the time and effort. This study owes to a part of Grant-in-Aid for Scientific Research (A) in 2004–2007 by Japan Society for the Promotion of Science (No. 16251005). I appreciate to Kyoichi KIKUCHI, Yoshio TAKEDA, Atsushi KURODA, Tohru DANHARA, Mitsuhisa WATANABE, Jun'ichi ITOH, Yoshikatsu NAKAMURA,,Youichi KONDO and Hisao BABA for their academic advices. Figures 15.3-6 were offered by Education School Board of Tono City.

References

Anthropology and Archaeology Research Group for Nojiri-ko Excavation. (2006). Results of the archaeological investigation of the 15th Nojiri-ko excavation and a reexamination of the Stone Artifacts from the Tategahana site. *Bulletin of the Nojiri-ko Museum, 14*, 31–53 (in Japanese).

Baba, H. (2001a). Pleistocene hominids of Japan. Long journey to prehistoric Japan, pp. 52–53. NHK and NHK Promotions (in Japanese).

Baba, H. (2001b). Where did the Minatogawa Hominids come from? Pleistocene hominids of Japan. Long journey to prehistoric Japan, pp. 54–55. NHK and NHK Promotions (in Japanese).

Baba, H. (2001c). Verification of unconfirmed fossils. Pleistocene hominids of Japan. Long journey to prehistoric Japan, pp. 104–107. NHK and NHK and Promotions (in Japanese).

Chen, Q.-J., & Li. Q.-T. (1994). A brief report on Xianrendong cave site, Jilin Province. *Acta Anthropologica Sinica, 8*(1), 12–19 (in Chinese).

Chen, Q.-J., & Cheng, X.-M. (1996). The first discovery of Paleolithic remains in Jilin District. In Y.-J. Lee (Ed.), *The Paleolithic culture in North East Asia* (pp. 247–257). Liaoning Provincial Archaeology Institute, Shenyang (in Chinese).

Danhara, T. (2002). *Preliminary report of tephra analysis in the Kanedori site*. Kyoto Fission-Track Co., Ltd. (unpublished).

Geology Research Group for the Nojiri-ko Excavation. (2004). Geology of the excavation site at Lake Nojiri, and its surrounding areas, northern part of Nagano Prefecture. On the redefinition of the Nojiri-ko formation. *Bulletin of the Nojiri-ko Museum, 12*, 1–13 (in Japanese).

Hasegawa, Y. (1980). Notes on vertebrate fossils from Late Pleistocene to Holocene of Ryukyu Islands, Japan. *The Quaternary Research, 18*(4), 263–267.

Itoh, J., Danhara,T., Iwama, H. (2007). Fission track age for Iwate-Yukiura Pumice, that is a lower unit in Iwate-Oide Black Ash group. Program and Abstracts for the Volcanological Society of Japan 2007. Fall Meeting 44, *Volcanological Society of Japan* (in Japanese).

JPRA Data Base Committee. (2007). An interim report on the Japanese Palaeolithic database project. *Palaeolithic Research, 3*, 147–152 (in Japanese).

Kawamura, Y. (2001). Mammalian. *Archaeology Quarterly, 74*, 36–40 (in Japanese).

Kidder, J. E. & Oda, S. (Eds.) (1975). *Nakazanya site. Occasional papers 1*. Archaeological Research Center of International Christian University (in Japanese).

Kikuchi, K. (Ed.) (1986). *Kanedori site : excavation report*. Education School Board of Miyamori Village (in Japanese).

Kizaki, Y. (Ed.) (1987). *Tanukidani site*. Education School Board of Kumamoto Prefecture (in Japanese).

Konishi, S., & Yoshikawa, S. (1999). Immigration times of the two proboscidean species, Stegodon orientaris and Palaeoloxodon naumanni, into the Japanese Islands and the formation of land bridge. *Earth Science, 53*, 125–134. in Japanese.

Kuroda, A. (Ed.) (2005). *Kanedori site: 2nd and 3rd excavation report.* Education School Board of Miyamori Village (in Japanese).

Machida, H. & Arai, F. (2003). *New Atlas of Tephra in and around Japan.* University of Tokyo Press, p. 148 (in Japanese).

Matsufuji, K. (1986). The Paleolithic man's cultures in Japan. In K. Mori (Ed.), *Early lives in Japan, Vol.4* (pp. 25–74). Shogakukan (in Japanese, Tokyo).

Matsufuji, K. (1999). Paleolithic mans crossed over channel. Collection of essays for the retirement of Professor Koichi Mori. *The Annual of Culturology, 48*, 1–25 (in Japanese).

Matsufuji, K. (2002). *The most ancient industry in Japan. Paleolithic archaeology in Northeast Asia. International seminar for commemorating the Chongokni Paleolithic site* (pp. 63–88). Yeoncheon County and Institute of Cutural Properties of Hanyang University and the Korean Paleolithic Society (in Japanese).

Matsufuji, K. (2004). The beginning of the Upper Palaeolithic culture in the Japanese Archipelago. In K. Matsufuji (Ed.), Study of the origin of the Upper Palaeolithic culture in the Japanese archipelago: The result report of the Aid for scientific research (C) (2) by the Ministry of Education, Culture, Sports, Science and Technology, pp. 79–118.

Matsu'ura, S., & Kondo, M. (2001). Chronologies of "Palaeolithic Hominid Remain" of Japan: issues of current research. Pleistocene hominids of Japan. Long Journey to Prehistoric Japan, pp. 104–107. NHK, NHK Promotions (in Japanese).

Narasaki, S., Baba, H., Matsu'ura, S., & Kondo, M. (2000). Palaeolithic Hominids of Japan. *Bulletin of Gunma Museum of Natural History, 4*, 23–46.

Oda, S., Itoh, F,. Keally, C. T., Shigezumi, Y. (1977). Takaido-Higashi site. The excavation team for the Takaido-Higashi site (in Japanese).

Otake, N. (Ed.) (2005). Takesa-Nakahara Site, Nagano. Industry of the Japanese Paleolithic Age. Ministry of Land, Infrastructure and Transport Chubu Regional Bureau and Archaeological Research Center of Nagano Prefecture (in Japanese).

Otsuka, T. (1987). Middle-Upper Pleistocene mammalian fossils in the Japanese archipelago. *Bulletin of the national Museum of Japanese History, 13*, 275–287 (in Japanese).

Satoh, R. (1987). A note about blade technique of Bisan-Seto type. *Hanazono Shigaku, 8*, 20–30 (in Japanese).

Shackleton, N. J. (1995). New data on the evolution of Pliocene climatic variability. In E.S. Vrba et al. (Eds.), *Paleoclimate and evolution with emphasis on human origins*. Yale University Press.

Shirai, M. (2001). Wide area tephra of the Middle Pleistocene found in the east Japanese Sea. *Chikyu Monthly, 267*, 600–604 (in Japanese).

Shiraishi, H. (2002). *The Paleolithic society and culture* (pp. 47–48). Yamakawa Press, (in Japanese), Tokyo.

Shiraishi, H., & Kato, C. (1996). Yoshioka Sites IX; Lithic Culturs before AT. Excavation Report of Kanagawa Archaeological Foundation 49 (in Japanese).

Soda, T. (2005). *Sediment and tephra at the Kanedori site. The report of the second and third excavations of the Kanedori site; 2nd and 3rd excavation report* (pp. 47–60). Board of Education of Miyamori Village (in Japanese).

Sudoh, T. (1999). *Happusan sites*. School board of Saku City (in Japanese).

Suzuki, H., & Hanihara, K. (Eds.) (1982). *The Minatogawa man*. Tokyo University Press.

Tachibana, M., Sato, H., Yamada, T. (Eds.) (2002). *Ushiromuta site*. The excavation group of Ushiromuta Site and Education School Board of Kawaminami Town (in Japanese).

Terui, K. (2005). Geological examination the Paleolithic raw materials and natural pebbles from the Kanedori site. In A. Kuroda (Ed.), *Kanedori site; 2nd and 3rd excavation report* (pp. 75–81). Education School Board of Miyamori Village (in Japanese).

Watanabe, M. (1991). Chronology of the fluvial Terrace surfaces in the Kitakami Lowland, Northeast Japan and fluctuations of debris supply in the Late Pleistocene. *Quaternary Research, 30*(1), 19–42 (in Japanese).

Watanabe, M., Danhara, T., Fujiwara, O. (2005). Alluvial fan terraces in Kitakami Lowland. Abstracts of spring meeting of Japanese geographical society, 111 (in Japanese).

Site Index

Subject Index

MIX
Papier aus verantwortungsvollen Quellen
Paper from responsible sources
FSC® C105338

If you have any concerns about our products,
you can contact us on
ProductSafety@springernature.com

In case Publisher is established outside the EU,
the EU authorized representative is:
Springer Nature Customer Service Center GmbH
Europaplatz 3, 69115 Heidelberg, Germany

Printed by Libri Plureos GmbH
in Hamburg, Germany